INORGANIC
SYNTHESES
Volume XII

Editor-in-Chief

ROBERT W. PARRY

University of Utah, Salt Lake City, Utah

●●

INORGANIC SYNTHESES

Volume XII

McGRAW-HILL BOOK COMPANY

New York   St. Louis   San Francisco   Düsseldorf
London   Mexico   Panama   Sydney   Toronto

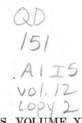

INORGANIC SYNTHESES, VOLUME XII

To **WALTER C. SCHUMB**
(1892–1967)

Dedicated teacher, researcher, and scholar
whose many contributions to inorganic
chemistry included distinguished service to
INORGANIC SYNTHESES.

PREFACE

This volume follows the organizational patterns initiated in Volumes X and XI. Syntheses are grouped according to interest area, using this criterion, we have placed in Vol. XII a chapter on Metal Complexes of Molecular Nitrogen and one on Significant Solids. More traditional chapter headings such as Nonmetal Systems and Boron Compounds also fit appropriately under the criterion of interest areas.

Inorganic Syntheses, Inc., is a nonprofit organization dedicated to the selection and presentation of tested procedures for the preparation of compounds of more than routine interest. The Editorial Board seeks the cooperation of the entire scientific community in realizing these goals. The editor for Volume XIII is F. A. Cotton; for Volume XIV, J. K. Ruff; and for Volume XV, Aaron Wold. Directions for submitting preparations are included in the section which follows this one.

The editor of Volume XII is particularly indebted to a number of people who have contributed significantly to this volume. Dr. George Parshall, Prof. A. D. Allen, Dr. Frank Bottomley, and their associates made particularly valuable contributions to Chapter One. Professor John Fackler assisted in the accumulation of syntheses on pentanedionato complexes; Prof. Aaron Wold assisted with the work on solids, and Prof. John Ruff assisted with the section on fluorine chemistry. The assistance of these men is gratefully acknowledged. Finally, every volume of INORGANIC SYNTHESES owes a large debt to the many scientists who willingly contribute their time and materials to the checking operation.   We are particularly grateful for such help.   It is a real pleasure to thank Janet Scott who has been of great help in the editing and indexing, as well as Dr. J. Richard Weaver and Mrs. Jeanne Gordus who helped willingly at crucial points in the preparation of this volume.   The generous assistance of other members of Inorganic Syntheses, Inc., is also gratefully acknowledged.

R. W. Parry

NOTICE TO CONTRIBUTORS

The INORGANIC SYNTHESES series is published to provide all users of inorganic substances with detailed and foolproof procedures for the preparation of important and timely compounds. Thus the series is the concern of the entire scientific community. The Editorial Board hopes that all chemists will share in the responsibility of producing INORGANIC SYNTHESES by offering their advice and assistance both in the formulation and laboratory evaluation of outstanding syntheses. Help of this type will be invaluable in achieving excellence and pertinence to current scientific interests.

There is no rigid definition of what constitutes a suitable synthesis. The major criterion by which syntheses are judged is the potential value to the scientific community. An ideal synthesis is one which presents a new or revised experimental procedure applicable to a variety of related compounds, at least one of which is critically important in current research. However, syntheses of individual compounds that are of interest or importance are also acceptable.

The Editorial Board lists the following criteria of content for submitted manuscripts. Style should conform with that of previous volumes of INORGANIC SYNTHESES. The *Introduction* should include a concise and critical summary of the available procedures for synthesis of the product in question. It should also include an estimate of the time required for the synthesis, an indication of the importance and utility of the product, and an admonition if any potential hazards are associated with the procedure. The *Procedure* should present detailed and unambiguous laboratory directions and be written so that it anticipates possible mistakes and misunderstandings on the part of the person who attempts to duplicate the procedure. Any unusual equipment or procedure should be clearly described. Line drawings should be included when they can be helpful. All safety measures should be clearly stated. *Sources of unusual starting materials must be given*, and, if possible, minimal standards of purity of reagents and solvents should be stated. The scale should be reasonable for normal laboratory operation, and any problems involved in scaling the procedure either up or down should be discussed. The criteria for judging the purity of the final product should be clearly delineated. The section on *Properties* should list and discuss those physical and chemical characteristics that are relevant to judging the purity of the product and to permitting its handling and use in an intelligent manner. Under *References*, all pertinent literature citations should be listed in order.

The Editorial Board determines whether submitted syntheses meet the general specifications outlined above. Every synthesis must be satisfactorily reproduced in a different laboratory other than that from which it was submitted.

Each manuscript should be submitted in duplicate to the Secretary of the Editorial Board, Professor Stanley Kirschner, Department of Chemistry, Wayne State University, Detroit, Michigan 48202, U.S.A. The manuscript should be typewritten in English. Nomenclature should be consistent and should follow the recommendations presented in "The Definitive Rules for Nomenclature of Inorganic Chemistry," *J. Am. Chem. Soc.*, **82**, 5523 (1960). Abbreviations should conform to those used in publications of the American Chemical Society, particularly *Inorganic Chemistry*.

CONTENTS

INORGANIC
SYNTHESES
Volume XII

Chapter One

METAL COMPLEXES OF MOLECULAR
NITROGEN AND RELATED SPECIES

(Assembled through the cooperation of Dr. George W. Parshall.)

Metal complexes containing coordinated carbon monoxide have
been known for years. Some speculation about the existence
of related compounds containing the isoelectronic nitrogen
molecule appeared in the earlier literature, but all attempts at
the synthesis of these nitrogen analogs were unsuccessful until
1965. In December of that year, a highly significant communication by A. D. Allen and C. V. Senoff (see reference 1, synthesis
1) described the first successful synthesis of a family of nitrogen
complexes of the form $[Ru(NH_3)_5N_2]X_2$ (where X^- was Br^-,
I^-, BF_4^-, and PF_6^-). Since the initial report of Allen and
Senoff, interest in these compounds has been high, partly
because of the apparent relationship between these substances
and the nitrogen-fixation process. A number of significant
compounds containing coordinated molecular nitrogen or a
closely related molecule are described herein.

Synthesis 4 describes a well-characterized substance which will
pick up elementary nitrogen under carefully controlled conditions. Procedures described in these pages have all been verified
in at least two independent laboratories. In every case, however, checkers reported that *all details* of each synthesis had to be

1

carried out as indicated. Care is required. ■ *Caution. In cases where azides are used, the explosive nature of heavy metal azides should always be kept in mind.*

1. PENTAAMMINE(NITROGEN)RUTHENIUM(II) SALTS AND OTHER AMMINES OF RUTHENIUM

$$RuCl_3 + 4H_2NNH_2 \xrightarrow{H_2O} [Ru(NH_3)_5N_2]^{2+} + \cdots$$

$$[Ru(NH_3)_5N_2]^{2+} + HCl_{(conc.)} \xrightarrow[\Delta]{H_2O} [Ru(NH_3)_5Cl]Cl_{2(s)} + \cdots$$

$$[Ru(NH_3)_5Cl]Cl_{2(s)} + NaN_3 \xrightarrow[\substack{CH_3SO_3H \\ NaCl}]{NH_3} [Ru(NH_3)_5N_2]Cl_2 + \cdots$$

Submitted by A. D. ALLEN,* F. BOTTOMLEY,* R. O. HARRIS,*
V. P. REINSALU,* AND C. V. SENOFF*
Checked by DAVID W. BADGER†

Solutions of pentaammine(nitrogen)ruthenium(II) have been prepared from ruthenium(III) chloride and hydrazine hydrate.[1,2] These solutions have been used to prepare pentaammine-haloruthenium(III) salts: $[Ru(NH_3)_5X]X_2$ (X = Cl, Br, I). $[Ru(NH_3)_5Cl]Cl_2$ has been converted to pure pentaammine-(nitrogen)ruthenium(II) salts—$[Ru(NH_3)_5N_2]X_2$ ($X^- = Cl^-$, Br^-, I^-, BF_4^-, PF_6^-)—via the reaction between azide ion and aquopentaammineruthenium(III).[2] Hexaammineruthenium(III) salts—$[Ru(NH_3)_6]X_3$ (X = I^-, BF_4^-)—have been prepared by the reaction between pentaamminechlororuthenium-(III) chloride and hydrazine monohydrochloride.

Syntheses of the ammine complexes of ruthenium(III) have appeared in the literature,[3] but the methods given here offer advantages in time and yield over the earlier procedures.

* Lash-Miller Chemical Laboratories, University of Toronto, Toronto 5, Ontario, Canada.
† Department of Chemistry, Geneva College, Beaver Falls, Pa. 15010.

A. SOLUTIONS CONTAINING PENTAAMMINE(NITROGEN)-RUTHENIUM(II) CATION (IMPURE PRODUCT)

$$RuCl_3 + 4H_2NNH_2 \xrightarrow{H_2O} [Ru(NH_3)_5N_2]^{2+} + \cdots$$

Procedure

Ruthenium(III) chloride* (2.0 g., 9 mmoles) (free from nitrosyl impurities) is dissolved in water (10 ml.) in a 200-ml. beaker. Hydrazine hydrate (85%, 20 ml.) is added carefully to the well-stirred solution. The initial reaction is exothermic, and large quantities of gas are evolved. The solution is stirred for about 12 hours and filtered by gravity. Because of the vigorous evolution of gas and the highly exothermic nature of the initial reaction, it is *not* recommended that the scale of the reaction be increased. Chloro complexes such as $(NH_4)_2[RuCl_6]$ or $K_2[RuCl_5H_2O]$ may be substituted for ruthenium(III) chloride.

The solution obtained as a product contains a mixture of hexaammineruthenium(II) and pentaammine(nitrogen)ruthenium-(II) ions.

B. PENTAAMMINECHLORORUTHENIUM(III) CHLORIDE, $[Ru(NH_3)_5Cl]Cl_2$

$$[Ru(NH_3)_5N_2]^{2+}_{(soln.)} + HCl_{(conc.)} \xrightarrow{H_2O}$$
$$[Ru(NH_3)_5Cl]^{2+} + \cdots$$
$$\xrightarrow[Cl^-]{} [Ru(NH_3)_5Cl]Cl_{2(s)}$$

Procedure

To the filtered solution, prepared as in Sec. A, is carefully added *concentrated hydrochloric acid* until the pH is 2 (Alkacid

---

* Ruthenium trichloride can be purchased from Engelhard Industries, 113 Astor St., Newark, N.J. 07114.

paper). Vigorous gas evolution takes place. The solution is *boiled, with stirring, and becomes yellow, slowly precipitating the yellow product.* When no further precipitation is observed, the mixture is cooled to room temperature, and the crude pentaamminechlororuthenium(III) chloride is collected by filtration, washed once with 6 M hydrochloric acid, alcohol, and acetone, and then air-dried. The yield of crude product is 3 g., 46% based on $RuCl_3$.

The product is recrystallized by heating an aqueous slurry (3 g. in 10 ml. of water) to 60°C. and adding concentrated ammonia dropwise until the yellow complex dissolves to give a wine-red solution. The solution is *filtered hot* and then cooled in an ice bath. To the cold solution concentrated hydrochloric acid is added dropwise to reprecipitate the mustard-yellow pentaamminechlororuthenium(III) chloride. The product is collected by filtration, washed once with 6 M hydrochloric acid, and then quickly washed with water, alcohol, and acetone. Finally it is dried at 78°C. over P_2O_5 *in vacuo.* Yield is 1.1 g., 39%. *Anal.* Calcd. for $Cl_3H_{15}N_5Ru$: H, 5.19; N, 23.98; Cl, 36.35. Found: H, 5.30; N, 24.53; Cl, 34.84.

C. PENTAAMMINEIODORUTHENIUM(III) IODIDE, [Ru(NH₃)₅I]I₂, AND PENTAAMMINEBROMORUTHENIUM(III) BROMIDE, [Ru(NH₃)₅Br]Br₂

These complexes are prepared from the solution prepared in Sec. A by methods similar to those described for pentaamminechlororuthenium(III) chloride in Sec. B, using *hydriodic* or *hydrobromic acid* as a reagent, and 6M HI or HBr as a wash liquid. Yields are [Ru(NH₃)₅I]I₂, 2.7 g., 49%; [Ru(NH₃)₅Br]Br₂, 1.25 g., 31%; based on $RuCl_3$. *Anal.* Calcd. for $H_{15}I_3N_5Ru$: H, 2.67; I, 67.19; N, 12.35. Found: H, 2.95; I, 67.60; N, 11.80. Calcd. for $Br_3H_{15}N_5Ru$: Br, 56.33; H, 3.88; N, 16.44. Found: Br, 56.10; H, 3.50; N, 16.92.

D. PENTAAMMINE(NITROGEN)RUTHENIUM(II) SALTS, [Ru(NH₃)₅N₂]X₂ (X = Cl⁻, Br⁻, I⁻, BF₄⁻, PF₆⁻)

D. PENTAAMMINE(NITROGEN)RUTHENIUM(II) SALTS, $[Ru(NH_3)_5N_2]X_2$ (X = Cl⁻, Br⁻, I⁻, BF₄⁻, PF₆⁻)

$$[Ru(NH_3)_5Cl]Cl_{2(s)} + NaN_3 \xrightarrow[NH_3]{CH_3SO_3H} [Ru(NH_3)_5N_2]Cl_2 + \cdots$$
yellow   yellow*

Procedure

A slurry of pentaamminechlororuthenium(III) chloride (1.0 g., 3.4 mmoles), in water (25 ml.) is heated to 60°C. Concentrated ammonia is added *dropwise*, and the temperature is kept at 60°C. until all the yellow complex dissolves to give a wine-red solution. This solution is cooled in an ice bath; it is then stirred vigorously while methanesulfonic acid is added dropwise to the ice-cold solution until the pH is 2. The solution becomes colorless to pale pink.†

To the solution is added sodium azide, 5 g. (77 mmoles).

■ *Caution. From the reaction between azide and aquopentaammineruthenium(III), it is possible to precipitate a red solid*

* *Editor's note:* The equations shown below are proposed by the editor to represent the various color changes reported.

a. $[Ru(NH_3)_5Cl]Cl_2 + OH^-_{(conc.)} \rightarrow [Ru(NH_3)_5OH]^{2+} + 3Cl^-$
   yellow solid   wine-red

b. $[Ru(NH_3)_5OH]^{2+} + CH_3SO_3H \xrightarrow{H_2O} [Ru(NH_3)_5H_2O]^{3+} + CH_3SO_3^-$
   wine-red   colorless to pink

c. $[Ru(NH_3)_5H_2O]^{3+} + NaN_3 \xrightarrow{H_2O} [Ru(NH_3)_5N_2]^{2+} + \cdots$
   colorless   yellow

d. $[Ru(NH_3)_5N_2]^{2+} + 2Cl^- \rightarrow [Ru(NH_3)_5N_2]Cl_2$
                             nearly white solid

■ *Warning. If the azide group is coordinated in step c without an accompanying oxidation-reduction reaction, the resulting red azide complex is dangerously explosive.* Equations e and f represent this process:

e. $[Ru(NH_3)_5H_2O]^{3+} + NaN_3 \rightarrow [Ru(NH_3)_5N_3]^{2+} + \cdots$

f. $[Ru(NH_3)_5N_3]^{2+} + 2X^- \rightarrow [Ru(NH_3)_5N_3]X_2$
                             red solid

† If, on adding methanesulfonic acid to the wine-red solution of $[Ru(NH_3)_5OH]^{2+}$, a fluffy, white precipitate is produced, ice-cold water may be added to redissolve it. If a yellow precipitate which is insoluble in water is produced, hydrochloric acid (6 *M*) should be added, the mixture refluxed for 10 minutes, the precipitated pentaamminechlororuthenium(III) chloride recovered, and the procedure started again.

believed to contain $[Ru(NH_3)_5N_3]^{2+}$. *This red compound is violently and unpredictably explosive. The reaction should be carried out under a good hood, and the precipitating agent should not be added until the solution has become yellow. If any red solid is observed in the reaction mixture, it should not be removed, but the mixture should be allowed to stand until decomposition to* $[Ru(NH_3)_5N_2]^{2+}$ *has taken place.*

The solution becomes orange-red, and gas is evolved. Methanesulfonic acid is added to keep the pH at 2 to 3. The solution is allowed to stand, with additions of methanesulfonic acid to keep the pH at 2 to 3, until the solution becomes yellow. Solid sodium chloride is added until precipitation of $[Ru(NH_3)_5N_2]Cl_2$ is complete. The crude product is removed by filtration, washed with alcohol and acetone, and air-dried. It is recrystallized by dissolving in water, filtering the solution, and reprecipitating the product with solid sodium chloride. Two recrystallizations by this method yield a product which is almost white (very pale yellow). The other salts are prepared by dissolving the purified chloride salt in water and adding saturated, filtered solutions of potassium iodide or bromide, sodium tetrafluoroborate, or ammonium hexafluorophosphate until precipitation of the desired salt is complete. The products are collected by filtration, washed with a little water, alcohol, and acetone, and then air-dried. Yield is 0.5 g., 33% for the pale yellow iodide salt. *Anal.* Calcd. for $[Ru(NH_3)_5N_2]I_2$: NH_3, 18.19; N_2, 5.98; I, 54.27. Found: NH_3, 18.04; N_2, 5.35; I, 54.05.

Conventional microanalysis for the percentage nitrogen in pentaammine(nitrogen)ruthenium(II) salts gives variable results. Ammonia may be determined by Kjeldahl methods, and N_2 either by dry heating of the compound *in vacuo* and determining the pressure of the evolved gas[2] or by decomposing the compound with excess cerium(IV) and measuring the volume of gas evolved.[8]

E. HEXAAMMINERUTHENIUM(III) SALTS,
$[Ru(NH_3)_6]X_3$ (X = I, BF_4^-)

$$2[Ru(NH_3)_5Cl]Cl_{2(\text{slurry})} + [H_3NNH_2]^+Cl^- \xrightarrow{\Delta} 2[Ru(NH_3)_6]^{3+} + \cdots$$

Procedure

Pentaamminechlororuthenium(III) chloride (0.5 g., 1.7 mmoles) is made into a slurry with water (5 ml.). Hydrazine monohydrochloride (1.0 g.), prepared as instructed in the footnote,* is added and the mixture gently heated. Vigorous gas evolution is observed, and the solution becomes almost colorless.† It is cooled to room temperature, filtered, cooled in ice, and to it is added a saturated filtered solution of potassium iodide or sodium tetrafluoroborate. The resultant precipitate (chocolate brown in the case of the iodide salt) is separated by filtration, washed with alcohol and acetone, and then air-dried.‡ Yield for $[Ru(NH_3)_6]I_3$ is 0.7 g., 70%. *Anal.* Calcd. for $H_{18}I_3N_6Ru$: H, 3.11; I, 65.19; N, 14.39. Found: H, 3.42; I, 65.01; N, 15.42.

Properties

Pentaammine(nitrogen)ruthenium(II) salts show a strong, sharp band in the infrared spectrum in the region 2100–2169

* Hydrazine monohydrochloride may be purchased commercially. Alternatively, it may be prepared by adding concentrated hydrochloric acid dropwise, with stirring, into a well-cooled hydrazine hydrate solution until the pH is about 7 (Alkacid paper). The solution is evaporated to the point of crystallization and cooled. The product which separates is collected by filtration. The resulting solid has a high water of crystallization but is perfectly suitable for the preparation of hexaammineruthenium(III) salts.

† The checker found that in some cases the solution did not become completely colorless; in these cases it was permissible to filter off the small amount of yellow precipitate and proceed.

‡ The checker found that the product could be recrystallized from dilute ammonia.

cm.$^{-1}$, which is assigned to ν_{N-N}; the exact position depends on the counteranion.[2] The chloride salt shows one band in the ultraviolet spectrum at 221 mμ, $\epsilon = 1.5 \times 10^4$. Other physical and chemical properties of these compounds are given in the literature,[2] as are the properties of the ammineruthenium(III) complexes.[3-7]

References

1. A. D. Allen and C. V. Senoff, *Chem. Commun.*, 621 (1965).
2. A. D. Allen, F. Bottomley, R. O. Harris, V. P. Reinsalu, and C. V. Senoff, *J. Am. Chem. Soc.*, **89**, 5595 (1967).
3. K. Gleu and K. Rehm, *Z. Anorg. Allgem. Chem.*, **227**, 237 (1936); **235**, 346 (1938); **237**, 89 (1938).
4. H. Hartmann and C. Buschbeck, *Z. Physik. Chem. (Frankfurt)*, **11**, 120 (1957).
5. A. D. Allen and C. V. Senoff, *Can. J. Chem.*, **45**, 1337 (1967).
6. W. P. Griffith, *J. Chem. Soc.*, 899 (1966).
7. J. A. Broomhead, F. Basolo, and R. G. Pearson, *Inorg. Chem.*, **3**, 826 (1964).
8. D. E. Harrison and H. Taube, *J. Am. Chem. Soc.*, **89**, 5706 (1967).

2. *trans*-CHLORO(NITROGEN)-BIS(TRI-PHENYLPHOSPHINE)IRIDIUM(I)

$$[IrCl(CO)\{P(C_6H_5)_3\}_2] + p\text{-}NO_2C_6H_4CON_3 + C_2H_5OH \xrightarrow{CHCl_3}$$
$$[IrCl(N_2)\{P(C_6H_5)_3\}_2] + p\text{-}NO_2C_6H_4CONHCO_2C_2H_5$$

Submitted by J. P. COLLMAN,* N. W. HOFFMAN,* and J. W. HOSKING*
Checked by GEORGE W. PARSHALL†

trans-Chloro(nitrogen)-bis(triphenylphosphine)iridium(I) is of interest as a complex of molecular nitrogen and is valuable as an intermediate in the synthesis of other iridium(I) complexes. This compound is prepared by treating carbonyl(chloro)-bis-(triphenylphosphine)iridium(I) with acid azides in the presence

* Department of Chemistry, Stanford University, Stanford, Calif. 94305.
† Central Research Department, Experimental Station, E. I. du Pont de Nemours & Company, Wilmington, Del. 19898.

of an alcohol which acts to intercept the intermediate acyl isocyanate by forming a carbamic ester.[2]   In the absence of such a trapping reagent, the acyl isocyanate reacts with the nitrogen complex, liberating nitrogen and forming an acyl isocyanate complex.[3,4]   When $[IrCl(CO)\{P(C_6H_5)_3\}_2]$ is treated with a slight excess of *p*-nitrobenzoyl azide in chloroform containing ethanol at 0°C. in the absence of oxygen, yields of 68–82% of the nitrogen complex are obtained.

Procedure

A magnetic stirring bar and a 0.39-g. (0.50-mmole) sample of $[IrCl(CO)\{P(C_6H_5)_3\}_2]$ are placed in a Schlenk apparatus (Fig. 1) which is flushed with nitrogen by alternately applying a vacuum and refilling with nitrogen.*   The apparatus is cooled in an ice-water bath, and 10 ml. of ice-cold chloroform is injected through the stopcock of bulb *A* (Fig. 1, position 1)† by means of a hypodermic syringe.   The magnetic stirrer (beneath the ice bath) is turned on, and a precooled solution of *p*-nitrobenzoyl azide[5] (105 mg., 0.55 mmol), in 2.5 ml. of chloroform is injected into tube *A*.‡   Within 2 to 3 minutes the yellow suspension becomes a clear yellow solution.   After a total reaction time of 5 minutes, the apparatus is inverted, and the solution is filtered into bulb *B* through the medium-porosity frit in tube *C*.   The filtration is accomplished by applying a slight vacuum to the stopcock in bulb *B*.   Fifty milliliters of cold methanol is immediately injected through the stopcock in bulb *B* to precipi-

---

* The Schlenk-tube apparatus (Fig. 1) is a relatively convenient system to exclude oxygen.   In all steps of the synthesis, air can be kept out of the apparatus by flushing with nitrogen when solvents are injected.

† All solvents used should be degassed by passing a stream of nitrogen through the solvent with a fritted bubbling tube for half an hour and then cooling to 0°C. Reagent-grade chloroform containing ethanol is used.

‡ The checker used 75 mg. of furoyl azide instead of the 105 mg. of *p*-nitro-benzoyl azide.   Reference 3 indicates that furoyl azide is equally active.   With a reaction time of 6 minutes, 0.32 g. of pure yellow nitrogen complex was obtained. The yield was 80%.

Position 1   Position 2

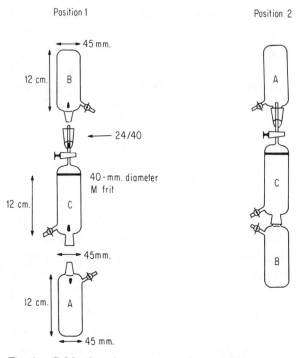

Fig. 1. *Schlenk-tube apparatus for carrying out the azide reactions under an inert atmosphere.*

tate the nitrogen complex.* While nitrogen is passing through the stopcock in bulb *B*, tube *C* is removed, the frit washed, and the apparatus reassembled so that the orientation of tube *C* has been inverted (Fig. 1, position 2) (at least 10 minutes should be allowed to complete the precipitation). After tube *C* and bulb *A* have been thoroughly flushed with nitrogen, the apparatus is again inverted and the nitrogen complex collected on the filter. A 20-ml. portion of cold methanol is introduced through the stopcock in tube *C* to wash the precipitate. Bulb *A* is replaced by a fresh bulb and vacuum applied at the stopcock in the new

---

* If the solution is allowed to stand for a longer period before the nitrogen complex is precipitated, the complex may react with the excess azide, reducing the yield.

bulb A.   The product is dried *in vacuo* on the frit for one hour.*
The apparatus is then disassembled, and the yellow microcrystalline product is removed with a spatula.   (When dry, it is
stable in the atmosphere for a few hours.)   The yield is approximately 0.29 g. (75%).   The purity of the product can be determined by measuring its infrared spectrum.   The carbonyl band
of the starting material at 1970 cm.$^{-1}$ should be absent in the
product.   The complex should be stored in a refrigerator under
nitrogen.   For analysis, see references 2 and 4.

Properties

trans-Chloro(nitrogen)-bis(triphenylphosphine)iridium(I) is a
bright yellow crystalline solid ($\nu_{N\equiv N}$ = 2105 cm.$^{-1}$ in chloroform)
which is moderately stable in air but which takes up oxygen
rapidly in solution.   The crystals decompose at 150–152°C.,
but in chloroform solution the nitrogen complex is unstable
above 25°C. even in the absence of air.   It is soluble in chloroform, dichloromethane, and benzene and is insoluble in ether,
hexane, and methanol.

The nitrogen complex reacts[2] with disubstituted acetylenes,
$(C_6H_5)_3P$ and CO, with evolution of nitrogen, to form four-
coordinate complexes in which nitrogen is replaced by a ligand.
It reacts with diethyl maleate to afford a five-coordinate complex
containing both nitrogen and the olefin.[2,4]

References

1. K. Vrieze, J. P. Collman, C. T. Sears, Jr., and M. Kubota, *Inorganic Syntheses*, **11**, 101 (1968).
2. J. P. Collman and J. W. Kang, *J. Am. Chem. Soc.*, **88**, 3459 (1966).
3. J. P. Collman, M. Kubota, J. Y. Sun, and F. D. Vastine, *ibid.*, **89**, 169 (1967).
4. J. P. Collman, M. Kubota, F. D. Vastine, J. Y. Sun, and J. W. Kang, *ibid.*, **90**, 5430 (1968).
5. P. A. S. Smith, in "Organic Reactions," R. Adams (ed.), Coll. Vol. 3, pp. 366–375, John Wiley & Sons, Inc., New York, 1946.

* In the solid state, the nitrogen complex slowly picks up oxygen from the air. In solution, it is rapidly attacked by oxygen to give a greenish color.

3. *trans*-HYDRIDO(NITROGEN)-TRIS(TRI-PHENYLPHOSPHINE)COBALT(I)

$$Co(acac)_3{}^* + 3P(C_6H_5)_3 + AlR_3 + N_2 \rightarrow$$
$$[CoH(N_2)\{P(C_6H_5)_3\}_3] + \cdots$$

Submitted by AKIRA MISONO†
Checked by KENNETH RUBINSON,‡ HANS BRINTZINGER,‡
V. P. REINSALU,§ and A. D. ALLEN§

The synthesis of nitrogen complexes *from molecular nitrogen* was reported almost simultaneously by three research groups.[1-3] Two of the groups[1,2] used a procedure based on the reaction between tris(2,4-pentanedionato)cobalt(III) (i.e., acetylacetonate), an organoaluminum compound, triphenylphosphine, and nitrogen gas. The formula of the complex was first reported as $[Co(N_2)\{P(C_6H_5)_3\}_3]$.[1,2] The third group used a procedure based on the reaction between trihydrido-tris(triphenylphosphine)-cobalt(III), $[CoH_3\{P(C_6H_5)_3\}_3]$, and elementary nitrogen gas. The formula of the product was reported as $CoH(N_2)\{P(C_6H_5)_3\}_3$. Recent data indicate that the two products are identical; both are correctly formulated as $[CoH(N_2)\{P(C_6H_5)_3\}_3]$. Recent x-ray data of Ibers and his co-workers suggest a distorted trigonal bipyramidal structure with the hydride and nitrogen in axial positions.[4] This model is also consistent with n.m.r. and i.r. data obtained in this laboratory.[5] The complex is of particular interest because it is made by using molecular nitrogen gas, N_2, as a reactant.

Procedure

(Read checkers' Note II at the end of this synthesis before undertaking laboratory work.)

* acac = acetylacetonate(2,4-pentanedione derivative).
† Department of Industrial Chemistry, University of Tokyo, Hongo, Tokyo, Japan.
‡ Department of Chemistry, University of Michigan, Ann Arbor, Mich. 48104.
§ Lash-Miller Chemical Laboratories, University of Toronto, Toronto 5, Ontario, Canada.

A 3.56-g. (0.01-mole) sample of cobalt(III) acetylacetonate,* a 7.86-g. (0.03-mole) portion of triphenylphosphine, and 40 ml. of toluene† are charged into a 250-ml. four-necked flask (standard-taper necks are preferred if available). One neck of the flask is equipped with a nitrogen inlet; the second accommodates a nitrogen-bubbling tube; the third is connected to an efficient condenser; the fourth accommodates a magnetic stirrer. All air is flushed from the system by a nitrogen stream. The mixture in the flask is cooled to a temperature of −30°C. to −40°C., and a 4.6-ml. sample of triisobutylaluminum (Alfa Inorganics) is added by means of a hypodermic syringe. (See Note I by Reinsalu and Allen at the end of this synthesis before carrying out this operation.)

The mixture is stirred with a magnetic stirrer while N_2 gas is bubbled in. The temperature is held at −20°C. for the first hour, then raised to 10°C., and held for an additional 8 to 9 hours‡ with stirring. The mixture becomes reddish-brown, and all the reactants dissolve in toluene.

If the homogeneous product obtained is now allowed to stand undisturbed at a temperature of 0 to 10°C., some of the nitrogen complex precipitates. When a 100-ml. quantity of cold petroleum ether is added to the solution, an additional quantity of the solid orange complex is deposited. The solution is filtered under a nitrogen atmosphere by a Schlenk apparatus, as described in the preceding preparation (synthesis 2). The Schlenk filter is placed in the flask neck occupied by the con-

---

* The compound tris(2,4-pentanedionato)cobalt(III) is sold commercially by Alfa Inorganics as cobalt(III) acetylacetonate, which name will be used through the remainder of this report.

│ Toluene and petroleum ether (30–40°C. fraction) are purified by the usual methods and distilled from sodium wire in an atmosphere of nitrogen. The nitrogen gas is purified by passing it through an activated copper column at a temperature of 170°C.

‡ Checkers Reinsalu and Allen reported that the over-all time element for the reaction is important. If the reaction is stopped in 3–4 hours, a mixture of products is obtained. Free triphenylphosphine and some nitrogen complex are present. If the reaction is allowed to continue for more than 11 hours, extensive decomposition of the N_2 complex occurs.

denser while an N_2 stream is bubbled through the system. An alternative and highly effective vacuum-line filtration procedure has been described by checkers Rubinson and Brintzinger (see Note II at the end). The compound is very air-sensitive; hence air must be carefully avoided in all operations. The precipitate is washed under nitrogen with 20 ml. of petroleum ether. After the washing, the remaining petroleum ether is pulled off *in vacuo*. An 8.0-g. sample of orange complex is obtained (yield is 92%). The crude complex can be recrystallized under N_2 in a Schlenk apparatus, using the following procedure. Pure, dry tetrahydrofuran* is added from a syringe, inserted through a stopcock, until all the precipitate *just dissolves*. If cold petroleum ether is added carefully to this cold solution (through the stopcock), well-shaped crystals will be obtained. *Anal.* Calcd. for $CoH(N_2)\{P(C_6H_5)_3\}_3$ (that is, $C_{54}H_{46}N_2P_3Co$): C, 74.14; H, 5.30; N, 3.20; Co, 6.74. Found: C, 74.07; H, 5.76; N, 3.08; Co, 6.80. The checkers reported: C, 72.38; H, 5.16; N, 1.82.

Properties

The complex forms orange crystals which decompose in a few minutes to a gray-black residue in air. At *ca.* 77°C. *in vacuo* the complex evolves nitrogen gas. When sealed in an atmosphere of nitrogen, the complex retains its orange color and original i.r. spectrum for at least a week. It is soluble in tetrahydrofuran, benzene, toluene, ether, etc., but insoluble in water. A sharp intense band at 2088 cm.$^{-1}$ in the i.r. spectrum is assignable to the N≡N stretching mode of the coordinated nitrogen.† The nitrogen is readily exchanged reversibly by hydrogen, ethylene, and ammonia,[6] and irreversibly by carbon monoxide.[7]

---

* Tetrahydrofuran is purified in the usual manner and distilled from sodium before use. (See synthesis 19, Sec. A, and Appendix I.)

† Rubinson and Brintzinger (in accord with Enemark et al.[4]) reported two lines, one at 2085 cm.$^{-1}$ and the other at 2104 cm.$^{-1}$. They also obtained a sample with one line at 2090 cm^{-1}. This suggests the possibility of two crystalline forms.

Checkers' Special Notes

I. Notes by Reinsalu and Allen

    A. The preparation was run on a $\frac{1}{3}$ scale to facilitate filtering. The procedure was acceptable on this reduced scale. A Dry Ice–acetone bath was used to cool the mixture at the beginning of the reaction; then a cold-water bath was employed.

    B. Ethoxydiethylaluminum(III) was used and recommended by Reinsalu and Allen as a reducing agent instead of the more dangerous aluminum trialkyls. It is readily obtainable from the reaction of triethylaluminum with ethanol.

II. Procedural Notes by Rubinson and Brintzinger. These checkers reported as follows:

It was found that the procedure is carried out most successfully in a chemical vacuum system which allows handling of all solvents and solutions in the complete absence of air. All other procedural details, except for the details of the apparatus, were left unchanged. A standard filtration device for use on a vacuum manifold was employed (see Fig. 2). Yields of unrecrystallized product obtained in this way were in excess of 85%.

Recrystallization can also be carried out in the same system. The procedures follow closely the conventionally employed vacuum-line filtration techniques: The 3.56-g. sample of cobalt-(III) acetylacetonate (0.01 mole) and the 7.86-g. sample of triphenylphosphine (0.03 mole) are placed in flask *A*; then 40 ml. of toluene is distilled into the reactor *in vacuo* from the vacuum manifold by placing flask *A* into a Dry Ice–methanol cooling bath (−78°C.).

The system is filled with nitrogen gas to a slightly positive pressure and, with flask *A* still in the cooling bath at −78°C., a 4.6-ml. sample of triisobutylaluminum (Alfa Inorganics) is introduced with a syringe through opening 4. The nitrogen inlet tube is again placed in this opening, allowing nitrogen gas

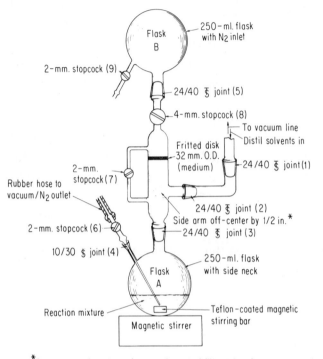

Fig. 2. *Vacuum-line filtration apparatus.*

to bubble slowly through the mixture, escaping through stopcock 9.   The cooling bath is allowed to warm to −20°C. and is kept at this temperature for one hour by continuous addition of small quantities of Dry Ice.   The temperature is then raised to 10°C. and is kept there for an additional 8 to 9 hours.*   The N₂ pressure is held at about 1 atmosphere during this period, with stopcock 9 closed.   The mixture becomes reddish-brown, and all the reactants dissolve in the toluene during the reaction period.

At this time the nitrogen inlet tube in joint 4 is replaced by a stopper; the liquid in the reaction flask is again cooled to −78°C.

* Proper timing is important to good yields.

with Dry Ice–methanol; the system is evacuated, and a 100-ml. sample of petroleum ether is distilled in. Fine orange crystals of the product precipitate. After the system is warmed to room temperature, filtration is accomplished by rotating the entire filtration assembly around the standard-taper joint (2). Nitrogen is then introduced into flask *A*, with stopcock 7 closed; the mixture is thus pushed through the filtering frit. After filtration is complete, stopcock 8 is closed; flask *A* is evacuated again, and a small (*ca.* 20- to 30-ml.) portion of fresh petroleum ether is distilled into *A* by cooling its walls with a swab soaked with cold methanol from a Dry Ice–methanol bath. When this operation is done properly, the condensing petroleum ether washes the product from the flask onto the filter disk. The solvent is then forced through the frit by again introducing nitrogen to flask *A* and opening stopcock 8. This procedure is repeated two to three times. After several washings, most of the product appears on the frit, and the washings run through clear. The product is dried *in vacuo* on the frit. Then, with a positive nitrogen pressure introduced through joint 1 from the line, flasks *A* and *B* are removed, cleaned, filled with nitrogen, and put back in their places on the filtering assembly (flask *A* with the nitrogen inlet tube in opening 4 again). Before flask *A* is replaced, the filter cake, always protected by a stream of nitrogen through opening 1, is carefully loosened from the frit and broken up with a long spatula. After the flasks are replaced, all traces of air which might have been introduced are removed by completely evacuating the assembly through stopcocks 6 and 9 and then filling the system with fresh nitrogen from the line through opening 1. After the assembly is tipped back into its original position (flask *A* down), the product is dumped into flask *A* by tapping the filter tube lightly. The product can be stored by removing flask *A*, which is protected by a stream of nitrogen through inlet tube 6. After removal, the flask is closed rapidly with a standard-taper stopper. Yields are in excess of 85%.

Alternatively, the product collected in flask *A* can be recrystallized without again removing flask *A* from the assembly. Tetrahydrofuran is distilled in through opening 1 until the product just dissolves. If the filter tube is cooled with a swab, the condensing tetrahydrofuran washes most of the remaining product from the tube into flask *A*. A small quantity of petroleum ether (*ca.* 20–30 ml.) is then distilled in to initiate crystallization of the product, which is filtered off and washed in the same manner as above.

References

1. A. Yamamoto, S. Kitazume, L. S. Pu, and S. Ikeda, *Chem. Commun.*, 79 (1967).
2. A. Misono, Y. Uchida, and T. Saito, *Bull. Chem. Soc., Japan*, **40**, 700 (1967); *Chem. Commun.*, 419 (1967).
3. A. Sacco and M. Rossi, *Chem. Commun.*, 316 (1967); *Inorg. Chim. Acta*, **2**, 127 (1968).
4. J. H. Enemark, B. R. Davis, J. A. McGinnety, and J. A. Ibers, *Chem. Commun.*, 96 (1968).
5. A. Misono, Y. Uchida, M. Hidai, and M. Araki, *ibid.*, 1044 (1968).
6. A. Yamamoto, L. S. Pu, S. Kitazume, and S. Ikeda, *J. Am. Chem. Soc.*, **89**, 3071 (1967).
7. A. Misono, Y. Uchida, M. Hidai, and T. Kuse, *Chem. Commun.*, 981 (1968).

4. TRIHYDRIDO-TRIS(TRIPHENYLPHOSPHINE)-COBALT(III) AND HYDRIDO(NITROGEN)-TRIS-(TRIPHENYLPHOSPHINE)COBALT(I)

$$CoCl_2 + P(C_6H_5)_3 \xrightarrow[\text{EtOH}]{\text{NaBH}_4} [CoH_3\{P(C_6H_5)_3\}_3] \xrightarrow{N_2}$$

$$[CoH(N_2)\{P(C_6H_5)_3\}_3]$$

Submitted by A. SACCO* and M. ROSSI*
Checked by FRANK BOTTOMLEY† and A. D. ALLEN†

Molecular nitrogen complexes of cobalt have been reported by three research groups.[1–4] Preparation of the compound—now

* Istituto di Chimica Generale e Inorganica, Universita di Bari, Italy.

† Lash-Miller Chemical Laboratories, University of Toronto, Toronto 5, Ontario, Canada.

known to be $[CoH(N_2)\{P(C_6H_5)_3\}_3]$—by the reaction between cobalt(III) acetylacetonate, triphenylphosphine, an organo-aluminum compound, and atmospheric N_2 has been described in the preceding synthesis. The present synthesis describes the preparation of this same compound, hydrido(nitrogen)-tris(tri-phenylphosphine)cobalt(I), by the reaction between trihydrido-tris(triphenylphosphine)cobalt(III) and elementary N_2. This reaction represents the first example of a well-characterized compound which is able to react with molecular nitrogen in either solid or solution phases.

A. TRIHYDRIDO-TRIS(TRIPHENYLPHOSPHINE)COBALT(III)

$$CoCl_2 + 3P(C_6H_5)_3 + 2NaBH_4 + 6C_2H_5OH \rightarrow$$
$$[CoH_3\{P(C_6H_5)_3\}_3] + H_2 + \cdots \text{ or } + \text{ other products}$$

Trihydrido-tris(triphenylphosphine)cobalt(III) has been obtained thus far only by the following procedure.[1] Since the compound is very sensitive to air, it must be prepared and handled in an atmosphere of hydrogen or argon by standard vacuum-line techniques. The vacuum filtration apparatus, shown in Fig. 3, is used in conjunction with the vacuum line. It is important that an excess of sodium tetrahydroborate be used; otherwise some chloro-tris(triphenylphosphine)cobalt(I)[2] will be obtained also.

Procedure

A 250-ml. flask (*A*), containing a magnetic stirring bar, is charged in a countercurrent of hydrogen or argon with a mixture of 0.450 g. (11.8 mmoles) of sodium tetrahydroborate (NaBH₄) in 15 ml. of 95% ethanol.* Cobalt(II) chloride hexahydrate (0.42 g., 1.76 mmoles) is dissolved in 20 ml. of boiling 95% ethanol and added to a solution made by dissolving triphenyl-phosphine (1.50 g., 5.7 mmoles) in 40-ml. of hot ethanol.

---

* All solvents used in this preparation were deoxygenated by bubbling pure N_2 slowly through the system for about 2 hours.

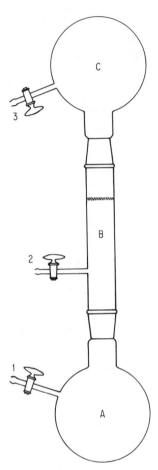

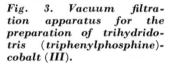

*Fig. 3. Vacuum filtra-
tion apparatus for the
preparation of trihydrido-
tris (triphenylphosphine)-
cobalt (III).*

After cooling to about 50°C., this solution is added, in a
countercurrent of inert gas and under stirring, to the flask
containing the sodium tetrahydroborate. The NaBH₄ solution
is maintained at about 0°C. After the addition is completed
(about 2 minutes), the flask is closed* and the stirring is continued for one hour at 0°C. The stirring is then discontinued,
and the reaction mixture is left one hour at 0°C.

---

* The H_2 evolved is allowed to escape through stopcock 1, which is connected
to the vacuum manifold. The checker noted that a brown precipitate formed on
addition of NaBH₄. This slowly turned yellow-green.

The vacuum filtration apparatus (*B* and *C* of Fig. 3), previously filled with inert gas,* is connected to flask *A* and the apparatus is rotated, so that the reaction mixture flows onto the filter plate. Receiver flask *C* is partially evacuated through stopcock 3. After the filtration is finished, the precipitate is washed three times with ethanol, three times with deoxygenated water, six times with ethanol, and three times with *n*-pentane, using 15 ml. of solvent, added through stopcock 2 each time. Receiver flask *C* is now removed, and an empty flask, previously flushed with the inert gas, is substituted. After drying *in vacuo*, 1.1–1.4 g. of crude product is obtained (70–90%).†

Properties

The yellow compound, which decomposes at 80°C., releasing hydrogen, is soluble in benzene, toluene, and tetrahydrofuran. It reacts with a wide variety of inorganic and organic compounds and with many simple molecules such as oxygen, nitrogen, and carbon monoxide.

The infrared spectrum, run in Nujol mull, shows two Co—H stretching bands at 1933(m) and at 1745(s) cm.$^{-1}$.

B. HYDRIDO(NITROGEN)-TRIS(TRIPHENYLPHOSPHINE)-COBALT(I)

The crude trihydrido-tris(triphenylphosphine)cobalt(III) prepared above will pick up nitrogen directly to form the nitrogen complex.

---

* Stopcocks 2 and 3 are connected to the manifold.

† The authors recommended the following procedure for recrystallizing the product, but the checkers were unsuccessful in their attempts at recrystallization. *The crude product works well, however, in the synthesis of the nitrogen complex described in the next section.*

"The product is recrystallized in the same apparatus by dissolving in 20 ml. of pure benzene, filtering, concentrating under vacuum to about 8 ml., and adding ethanol or light petroleum ether. After filtering, washing with ethanol and drying *in vacuo*, 0.9–1.0 g. (50–60% of over-all yield) of pure, yellow, dry product is obtained."

Procedure

An impure sample of the hydrido(nitrogen)-tris(triphenyl-phosphine)cobalt(I) can be obtained directly by carrying out the above procedure in an atmosphere of nitrogen (50–60% yield). The pure product can be obtained as described below.

A 2.8-g. sample of the crude $CoH_3[P(C_6H_5)_3]_3$ is dissolved under nitrogen in 40 ml. of toluene; the solution is filtered, concentrated under vacuum to about 15 ml., and mixed with ethanol or light petroleum ether. The product precipitates. After filtering, washing with ethanol, and drying *in vacuo*, 1.3–1.4 g. (42–45% of over-all yield) of pure orange-red product is obtained.

Properties

The orange-red compound, which decomposes at 80°C., releasing hydrogen and nitrogen, is soluble in benzene, toluene, tetrahydrofuran, and diethyl ether. In the crystalline form it can be handled in the air for a very short time without decomposition. Both in solution and in the solid state, it reacts with hydrogen, giving back the trihydrido complex. The infrared spectrum, run in Nujol mull, shows one $N{\equiv}N$ stretching band at 2096 cm.$^{-1}$ when the compound is recrystallized from benzene, and two $N{\equiv}N$ stretching bands at 2102 cm.$^{-1}$ and at 2087 cm.$^{-1}$ when recrystallized from tetrahydrofuran.

References

1. A. Sacco and M. Rossi, *Chem. Commun.*, 316 (1967); *Inorg. Chim. Acta*, **2**, 127 (1968).
2. A. Sacco and M. Rossi, "Proceedings of the 10th International Conference on Coordination Chemistry," p. 125, The Chemical Society of Japan, Tokyo, 1967.
3. A. Yamamoto, S. Kitazume, L. S. Pu, and S. Ikeda, *Chem. Commun.*, 79 (1967).
4. A. Misono, Y. Uchida, and T. Saito, *ibid.*, 419 (1967); *Bull. Chem. Soc. Japan*, **40**, 700 (1967).
5. B. R. Davis, N. C. Payne, and J. A. Ibers, *J. Am. Chem. Soc.*, **91**, 1240 (1969); J. H. Enemark, B. R. Davis, J. A. McGinnety, and J. A. Ibers, *Chem. Commun.*, 96 (1968).

5. AZIDO(NITROGEN)-BIS(ETHYLENEDIAMINE)-RUTHENIUM(II) HEXAFLUOROPHOSPHATE

cis-[Ru(en)$_2$*Cl$_2$]Cl·H$_2$O + 3Ag(tps)† →
$$cis\text{-[Ru(en)}_2(\text{H}_2\text{O})_2]\text{(tps)}_3† + 3\text{AgCl}$$
cis-[Ru(en)$_2$(H$_2$O)$_2$](tps)$_3$ + 2NaN$_3$ + NaPF$_6$ →
$$cis\text{-[Ru(en)}_2(\text{N}_3)_2]\text{PF}_6 + 3\text{Natps} + 2\text{H}_2\text{O}$$
$$cis\text{-[Ru(en)}_2(\text{N}_3)_2]\text{PF}_6 \xrightarrow{\Delta} cis\text{-[Ru(en)}_2(\text{N}_2)(\text{N}_3)]\text{PF}_6 + \tfrac{1}{2}\text{N}_2$$

Submitted by LEON A. P. KANE-MAGUIRE,‡ PETER S. SHERIDAN,‡
and FRED BASOLO‡
Checked by FRANK BOTTOMLEY§ and A. D. ALLEN§

Recent studies[1,2] have shown the azido-pentaammineruthenium(III) species to be very unstable toward decomposition to pentaammine(nitrogen)ruthenium(II). Unlike [Ru(NH$_3$)$_5$-(N$_3$)]$^{2+}$, the cis-[Ru(en)$_2$(N$_3$)$_2$]$^+$ cation is moderately stable at room temperature, and its isolation is described below. Heating cis-[Ru(en)$_2$(N$_3$)$_2$]PF$_6$ in the solid state provides a rapid and efficient preparation of cis-[Ru(en)$_2$(N$_2$)(N$_3$)]PF$_6$. ■ *Caution. All azides are potentially explosive and should be handled with care.*

Procedure

cis-[Ru(en)$_2$Cl$_2$]Cl·H$_2$O [3,4] (0.200 g., 0.58 mmole) is mixed with silver toluene-p-sulfonate (0.483 g., 1.74 mmoles) and 5 ml. of water is added. The suspension is stirred near the boiling point until no further silver chloride precipitates, leaving a clear, red supernatant liquid. This generally requires 5 to 6 minutes of heating. The silver chloride precipitate is then separated by filtration. The precipitate is washed with water (1 ml.) and the washings combined with the filtrate. The filtrate contains

* en = ethylenediamine.
† (tps) = toluene-p-sulfonate.
‡ Department of Chemistry, Northwestern University, Evanston, Ill. 60201.
§ Lash-Miller Chemical Laboratories, University of Toronto, Toronto 5, Ontario, Canada.

cis-[Ru(en)$_2$(H$_2$O)$_2$]$^{3+}$.   The red color is thought to be due to a small amount of an intensely colored impurity, since *cis*-[Ru(en)$_2$(H$_2$O)$_2$]$^{3+}$ is almost colorless.   The red solution is cooled and added to solid sodium azide (0.72 g., 11.0 mmoles) and the mixture is stirred in a beaker in an ice bath.   ■ *Caution.* *If a precipitate settles out at this stage, it is probably silver azide from an excess of silver toluene-p-sulfonate, and it should be removed immediately by filtration.   Silver azide is explosive.   It can be destroyed after removal by treatment with dilute bromine water.   This gives silver bromide and gaseous nitrogen.   It will also decompose if allowed to stand in the light.   Solid silver metal and gaseous nitrogen are products.*   Slow evolution of gas is observed.   After 5 minutes, an ice-cold filtered aqueous solution of sodium hexafluorophosphate (saturated, 6 ml.) is added. [The sodium salt (NaPF$_6$) must be used because other PF$_6^-$ salts have too low a solubility.]   Deep purple-red microcrystals of *cis*-[Ru(en)$_2$(N$_3$)$_2$]PF$_6$ slowly form.   Small explosions were observed when [Ru(en)$_2$(N$_3$)$_2$]PF$_6$ crystals were scratched with a metal spatula.   The product is collected by filtration after 30 minutes and washed twice with small volumes of ice-cold ethanol (yield is 0.095 g.).   A further crop of crystals is obtained from the filtrate after an additional 40 minutes at ice-bath temperatures.   Total yield is 0.14 g. (54%).   *Anal.* Calcd. for C$_4$H$_{16}$N$_{10}$PF$_6$Ru: C, 10.7; H, 3.6; N, 31.1.   Found: C, 10.7; H, 3.6; N, 30.0.

The volumes employed above should be followed closely for satisfactory results.   If the volumes are reduced, a white impurity separates with the product, whereas the use of larger volumes results in a sharp decrease in the product yield.

Solid *cis*-[Ru(en)$_2$(N$_3$)$_2$]PF$_6$ (0.10 g., 0.22 mmole) is placed in the bottom of a test tube, which is partially immersed in a water bath at 65°C. and maintained at this temperature for 25 minutes. This completes the redox process, producing almost pure *cis*-[Ru(en)$_2$(N$_2$)(N$_3$)]PF$_6$ in quantitative yield.   *Anal.* Calcd. for C$_4$H$_{16}$N$_9$PF$_6$Ru: C, 11.0; H, 3.6; N, 28.8.   Found: C, 11.0;

H, 3.6; N, 27.4. The heating time should not be extended beyond 30 minutes. Samples heated for longer periods experience a further slow loss of molecular nitrogen, and their i.r. spectra indicate the formation of a nitrosyl ruthenium species.

On a microscale (0.010 g.), the conversion of *cis*-[Ru(en)$_2$-(N$_3$)$_2$]PF$_6$ to *cis*-[Ru(en)$_2$(N$_2$)(N$_3$)]PF$_6$ may be performed conveniently on a Faraday magnetic balance. The sample is kept at 65°C. by passing a stream of hot air through a jacket surrounding the sample pan. Heating is stopped as soon as the sample becomes diamagnetic (about 30 minutes). The weight loss at zero magnetic field during the reaction corresponds closely ($\pm 5\%$) to the calculated weight loss for the release of one atom of nitrogen per molecule of original *cis*-[Ru(en)$_2$(N$_3$)$_2$]PF$_6$. Mass spectral analysis of the gas released shows it to be greater than 99% nitrogen.

Properties

Freshly prepared *cis*-[Ru(en)$_2$(N$_3$)$_2$]PF$_6$ is paramagnetic ($\mu_{\text{eff.}} = 2.0$ B.M.). In the air at room temperature the purple-red crystals decompose over a few days to a dark brown mixture of *cis*-[Ru(en)$_2$(N$_2$)(N$_3$)]PF$_6$ and unknown nitrosyl ruthenium species. It may be stored in the dark at -15°C. for several weeks without significant change. Its i.r. spectrum shows a characteristic coordinated azide-ion band at 2050 cm.$^{-1}$. Most samples also show a sharp weak peak at 2130 cm.$^{-1}$, which is due to the presence of small amounts of *cis*-[Ru(en)$_2$(N$_2$)(N$_3$)]PF$_6$. *cis*-[Ru(en)$_2$(N$_3$)$_2$]PF$_6$ is readily soluble in water and dimethyl sulfoxide, in which it is initially a typical 1:1 electrolyte. Subsequent complex chemical changes occur in these solvents. In acetone, methanol, and acetonitrile, however, *cis*-[Ru(en)$_2$(N$_3$)$_2$]$^+$ converts cleanly and quantitatively to *cis*-[Ru(en)$_2$(N$_2$)(N$_3$)]$^+$.

cis-[Ru(en)$_2$(N$_2$)(N$_3$)]PF$_6$ is a brown diamagnetic solid. It loses molecular nitrogen very slowly at room temperature, and it may be stored at -15°C. in the dark for long periods without change.   It is readily characterized by its i.r. spectrum, which contains a strong, sharp band at 2130 cm.$^{-1}$, attributed to the coordinated molecular nitrogen, and a strong band at 2050 cm.$^{-1}$ associated with the coordinated azide group.   It is readily soluble in water and dimethyl sulfoxide, in which it is initially a typical 1:1 electrolyte.   The compound undergoes further reaction (e.g., ligand substitution) in these solvents.   Solutions in acetone and methanol are relatively stable for a few hours.

Reaction of both solid cis-[Ru(en)$_2$(N$_3$)$_2$]PF$_6$ and cis-[Ru(en)$_2$-(N$_2$)(N$_3$)]PF$_6$ with a sulfuric acid solution saturated with cerium-(IV) ion releases both coordinated azide and coordinated nitrogen as nitrogen gas.   cis-[Ru(en)$_2$(N$_3$)$_2$]PF$_6$ releases 3.0 moles of gas, and cis-[Ru(en)$_2$(N$_2$)(N$_3$)]PF$_6$ releases 2.5 moles of nitrogen gas for each mole of original complex.

References

1. A. D. Allen, F. Bottomley, R. O. Harris, V. P. Reinsalu, and C. V. Senoff, *J. Am. Chem. Soc.*, **89**, 5595 (1967).
2. H. Tobias, private communication.
3. J. A. Broomhead and L. A. P. Kane-Maguire, *J. Chem. Soc.*, A, 546 (1967).
4. J. K. Witschy and J. K. Beattie, *Inorg. and Nucl. Chem. Letters*, **5**, 969 (1969).

6. AZO, DIIMIDE, AND HYDRAZINE COMPLEXES OF PLATINUM

$$[PtCl_2\{P(C_2H_5)_3\}_2] \xrightarrow[H_2O]{N_2H_4} [PtHCl\{P(C_2H_5)_3\}_2] \xrightarrow{[p\text{-}FC_6H_4N_2]BF_4}$$
$$[(p\text{-}FC_6H_4N{=}NH)PtCl\{P(C_2H_5)_3\}_2]BF_4$$

Submitted by GEORGE W. PARSHALL*
Checked by M. D. CURTIS† and R. C. JOB†

The reactions of arenediazonium salts with *trans*-chlorohydrido-bis(triethylphosphine)platinum(II)[1] give aryldiimide complexes which are precursors to new arylazoplatinum compounds[2] and

* Central Research Department, Experimental Station, E. I. du Pont de Nemours & Company, Wilmington, Del. 19898.
† Department of Chemistry, University of Michigan, Ann Arbor, Mich. 48104.

arylhydrazine complexes.[3] These compounds have been useful models for the chemistry of the nitrogenase enzyme[3,4] and have also provided a new route to arylplatinum compounds.[2] The preparation of this family of compounds is illustrated below for the *p*-fluorophenyl derivatives which are, in general, the most stable and tractable.

The syntheses of *cis*-[PtCl$_2${P(C$_2$H$_5$)$_3$}$_2$] and *trans*-[PtHCl{P(C$_2$H$_5$)$_3$}$_2$] are included because these compounds are widely used as chemical intermediates and because the purity of these starting materials is critical to the success of later reactions. The reaction of triethylphosphine with K$_2$PtCl$_4$ is like that of tributylphosphine,[5] but the procedure for purification of the product is somewhat different. The synthesis of [PtHCl{P(C$_2$H$_5$)$_3$}] by reduction of *cis*-[PtCl$_2${P(C$_2$H$_5$)$_3$}$_2$] with hydrazine hydrate is particularly convenient, although Chatt and Shaw, who developed this procedure, also evaluated many other reducing agents.[1]

A. *cis*-DICHLORO-BIS(TRIETHYLPHOSPHINE)PLATINUM(II)

$$2(C_2H_5)_3P + K_2PtCl_4 \rightarrow [PtCl_2\{P(C_2H_5)_3\}_2] + 2KCl$$

A solution of 20.8 g. (55 mmoles) of potassium tetrachloroplatinate(II) in 100 ml. of water is prepared in a 250-ml. suction flask with a nitrogen flow through the sidearm to provide an inert atmosphere. (The procedures work equally well at half-scale.) Triethylphosphine* (15 ml., 100 mmoles) is added all at once and the mixture is stirred with a magnetic stirrer at room temperature for one hour. A pink-tan precipitate of tetrakis-(triethylphosphine)platinum(II)tetrachloroplatinate(II) forms. The mixture is heated on a steam bath for one hour. The pink salt dissolves, and a layer of pale yellow dichloro-bis(triethylphosphine)platinum forms on the liquid. The solid is filtered, washed with water, crushed in a mortar, and dried under

---

* Strem Chemicals, Inc., 150 Andover Street, Danvers, Mass. 01923, or Orgmet, Inc., Hampstead, N.H. 03841.

vacuum. The dry solid is suspended in 150 ml. of pentane containing 2 drops of triethylphosphine and is stirred until the color changes from yellow to pure white (about 10 minutes) as a result of the isomerization of the trans isomer to the less soluble cis form. The mixture is filtered, and the white solid is dried under vacuum to remove triethylphosphine. Recrystallization from acetonitrile (about 130 ml.) gives large white crystals of *cis*-dichloro-bis(triethylphosphine)platinum(II); m.p. 195–197°C.;* yield 17–23 g., 68–91%. The checkers reported a yield of 85% crude product.

Properties

The cis isomer of dichloro-bis(triethylphosphine)platinum(II) is a white crystalline solid with low solubility in benzene, ethanol, and acetone. It may be isomerized to an equilibrium mixture with the yellow trans isomer by heating to 180°C. or by exposure to sunlight in benzene solution.[6] The trans isomer (m.p. 140–142°C.) is readily soluble in nonpolar organic solvents and can easily be separated from the equilibrium mixture by extraction with ether.

B. *trans*-CHLOROHYDRIDO-BIS(TRIETHYLPHOSPHINE)-PLATINUM(II)

$$cis\text{-}[PtCl_2\{P(C_2H_5)_3\}_2] \xrightarrow{N_2H_4} trans\text{-}[PtHCl\{P(C_2H_5)_3\}_2]$$

It is desirable that pure *cis*-dichloro complex be used in this preparation because the trans isomer is not reactive under these conditions and is difficult to separate from the hydride.

Procedure

A 500-ml. suction flask is charged with 20 g. of *cis*-dichloro-bis(triethylphosphine)platinum, 10 ml. of hydrazine hydrate,

---

* The checkers noted that the melting point depends strongly on rate of heating, since the equilibrium indicated below occurs at 180°C.:

$$cis\text{-}[PtCl_2\{P(C_2H_5)_3\}_2] \rightleftharpoons trans\text{-}[PtCl_2\{P(C_2H_5)_3\}_2]$$
$$\text{m.p. } 195\text{–}197°C. \qquad\qquad \text{m.p. } 140\text{–}142°C.$$

and 200 ml. of water. A slow nitrogen stream is passed through the sidearm of the flask, and the mixture is heated on a steam bath for one hour. The solid compound dissolves, and an oil separates from the mixture. When the flask is cooled, the oil crystallizes and is collected by filtration. The filtrate may be neutralized with dilute hydrochloric acid to give some additional solid product.

The two solid products are recrystallized from a small quantity of methanol to give fine white crystals of *trans*-chlorohydrido-bis(triethylphosphine)platinum. Some additional material can be recovered by adding water to the methanol solution. The product is dried under vacuum and recrystallized from hexane, again giving white needles: m.p. 80–81°C.; yield 15 g., 81%. The checkers reported m.p. 82–83.5°C., yield 73%.

Properties

trans-Chlorohydrido-bis(triethylphosphine)platinum(II) is moderately air-stable, both in the solid state and in solution. It is very soluble in most organic solvents but can be conveniently recrystallized as described in the preparation. The platinum-hydrogen bond appears in the infrared spectrum as a sharp band of medium intensity at 2183 cm.$^{-1}$. In the proton n.m.r. spectrum in benzene solution, the Pt—H signal appears as a 1:4:1 triplet (J_{PtH} = 1276 Hz.) centered at $\tau 26.9$. The chloride ligand is easily displaced by nucleophilic anions.

C. *trans*-CHLORO-(*p*-FLUOROPHENYLDIIMIDE)-BIS(TRIETHYLPHOSPHINE)PLATINUM(II) TETRAFLUOROBORATE

$$[p\text{-}FC_6H_4N_2]BF_4 + trans\text{-}[PtHCl\{P(C_2H_5)_3\}_2] \rightarrow$$
$$[(p\text{-}FC_6H_4N{=}NH)PtCl\{P(C_2H_5)_3\}_2]BF_4$$

In this preparation it is desirable to use freshly prepared diazonium salt and thoroughly purified platinum hydride. The aryldiimide complex is very sensitive to nucleophilic reagents and should not be allowed to remain in contact with solvents any longer than is necessary.

Procedure

A solution of 1.92 ml. of freshly distilled p-fluoroaniline (b.p. 187–189°C.) and 10 ml. of 48% tetrafluoroboric acid is stirred at 0°C. in a 50-ml. beaker while a solution of 1.68 g. of sodium nitrite and 3 ml. of water is added over a period of 15 minutes. p-Fluorobenzenediazonium tetrafluoroborate precipitates as white crystals. The mixture is filtered, and the crystals are washed with 3 ml. of cold concentrated fluoroboric acid, 3 ml. of cold ethanol, and 30 ml. of ether. Yield is 3.8 g.; $\nu_{N \equiv N} = 2274$ cm.$^{-1}$.

A suspension of 2.10 g. of the freshly prepared diazonium salt and 50 ml. of ethanol is stirred in a 200-ml. Erlenmeyer flask and cooled to 0°C. while a solution of 4.7 g. of *trans*-chloro-hydrido-bis(triethylphosphine)platinum(II) in 25 ml. of ethanol is added over a period of 10 minutes. The mixture becomes yellow immediately and is stirred at 0°C. for 10 minutes. At this point, yellow crystals of the aryldiimide complex usually separate. The mixture is filtered, and the filtrate is cooled to $-78°$C. to give a second crop of the diimide complex. Occasionally crystallization does not occur immediately. In this situation, the mixture is cooled to $-78°$C. directly, and the crystals which form are filtered and washed with ether. In either case the total yield is about 4.9–5.4 g. (72–78%). (The checkers reported an 87% yield with a melting point of 105–108°C. with decomposition.) The purity of this material is adequate for most preparatory purposes. Further purification may be accomplished by conversion to the arylazoplatinum compound (Sec. D) followed by reprotonation with concentrated aqueous fluoroboric acid, HBF$_4$.

Properties

The p-fluorophenyldiimide complex is moderately stable at room temperature when pure but is best stored at $-78°C$. It melts with decomposition to a red liquid at 105–106°C. The NH proton is moderately acidic (pK_a 4.5)[3] and exchanges rapidly with D_2O and D_2.[4] The complex is readily soluble in alcohols, ketones, ethers, and chloroform and even has some solubility in benzene.

The analogous complexes in which the aromatic group is phenyl, m-fluorophenyl, and p-nitrophenyl are prepared in the same way and have similar stabilities.

D. *trans*-CHLORO-(p-FLUOROPHENYLAZO)-BIS(TRIETHYLPHOSPHINE)PLATINUM(II)

$$[(p\text{-}FC_6H_4N{=}NH)PtCl\{P(C_2H_5)_3\}_2]^+ + CH_3CO_2^- \rightarrow$$
$$[(p\text{-}FC_6H_4N{=}N)PtCl\{P(C_2H_5)_3\}_2] + CH_3CO_2H$$

This product, like the diimide complex from which it is prepared, is very sensitive to nucleophilic reagents and should not be left in contact with solvents unnecessarily.

Procedure

A solution of 1.38 g. of *trans*-chloro-(p-fluorophenyldiimide)-bis(triethylphosphine)platinum(II) tetrafluoroborate in 10 ml. of methanol is stirred and cooled to 0°C. in a 50-ml. Erlenmeyer flask while a solution of 0.27 g. of sodium acetate trihydrate in 5 ml. of 50% aqueous methanol is added over a period of 1 minute. The mixture becomes red immediately, and a violet precipitate forms. The mixture is stirred for 5 minutes at 0°C. and then is filtered *promptly*. After drying under vacuum to remove methanol, recrystallization from hexane (about 30 ml.) gives rose-colored needles (m.p. 116–117°C.) of *trans*-chloro-

(*p*-fluorophenylazo)-bis(triethylphosphine)platinum(II).   Yield is 0.52–0.75 g. (44–63%).

Properties[2]

The arylazoplatinum complexes are colored, air-stable solids with good solubility and moderate stability in hydrocarbon solvents.   Treatment of ethereal solutions with aqueous fluoroboric acid regenerates the yellow aryldiimide complexes in high yield and purity.[4]   The phenylazo and *m*-fluorophenylazo complexes are violet solids that melt at 109.5–110.5°C. and 38°C., respectively.   The *p*-nitrophenylazo compound is green and melts at 92–94°C. with gas evolution.   The arylazo compounds decompose with loss of nitrogen to give the corresponding *trans*-arylchloro-bis(triethylphosphine)platinum(II)   complexes in variable yields on treatment with active alumina or with copper salts.   The N=N stretching frequencies[3] fall in the range 1545–1565 cm.$^{-1}$.

E. *trans*-CHLORO-(*p*-FLUOROPHENYLHYDRAZINE)-BIS(TRIETHYLPHOSPHINE)PLATINUM(II) TETRAFLUOROBORATE

$$[p\text{-}FC_6H_4N_2]BF_4 + [PtHCl\{P(C_2H_5)_3\}_2] + H_2 \rightarrow$$
$$[(p\text{-}FC_6H_4NHNH_2)PtCl\{P(C_2H_5)_3\}_2]BF_4$$

Procedure

A solution of 4.7 g. of *trans*-chlorohydrido-bis(triethylphosphine)platinum(II) in 25 ml. of ethanol is added to a stirred, chilled suspension of 2.1 g. of *p*-fluorobenzenediazonium tetrafluoroborate in 50 ml. of ethanol, as in Sec. C.   The mixture is then allowed to warm to room temperature over a period of 20 minutes.   A 0.10-g. portion of 5% platinum-on-carbon catalyst is added, and hydrogen is bubbled through the mixture for 2 hours.   The mixture is cooled to 0°C. and is filtered to collect the white crystalline precipitate (5.9 g.). The precipitate is dissolved in a minimum volume of methanol and filtered to remove the residual catalyst. On cooling, fine white needles of trans-chloro-(*p*-fluorophenylhydrazine)-bis(triethylphosphine)platinum tetrafluoroborate separate; yield is 3.4 g. (50%), m.p. 180–183°C., with decomposition to a red liquid. Subsequent recrystallization for analysis may be done in a 1:2 methanol–ethanol mixture. *Anal.* Calcd. for $C_{18}H_{37}BClF_5N_2P_2Pt$: C, 31.79; H, 5.49; N, 4.12; Pt, 28.7. Found: C, 31.65; H, 5.70; N, 4.29; Pt, 28.8. The checkers reported a 35% yield with a melting point of 182–185°C. with decomposition.

Properties

The *p*-fluorophenylhydrazine complex is stable to air and moisture, but the hydrazine ligand is easily displaced by nucleophiles such as chloride ion. The N-bonded protons readily exchange with D_2O and with D_2 but are much less acidic than that of the *p*-fluorophenyldiimide complex.[4] Prolonged hydrogenation cleaves the N—Pt bond to give the arylhydrazine and $[PtHCl\{P(C_2H_5)_3\}_2]$.

References

1. J. Chatt and B. L. Shaw, *J. Chem. Soc.*, 5075 (1962).
2. G. W. Parshall, *J. Am. Chem. Soc.*, **87**, 2133 (1965).
3. G. W. Parshall, *ibid.*, **89**, 1822 (1967).
4. E. K. Jackson, G. W. Parshall, and R. W. F. Hardy, *J. Biol. Chem.*, **243**, 4952 (1968).
5. G. B. Kauffmann and L. A. Teter, *Inorganic Syntheses*, **7**, 245 (1963).
6. P. Haake and T. A. Hylton, *J. Am. Chem. Soc.*, **84**, 3774 (1962).

Chapter Two

ORGANOMETALLIC COMPOUNDS

I. GENERAL ORGANOMETALLICS

7. SUBSTITUTED IRON DICARBONYL CATIONS

$(C_5H_5)_2Fe(CO)_2X + L \rightarrow [(C_5H_5)_2Fe(CO)_2L]^+ + X^-$
(L = electron-donor molecule; X^- = halide ion)

Submitted by E. O. FISCHER* and E. MOSER*
Checked by P. M. TREICHEL† and J. P. STENSON†

In recent years more and more emphasis has been laid on the chemistry of cyclopentadienyl metal carbonyls. In this field the reactions of cyclopentadienyl metal carbonyl halides with electron donors deserve special attention. Many of these reactions yield cations by replacement of halogen. In the following report, general procedures will be described by which substituted cyclopentadienyliron dicarbonyl cations of the type $[C_5H_5Fe(CO)_2L]^+$ can be prepared.

The starting materials, the cyclopentadienyliron dicarbonyl

* Anorganisch-Chemisches Laboratorium der Technischen Hochschule München, 8 München, West Germany.
† Department of Chemistry, University of Wisconsin, Madison, Wis. 53706.

halides, are prepared from commercially available bis(cyclopentadienyliron dicarbonyl).*

Oxidation of $[C_5H_5Fe(CO)_2]_2$ with air in the presence of hydrochloric acid yields $C_5H_5Fe(CO)_2Cl$.[2,3]† Cyclopentadienyliron dicarbonyl bromide [dicarbonylcyclopentadienyliron bromide] may be obtained by the same procedure,[2] but it is more conveniently prepared by cleavage of the iron-iron bond with bromine.[1,3] The reaction of $[C_5H_5Fe(CO)_2]_2$ with iodine yields cyclopentadienyliron dicarbonyl iodide[5,6] [*Inorganic Syntheses*, 7, 110 (1963)] which also has been prepared by exchange with iodide ion from $C_5H_5Fe(CO)_2Cl$[2,3] and from $Fe(CO)_4I_2$ by treating iron pentacarbonyl with iodine and subsequently with the sodium derivative of cyclopentadiene.[7] Whereas the cyclopentadienyliron dicarbonyl chloride seems to be the most reactive compound in halogen substitution reactions, the iodide is the most easily accessible of these materials.

Some General Comments on Preparations

Most cyclopentadienyl metal carbonyl compounds are to some extent sensitive to air. The compounds described herein can be exposed to air for a short time; however, it is recommended that they should be handled in an inert-gas atmosphere in order to obtain pure samples. Even if there is no decomposi-

---

* Bis(cyclopentadienyliron dicarbonyl) is available commercially from several sources, for example, Strem Chemicals, Inc., 150 Anders St., Danvers, Mass. 01923. If the commercial source is not used, $[C_5H_5Fe(CO)_2]_2$ can be prepared by treatment of $Fe(CO)_5$ with cyclopentadiene in an autoclave or, better, with dicyclopentadiene in an open flask. See *Inorganic Syntheses*, 7, 110 (1963).[1,3,5]

† The following procedure for the synthesis of $C_5H_5Fe(CO)_2Cl$ was recommended by the checkers. In a mixture made up of 250 ml. of C_2H_5OH, 50 ml. of chloroform, and 7.5 ml. of concentrated HCl is dissolved 3.5 g. of $[C_5H_5Fe(CO)_2]_2$. Air is bubbled through this mixture for 3 hours. The solution is evaporated to dryness in vacuum, and the residue is extracted with 300 ml. of water. The resulting red solution is poured through a filter and extracted with chloroform. The chloroform extract is dried, and the product is crystallized from a mixture made by adding petroleum ether to the chloroform (up to 15% by volume). Yield is 75%. The product is made up of well-defined red crystals, m.p. 87°C.

tion visible in the solid, solutions may decompose more easily. The solvents used should be dried by normal procedures in order to obtain reproducible conditions and because some of the compounds react with water, possibly with formation of a $[C_5H_5Fe(CO)_2H_2O]^+$ cation.[1,7,19,20]

A. AMMINE(CYCLOPENTADIENYL)IRON DICARBONYL TETRAPHENYLBORATE—THE REACTION OF CYCLOPENTADIENYLIRON DICARBONYL HALIDES WITH DONOR MOLECULES

{Amminedicarbonylcyclopentadienyliron Tetraphenylborate}

Some donor molecules, e.g., ammonia, hydrazine, and tri-ethylphosphine, react with cyclopentadienyliron dicarbonyl halides instantaneously under mild conditions to give cations of the type $[C_5H_5Fe(CO)_2L]^+$ [L = NH_3, N_2H_4, $P(C_2H_5)_3$],[8] whereas the formation of the corresponding cations with triphenyl-phosphine[9] and tris(dimethylamino)phosphine[10] takes place in refluxing tetrahydrofuran and benzene, respectively. Under pressure, substitution of the halogen by triphenylarsine, tri-phenylstibine, and carbon monoxide has also been achieved, yielding the cations:

$$[C_5H_5Fe(CO)_2\{As(C_6H_5)_3\}]^+, [C_5H_5Fe(CO)_2\{Sb(C_6H_5)_3\}]^+,[9]$$
$$\text{and } [C_5H_5Fe(CO)_3]^+ \text{ [9,11]}$$

A typical reaction is:

$$C_5H_5Fe(CO)_2Cl + NH_3 \rightarrow [C_5H_5Fe(CO)_2NH_3]Cl$$

The tetraphenylborate salt can be precipitated:

$$[C_5H_5Fe(CO)_2NH_3]Cl + NaB(C_6H_5)_4$$
$$\rightarrow [C_5H_5Fe(CO)_2NH_3][B(C_6H_5)_4] + NaCl$$

Procedure

A 210-mg. (1-mmole) sample of $C_5H_5Fe(CO)_2Cl$ is placed in a small Schlenk tube (see Fig. 1, page 10) of about 100-ml. volume.

About a 50-ml. quantity of dry ammonia is condensed into the Schlenk tube which is cooled with a Dry Ice–methanol bath. The reaction mixture is stirred with a magnetic stirrer. The Schlenk tube containing the red solution may be left conveniently in the bath overnight. The bath warms up very slowly, and the ammonia can evaporate without further precautions. After removal of ammonia, the yellow solid residue is dissolved in about 30 ml. of water and the solution filtered into an excess of a concentrated solution of sodium tetraphenylborate. The yellow precipitate is collected and dried under high vacuum. The crude product is stirred with several portions of dichloromethane until the filtrate is almost colorless and only a small quantity of white ammonium tetraphenylborate remains on the filter. Pentane is then added dropwise until no more precipitate separates. After the purification is repeated and the precipitate is dried under high vacuum, the pure product is obtained in about 40% yield, based on $C_5H_5Fe(CO)_2Cl$. *Anal.* Calcd.: C, 72.55; H, 5.50; Fe, 10.88; N, 2.73. Found: C, 73.04; H, 6.05; Fe, 11.04; N, 2.68.

Properties

The yellow microcrystalline salt is stable in air up to 170°C. It is insoluble in water and nonpolar organic solvents but soluble in polar organic solvents such as $C_6H_5NO_2$ and H_2CCl_2. The cation can be precipitated as the reineckate* and hexafluorophosphate from aqueous solution; however, the latter is quite soluble in water.

B. CYCLOHEXENE(CYCLOPENTADIENYL)IRON DICARBONYL HEXAFLUOROPHOSPHATE—PREPARATIONS MADE WITH THE AID OF LEWIS ACIDS

{Dicarbonyl(cyclohexene)cyclopentadienyliron Hexafluorophosphate}

Olefins do not react with cyclopentadienyliron dicarbonyl halides by themselves. However, the preparation of cations $[C_5H_5Fe(CO)_2olefin]^+$, where olefin = ethylene, propylene, 1-

* Reineckate contains the anion of Reinecke's salt, $NH_4[Cr(NH_3)_2(SCN)_4]$.

octadecene, cyclohexene, cyclooctene, butadiene, or cyclo-hexadiene, has been achieved with the aid of Lewis acids, for example, $AlBr_3$, $InCl_3$, $TiCl_4$, $FeCl_3$, $ZnCl_2$.[12,13] The $[C_5H_5Fe(CO)_3]^+$ cation has also been obtained by the same method.[12,14] A typical reaction is:

$$C_5H_5Fe(CO)_2Br + C_6H_{10} + AlBr_3 \rightarrow [C_5H_5Fe(CO)_2C_6H_{10}]AlBr_4$$

The hexafluorophosphate salt can be precipitated:

$$[C_5H_5Fe(CO)_2C_6H_{10}]AlBr_4 + NH_4PF_6 + 3H_2O \rightarrow$$
$$[C_5H_5Fe(CO)_2C_6H_{10}]PF_6 + NH_4Br + \cdots$$

Procedure

A 275-mg. (1-mmole) sample of $C_5H_5Fe(CO)_2Br$ is dissolved in 20 ml. of dry cyclohexene. After addition of 800 mg. (3 mmoles) of $AlBr_3$, the red reaction mixture is stirred at room temperature for 4 hours. A red viscous oil is formed quite rapidly, and after 4 hours the solution is almost colorless. The colorless cyclohexene layer is decanted and the residue dried under high vacuum. About 20 ml. of ice-cold water is added while cooling with ice, and the orange solution is filtered into an excess of a concentrated solution of ammonium hexafluorophos-phate. The precipitate is collected and dried under high vacuum. For further purification, the yellow salt is dissolved in 10 ml. of acetone, and, after filtration, ether is added drop-wise until no more precipitate separates. After a second crystallization, the product is obtained in about 40% yield, based on $C_5H_5Fe(CO)_2Br$. *Anal.* Calcd.: C, 38.64; H, 3.74; F, 28.21; Fe, 13.8; P, 7.67. Found: C, 38.55; H, 3.75; F, 28.6; Fe, 13.8; P, 7.43.

Properties

The complex salt crystallizes in yellow platelets, is fairly stable in air, darkens at 129°C., is soluble in acetone, and is insoluble in water and ether.

C. μ-CHLORO-BIS(CYCLOPENTADIENYLIRON DICARBONYL) TETRAFLUOROBORATE

[μ-Chloro-bis(dicarbonylcyclopentadienyliron) Tetrafluoroborate]

If cyclopentadienyliron dicarbonyl halides are allowed to react with themselves in the presence of Lewis acids, cations are formed in which the new substituent is the cyclopentadienyliron dicarbonyl halide itself, for example, $[C_5H_5Fe(CO)_2\text{-X-}(CO)_2\text{-}FeC_5H_5]^+$ (X = Cl, Br, I). All three cations can be prepared best by treatment of the corresponding halides with boron trifluoride diethyl etherate; all are isolated as tetrafluoroborates.[15] The bromine complex can also be obtained by a more complicated procedure by the reaction between $C_5H_5Fe(CO)_2Br$ and $AlBr_3$ in liquid sulfur dioxide;[16] the iodine cation can be isolated from a melt of cyclopentadienyliron dicarbonyl iodide and aluminum chloride.[17] In the latter two cases the hexafluorophosphate salts can be obtained. These binuclear cations are of special interest, because they are cleaved by electron donors,[15–17] e.g., aniline, pyridine, benzonitrile, acetonitrile, acrylonitrile, with the formation of the corresponding $[C_5H_5Fe(CO)_2L]^+$ cations and the parent halide. Equations for preparation of the tetrafluoroborate are:

$$2C_5H_5Fe(CO)_2Cl + BF_3{\cdot}O(C_2H_5)_2 \rightarrow$$
$$[C_5H_5Fe(CO)_2\text{-Cl-}(CO)_2FeC_5H_5][BF_3Cl] + O(C_2H_5)_2$$
$$[C_5H_5Fe(CO)_2\text{-Cl-}(CO)_2FeC_5H_5]BF_3Cl + NaBF_4 + 3H_2O \rightarrow$$
$$NaCl + B(OH)_3 + 3HF + [C_5H_5Fe(CO)_2\text{-Cl-}(CO)_2FeC_5H_5]BF_4$$

Procedure

A 1-g. (4.75-mmole) sample of $C_5H_5Fe(CO)_2Cl$ is placed in a small Schlenk tube (see page 10) of about 50-ml. volume. A 7-ml. sample of freshly distilled $BF_3{\cdot}O(C_2H_5)_2$ is added, and the solution is heated to 40–50°C. for $1\frac{1}{2}$ hours. During that time the color changes from light red to deep purple. The reaction mixture is then frozen in liquid nitrogen, and 10 ml. of a saturated solution of sodium tetrafluoroborate is added. The mixture is slowly warmed to room temperature with continuous shaking. Ether is now removed under vacuum (water pump). The dark red precipitate which separates is collected on a filter, washed with about 10 ml. of water, and dried under high vacuum. The crude product is dissolved in 5 ml. of acetone; the solution is filtered, and ether is then added dropwise to the filtrate.* Dark red crystals separate first; the addition of ether is stopped as soon as a white fog appears. As the heavy red crystals sink much faster than the white material, the latter can be removed easily by taking the solvent into a pipet and then adding fresh ether. This procedure has to be repeated until no further white material is present. After the red crystals are collected on a filter and dried under high vacuum, the desired salt is obtained in about 35% yield, based on $C_5H_5Fe(CO)_2Cl$. *Anal.* Calcd.: C, 35.31; H, 2.12; B, 2.27; Cl, 7.45; F, 15.96; Fe, 23.46. Found: C, 35.50; H, 2.20; B, 2.3; Cl, 7.7; F, 15.4; Fe, 23.33.

Properties

The red crystalline salt decomposes in air and under vacuum at 105°C. It is insoluble in water and nonpolar organic solvents; it is soluble in acetone with decomposition. Therefore, the addition of ether to the acetone solution must be fairly rapid.

D. ACETONITRILE(CYCLOPENTADIENYL)IRON DICARBONYL TETRAFLUOROBORATE—REACTIONS OF μ-HALO-BIS-(CYCLOPENTADIENYLIRON DICARBONYL) CATIONS WITH DONOR MOLECULES

{Acetonitriledicarbonylcyclopentadienyliron Tetrafluoroborate}

As already mentioned, the binuclear cations react with several donor molecules with cleavage of the halogen bridge; however, only ligands that do not react with the cyclopentadienyliron dicarbonyl halides can be used in this reaction, as the latter are

---

* The product decomposes on standing in acetone. Therefore, filtration and reprecipitation should be rapid.

produced in the reaction and would lead to the formation of a salt with mixed anions. The acetonitrile(cyclopentadienyl)iron dicarbonyl cation has also been prepared by the method of Sec. B.[18] The corresponding cation with pyridine instead of acetonitrile, which had only been obtained by substitution of carbon monoxide in the $[C_5H_5Fe(CO)_3]^+$ cation,[18] is prepared very easily by the method described below.[18] The acetonitrile cation is prepared by the reaction:

$$[C_5H_5Fe(CO)_2\text{-}Cl\text{-}(CO)_2FeC_5H_5]BF_4 + NCCH_3 \rightarrow$$
$$C_5H_5Fe(CO)_2Cl + [C_5H_5Fe(CO)_2NCCH_3]BF_4$$

Procedure

A 100-mg. (0.2-mmole) sample of μ-chloro-bis(cyclopenta-dienyliron dicarbonyl) tetrafluoroborate is dissolved in 2 ml. of acetone, and 0.5 ml. (10 mmoles) of acetonitrile is added to the dark red solution. The reaction mixture is heated to 20–30°C. for $1\frac{1}{2}$ hours. During that time the color changes to red-orange. Solvent and excess acetonitrile are removed under vacuum. The oily residue is treated with ether, and a yellow solid forms. The yellow solid is removed by filtration and is dissolved in 2 ml. of acetone. The solution is filtered; then ether is added dropwise to the filtrate until no more precipitate separates. After the yellow crystalline product is dried under high vacuum, the yield is about 60%, based on $[\{C_5H_5Fe(CO)_2\}_2Cl]BF_4$. *Anal.* Calcd.: C, 29.78; H, 2.23; Fe, 15.39; N, 3.86. Found: C, 29.86; H, 2.46; Fe, 15.47; N, 3.48.

Properties

The yellow crystalline salt is fairly stable in air in the solid state; solutions, however, decompose quite rapidly.

References

1. B. F. Hallam and P. L. Pauson, *J. Chem. Soc.*, 3030 (1956).

2. B. F. Hallam, O. S. Mills, and P. L. Pauson, *J. Inorg. Nucl. Chem.*, **1**, 313 (1955).
3. T. S. Piper, F. A. Cotton, and G. Wilkinson, *ibid.*, **1**, 165 (1955).
4. R. B. King, in "Organometallic Syntheses," Vol. I, p. 114, Academic Press, Inc., New York, 1965.
5. T. S. Piper and G. Wilkinson, *J. Inorg. Nucl. Chem.*, **2**, 38 (1956).
6. F. A. Cotton, A. L. Liehr, and G. Wilkinson, *Ibid.*, **1**, 175 (1955).
7. T. S. Piper and G. Wilkinson, *ibid.*, **3**, 104 (1956).
8. E. O. Fischer and E. Moser, *J. Organometallic Chem.*, **5**, 63 (1966).
9. A. Davison, M. L. H. Green, and G. Wilkinson, *J. Chem. Soc.*, 3172 (1961).
10. R. B. King, *Inorg. Chem.*, **2**, 936 (1963).
11. R. B. King, *ibid.*, **1**, 964 (1962).
12. E. O. Fischer and K. Fichtel, *Chem. Ber.*, **94**, 1200 (1961).
13. E. O. Fischer and K. Fichtel, *ibid.*, **95**, 2063 (1962).
14. E. O. Fischer, K. Fichtel, and K. Öfele, *ibid.*, **95**, 249 (1962).
15. E. O. Fischer and E. Moser, *Z. Anorg. Allgem. Chem.*, **342**, 156 (1966).
16. E. O. Fischer and E. Moser, *J. Organometallic Chem.*, **3**, 16 (1965).
17. E. O. Fischer and E. Moser, *Z. Naturforsch.*, **20b**, 184 (1965).
18. P. M. Treichel, R. L. Shubkin, K. W. Barnett, and D. Reichard, *Inorg. Chem.*, **5**, 1177 (1966).
19. M. L. H. Green and P. L. I. Nagy, *Proc. Chem. Soc.*, 74 (1962).
20. M. L. H. Green and P. L. I. Nagy, *J. Chem. Soc.*, 189 (1963).

8. TRIS(HYDRIDOMANGANESE TETRACARBONYL)
(*Dodecacarbonyl Trihydridotrimanganese*)

$$(CO)_5Mn—Mn(CO)_5 \xrightarrow{KOH} \left[\begin{array}{c} \text{uncharacterized} \\ \text{green crystals} \end{array}\right] \xrightarrow[\text{aq.}]{H_3PO_4} [HMn(CO)_4]_3$$

Submitted by B. F. G. JOHNSON,* R. D. JOHNSTON,* J. LEWIS,* and B. H. ROBINSON*
Checked by MILTON OLAZAGASTI† and EARL MUETTERTIES†

Tris(hydridomanganese tetracarbonyl) has been prepared by the reaction of dimanganese decacarbonyl with sodium amalgam, followed by treatment with phosphoric acid; the use of sodium borohydride (sodium tetrahydroborate) in the first step was found to give higher yields.[1,2] A preparation using methylmanganese pentacarbonyl as starting material has also been reported,[3] but it is of no synthetic value. A more convenient

* Department of Chemistry, University College, London W.C. 1, England.
† Central Research Department, Experimental Station, E. I. du Pont de Nemours & Company, Wilmington, Del. 19898.

44   *Inorganic Syntheses*

synthesis involves reaction of dimanganese decacarbonyl with potassium hydroxide followed by acidification.[4,5]   The particular value of this method is that it is an example of a general route to polynuclear carbonyl hydride compounds of transition metals.

Procedure

A solution of 6.7 g. (0.12 mole) of potassium hydroxide dissolved in 8 ml. of distilled water is added to 0.5 g. of pure manganese carbonyl[6] (0.0013 mole) contained in a 100-ml. three-necked flask connected via a coarse sinter to a similar flask. The system is evacuated and the reaction mixture heated to 60°C.* and stirred vigorously, preferably with a magnetic stirrer, for about $2\frac{1}{2}$ hours.   At the end of the reaction the solution is an emerald-green color, and on cooling to 0°C. dark green crystals precipitate.   These crystals are separated by filtration through the sinter under a pressure of purified nitrogen† and are then quickly redissolved in 20 ml. of 2 N alkali solution.

Under a nitrogen atmosphere 85% phosphoric acid is added to the green solution, which changes first to purple; subsequently the product precipitates as an orange-red solid.   It is separated by filtration, washed with water until the washings are neutral, and dried *in vacuo*.   Any unreacted manganese carbonyl may be removed by sublimation under reduced pressure at room temperature (1 mm./20°C.).   The pure compound is then obtained as deep orange-red crystals by sublimation (1 mm./ 60°C.).   The yield is 0.3–0.35 g. (64–74%).   *Anal.* Found: C, 28.6; H, 0.79; O, 37.7; Mn, 32.1.   Calcd. for $H_3Mn_3(CO)_{12}$: C, 28.60; H, 0.56; O, 38.10; Mn, 32.70.

This preparation can be carried out on a larger scale without difficulty, the only limit being that for effective filtration through the sinter.

* Accurate control of this temperature is necessary.
† Care must be exercised to prevent any air entering the system.

Properties

Certain of the properties of tris(hydridomanganese tetra-carbonyl) have been described previously.[3,5,7] It is a red-orange solid, stable to air; it decomposes at 123–125°C. It has a characteristic four-band spectrum in the CO stretching region of the infrared: 2080(m), 2034(s), 2008(s), 1986(s). It is soluble without decomposition in most organic solvents. The mass spectrometric and infrared spectral data are consistent with a triangular structure involving four terminal CO groups per metal atom. The H atoms are equivalent and may be bridging between the metal atoms.

Despite repeated attempts to isolate the green and purple intermediates, their instability has prevented identification.

References

1. D. K. Huggins, dissertation, University of California, Los Angeles, 1963.
2. W. Fellmann, D. K. Huggins, H. D. Kaesz, abstracts of papers presented at the 8th International Conference on Coordination Chemistry, September, 1964, pp. 255–257.
3. E. O. Fischer and R. Aumann, *J. Organometallic Chem.*, **8**, P1 (1967).
4. B. F. G. Johnson, R. D. Johnston, J. Lewis, and B. H. Robinson, *Chem. Commun.*, 851 (1966).
5. B. F. G. Johnson, R. D. Johnston, J. Lewis, and B. H. Robinson, *J. Organometallic Chem.*, **10**, 105 (1967).
6. R. B. King and F. G. A. Stone, *Inorganic Syntheses*, **7**, 198 (1963).
7. J. R. Smith, K. Mehner, and H. D. Kaesz, *J. Am. Chem. Soc.*, **89**, 1759 (1967).

9. ORGANOTIN HYDRIDES

Submitted by ERNEST R. BIRNBAUM* and PAUL H. JAVORA*

Organotin hydrides have found much interest in recent years both as synthetic reagents and as the subjects of numerous theoretical investigations.

* St. John's University, Jamaica, N.Y. 11432.

Kraus and Greer[1] originally prepared a few organotin hydrides with considerable difficulty from the reaction between an organo-tin sodium compound and ammonium bromide or ammonium chloride in liquid ammonia.   Their procedure remained the only synthetic route to these compounds for some 25 years until 1947 when Finholt, Bond, Wilzbach, and Schlesinger[2] obtained the methyltin hydrides in a more facile manner by reducing a mixture of the methyltin chlorides with lithium aluminum hydride (lithium tetrahydroaluminate).

Fair to good yields of some organotin hydrides have been produced in one or two cases from the reduction of the organotin chlorides with magnesium and water,[3] from the thermal decom-position of organotin formates,[4] from the reduction of a bis-organotin oxide or of a bisorganotin compound with lithium aluminum hydride,[5] and from the hydrolysis of an organotin lithium compound.[6]

Although excellent yields of a variety of organotin hydrides result either from the reaction of the organotin methoxides with diborane[7] or from the reduction of the organotin chlorides with dialkylaluminum hydrides,[8] the necessity first to prepare the respective organotin methoxides or the dialkylaluminum hydride renders the application of these methods inconvenient.   As modified by van der Kerk, Noltes, and Luijten,[9] the reduction of organotin chlorides by lithium aluminum hydride has thus been the most general and satisfactory method for the preparation of organotin hydrides in high yield.

By reducing organotin chlorides (a large number of which are commercially available) with sodium borohydride as described below, rather than with lithium aluminum hydride, several previously required experimental operations (stirring the reaction mixture for an hour or two, slowly destroying the excess lithium aluminum hydride with water, and drying the ether extract with sodium sulfate) may be eliminated, resulting in a smoother preparation and a significant saving in time, the entire synthesis usually being accomplished in 4–6 hours.   The yields from the sodium borohydride reduction are generally similar to those from the lithium aluminum hydride reduction[10] and, for the specific examples given below, are substantially better.

It should be noted that the diborane (a highly toxic gas, spontaneously flammable in air) expected from the stoichiometric equation

$$R_{4-y}SnCl_y + yNaBH_4 \rightarrow R_{4-y}SnH_y + y/2B_2H_6{}^* + yNaCl$$

strongly complexes in solution with the excess sodium borohydride employed; it is *not* evolved during the reaction.

A. TRI-*n*-BUTYLTIN HYDRIDE
[Tri-*n*-butyl(hydrido)tin]

$$(C_4H_9)_3SnCl + NaBH_4 \rightarrow (C_4H_9)_3SnH + \tfrac{1}{2}B_2H_6{}^* + NaCl$$

Checked by M. AKHTAR† and H. C. CLARK†

Procedure

In a fume hood, 19.2 g. (0.059 mole) of tri-*n*-butyltin chloride and then 140 ml. of ethylene glycol dimethyl ether (monoglyme) which has been purified by distillation from calcium hydride at atmospheric pressure on an efficient fractionating column (fraction used boils at 85 ± 1°C.) are added to a pressure-equalized 250-ml. dropping funnel, and the funnel gently agitated to dissolve all the tributyltin chloride.

Also in the fume hood a 1-l. three-necked flask containing a magnetic stirring bar is clamped in an aluminum pan (which will subsequently serve to hold a cooling bath) placed on top of a magnetic stirrer. The two outer necks of the flask are fitted with gas inlet and outlet tubes. Dry, oxygen-free nitrogen (or helium) is passed continuously into one of the gas inlets during

* In solution.
† Department of Chemistry, University of Western Ontario, London, Ontario, Canada.

the remainder of the reaction. The gas outlet in the other neck of the flask is now connected to a mineral-oil bubbler (a piece of glass tubing immersed to a few centimeters in a beaker of mineral oil). After the nitrogen or helium stream has flowed for a few minutes, 6.2 g. (0.16 mole) of powdered sodium borohydride and then 230 ml. of purified monoglyme are added to the flask through the center neck. The dropping funnel containing the tributyltin chloride solution is then fitted in place in the central neck. The stirrer is turned on, and acetone and sufficient Dry Ice (or ice-salt) are now put into the aluminum pan to form a $-10°C.$ bath. Additional Dry Ice is added as necessary from time to time to maintain the bath at -10 to $-11°C.$

After the flask containing the sodium borohydride solution has cooled for about half an hour, the tributyltin chloride solution is added dropwise with rapid stirring over a 30-minute period. A white sodium chloride precipitate forms as each drop of the tributyltin chloride solution is added. The reaction mixture is allowed to stand at -10 to $-11°C.$ for 10–15 minutes after the last of the tributyltin chloride solution has been added. Without filtration, the entire reaction mixture is now transferred cold under nitrogen or helium into a 1-l., single-necked, round-bottomed flask; the flask is attached to a flash evaporator (Buchler Model PTFE-1G or equivalent) and immersed in a bath maintained at $0°C.$ The evaporator's receiving flask (also 1-l., single-necked, round-bottomed) is immersed in a $-80°C.$ bath of Dry Ice–acetone.

The evaporator is then connected to a standard high-vacuum system; with continuous pumping through a vacuum trap maintained at $-196°C.$ by a Dewar flask of liquid nitrogen, the reaction mixture is flash-evaporated. Diborane (■ *caution*) will collect in the $-196°C.$ trap during the flash evaporation, and it is imperative that this vacuum trap be kept at $-196°C.$ until the diborane is disposed of. When the evaporation is complete, the entire system is filled to atmospheric pressure with nitrogen or helium, and the diborane destroyed as follows: With a vigorous flow of nitrogen or helium through the $-196°C.$ trap, one gas-tight safety trap (200–500 cc.) and two gas absorption vessels, each containing 100–200 ml. of acetone, are connected in series with the $-196°C$. trap. The $-196°C$. trap is then allowed to warm slowly to room temperature, and the diborane is carried into the acetone by the nitrogen or helium stream. The resultant diisopropoxyborane is then harmlessly hydrolyzed to boric acid and hydrogen by flushing the acetone down the drain in running water.*

The evaporation flask is now removed from the flash evaporator, and the residue (which may appear to be dry) is extracted under nitrogen or helium with three 25-ml. portions of anhydrous, peroxide-free diethyl ether. The ether extract is poured through a medium sintered-glass frit under an atmosphere of nitrogen or helium. Finally, the ether is removed by "pumping" on the filtered extract at 0°C. for 30–40 minutes. A pressure of less than 1 mm. is maintained. The colorless liquid product is identified as tri-*n*-butyltin hydride from its liquid-phase infrared spectrum[11] and refractive index, n_D^{25} 1.4715, literature, n_D^{25} 1.4711 (extrapolated from n_D^{20} 1.4726[12] and n_D^{22} 1.4720[6]). The yield is 16.5 g. (0.057 mole, 96%).

B. TRIPHENYLTIN HYDRIDE

(Hydridotriphenyltin)

$$(C_6H_5)_3SnCl + NaBH_4 \rightarrow (C_6H_5)_3SnH + \tfrac{1}{2}B_2H_6 + NaCl$$

Checked by M. AKHTAR† and H. C. CLARK†

Procedure

Triphenyltin chloride, 7.49 g. (0.019 mole) in 150 ml. of purified monoglyme, is allowed to react with sodium borohydride (3.3 g., 0.087 mole) in 150 ml. of purified monoglyme

---

* See alternative procedure for destroying B_2H_6 by Bond and Pinsky in Sec. E.

† Department of Chemistry, University of Western Ontario, London, Ontario, Canada.

exactly as described above in Sec. A.   The resultant slightly yellow solid product is distilled at reduced pressure on a short-path apparatus (Kontes Glass No. K28480).   The fraction distilling at 156–158°C./0.15 mm. is collected and identified as triphenyltin hydride from its liquid-phase infrared spectrum[13] and its refractive index, n_D^{28} 1.6327, literature,[6] n_D^{28} 1.6322   The yield is 5.63 g. (0.016 mole, 82%).

C. DIMETHYLTIN DIHYDRIDE

(Dihydridodimethyltin)

$$(CH_3)_2SnCl_2 + 2NaBH_4 \rightarrow (CH_3)_2SnH_2 + B_2H_6 + 2NaCl$$

Checked by M. AKHTAR* and H. C. CLARK* (qualified approval)† and by A. C. BOND‡ and M. L. PINSKY‡

Procedure

This synthesis employs diethylene glycol dimethyl ether (diglyme) as the reaction medium.   To prevent the possible formation of methyl borate as an inseparable by-product, it is important that the diglyme be thoroughly purified by reflux over and distillation from calcium hydride at atmospheric pressure on an efficient fractionating column (collect fraction boiling at 162 ± 1°C.).   The apparatus is as described in Sec. A except that the aluminum pan may be omitted since the synthesis is to be carried out at room temperature.   Dimethyltin dichloride, 12.5 g. (0.057 mole) in 75 ml. of purified diglyme, is added dropwise with rapid stirring to sodium borohydride

* Department of Chemistry, University of Western Ontario, London, Ontario, Canada.
† Checkers Akhtar and Clark found an impurity in the product which could not be removed by fractionation.   On the basis of infrared spectra, Birnbaum and Javora identified the impurity as trimethyl borate.   These authors as well as checkers Bond and Pinsky reported that methyl borate contamination can be avoided by very careful purification of the polyether used as solvent and by carefully excluding moisture during reaction.   (See page 56.)
‡ Department of Chemistry, Rutgers University, New Brunswick, N.J. 08903.

(11.5 g., 0.30 mole) in 350 ml. of purified diglyme. The reaction mixture is allowed to stand for 10–15 minutes after the last of the dimethyltin dichloride solution has been added; it is then filtered (nitrogen or helium atmosphere) through a medium sintered-glass frit.* The diglyme filtrate is caught in, or transferred (nitrogen or helium atmosphere) into, a 1-l., single-necked, round-bottomed flask, and the flask is placed on a flash evaporator (Buchler Model PTFE-1G or equivalent). The evaporator receiving flask, also a 1-l., single-necked, round-bottomed flask, is immersed in a 0°C. bath. The evaporator is then attached to a standard, all-glass, high-vacuum fractionation train. The first vacuum trap in the fractionation train immediately following the evaporator is kept at 0°C.; the next vacuum trap is held at −126°C. with a methylcyclohexane slush bath; and the third consecutive vacuum trap is maintained at −196°C. by liquid nitrogen. The entire system is now continuously "pumped on" through the last (−196°C.) trap, a 40°C. bath is placed around the flask containing the diglyme filtrate, and the diglyme filtrate is flash-evaporated at this temperature.

When the evaporation is complete, the 0°C. trap will contain a few milliliters of diglyme; the −196°C. trap will have trapped-out diborane (*caution*); and the −126°C. trap will have retained the dimethyltin dihydride together with traces of diglyme and diborane. The crude product is accordingly subjected to a second vacuum fractionation through clean, 0°C. and −126°C. vacuum traps into the same −196°C. trap. The diborane collected in the −196°C. trap is destroyed as described in Sec. A. The colorless, volatile liquid product is identified from its vapor-phase infrared spectrum[14] and from its refractive index, n_D^{20} 1.4475, literature,[2] n_D^{20} 1.4480, as dimethyltin dihydride. The yield is 8.25 g. (0.055 mole, 96%).

---

* *Editor's note:* The filtration under nitrogen can be effected by use of a Schlenk tube or a similar tube with a sealed sintered disk inserted in place of the gas exit tube. See Figs. 1–3, pages 10, 16, and 20.

D. TRIMETHYLTIN HYDRIDE

(Hydridotrimethyltin)

$$(CH_3)_3SnCl + NaBH_4 \rightarrow (CH_3)_3SnH + \tfrac{1}{2}B_2H_6 + NaCl$$

Checked by M. AKHTAR* and H. C. CLARK* (qualified approval)† and by
A. C. BOND‡ and M. L. PINSKY‡

Procedure

With the exception that $-80°C$. traps (Dry Ice–acetone) are used instead of the $-126°C$. traps, this synthesis is accomplished as detailed above in Sec. C. From 14.9 g. (0.075 mole) of trimethyltin chloride and 10.0 g. (0.26 mole) of sodium borohydride, an 11.4-g. (0.069-mole, 92%) sample of a colorless liquid product was obtained. The product was identified as trimethyltin hydride from its gas-phase infrared spectrum;[14] from its boiling point, 58°C./752 mm., literature,[2] 59°C./760 mm.; and from its refractive index, n_D^{18} 1.4472, literature,[2] n_D^{18} 1.4484.

Properties

The organotin hydrides are easily identified by their infrared spectra which contain very strong Sn—H absorptions in the 1800–1880 cm.$^{-1}$ region. Tri-*n*-butyltin hydride, trimethyltin hydride, and dimethyltin dihydride are colorless liquids with respective boiling points 65–67°C./0.6 mm., 59°C./760 mm.,

* Department of Chemistry, University of Western Ontario, London, Ontario, Canada.

† Checkers Akhtar and Clark found an impurity in the product which could not be removed by fractionation. On the basis of infrared spectra, Birnbaum and Javora identified the impurity as trimethyl borate. These authors as well as checkers Bond and Pinsky reported that methyl borate contamination can be avoided by very careful purification of the polyether used as solvent (see page 56) and by carefully excluding moisture during reaction.

‡ Department of Chemistry, Rutgers University, New Brunswick, N.J. 08903.

and 35°C./760 mm.   Although obtained as a colorless liquid distillate, triphenyltin hydride, upon standing at 0°C., slowly forms white needles melting at 26–28°C.

Gas-liquid chromatography affords another technique by which the purity of the organotin hydrides may be ascertained. Tri-*n*-butyltin hydride, for example, may be chromatographed in a Pyrex column packed with Gas-Chrom Z solid support containing a 5% dispersion of silicone oil.

The organotin hydrides decompose slowly on standing at room temperature and hence are best stored at 0°C. or below.   The decomposition is strongly catalyzed by air, silicone stopcock grease, metallic surfaces, and, in the case of triphenyltin hydride, by light.   Manipulations involving the organotin hydrides are thus usually done either *in vacuo* or in an inert atmosphere, using thoroughly cleaned glassware.   The thermal stability of the organotin hydrides increases as the number of alkyl or aryl groups is increased.   In the absence of the above impurities, the organotin monohydrides may be kept for a number of months, the organotin dihydrides for several weeks, and the organotin trihydrides for a few days without appreciable decomposition.

The alkyltin hydrides react slowly with water to form oxides or hydroxides.   The reaction is more rapid with aqueous alkali, and hydrogen is evolved in addition to the formation of the organotin oxide or hydroxide.

A very rapid and quantitative conversion to the organotin chloride and hydrogen occurs when alkyltin monohydrides are treated with dilute hydrochloric acid.   The organotin hydrides readily react with halogens to form the corresponding organotin halide and hydrogen halide.   Although organotin compounds have been used as biocides and are, in general, toxic, they do *not* present an unusual hazard provided normal precautions are taken to avoid their contact with the skin and inhalation of their vapors.

E. HIGH-VACUUM TECHNIQUES FOR THE PREPARATION OF DIMETHYLTIN DIHYDRIDE

Submitted by A. C. BOND* and MICHAEL L. PINSKY*

The previously described synthesis of dimethyltin dihydride (Sec. C) can also be carried out by using high-vacuum techniques. The high-vacuum procedures are particularly attractive if such an apparatus is available, but even if the apparatus has to be assembled we believe that the high-vacuum procedures offer sufficient advantages to make their use a worthy alternative.

Procedure

The schematic diagram of the apparatus is shown in Fig. 4. The apparatus is completely evacuated through stopcock S_1, with all other stopcocks open. Stopcock S_1 is then closed. Nitrogen gas is admitted through stopcock S_4 until it bubbles out the open-end manometer M. The liquid-adding funnel L is removed from the reaction flask F, which for this preparation is a 50-cc., two-necked, round-bottomed flask with 24/40 S.T. ground-glass joints. Sodium borohydride (0.4 g. 10.5 mmoles), 10 cc. of diglyme, and a magnetic stirring bar are introduced into the reaction flask. The liquid-adding funnel is replaced, the large stopcock on the funnel is closed, and a diglyme solution containing 1.24 g. (4.03 mmoles) of dimethyltin dibromide† in 8 cc. of diglyme is introduced into the funnel. The ground-glass plug to the funnel is replaced and stopcock S_4 is closed. Liquid nitrogen is placed around the U-tube, and an acetone bath at $-45°C$. is placed around trap T. The temperature of the acetone is maintained by periodically adding small amounts of Dry Ice. An alternative $-45°C$. bath

---

* Department of Chemistry, Rutgers University, New Brunswick, N.J. 08903.
† Dimethyltin dichloride can be used just as well.

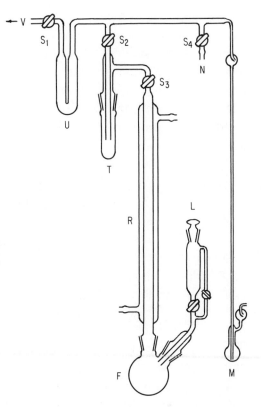

Fig. 4. (V), to vacuum; (S_{1-4}), 2- to 4-mm. oblique-bore high-vacuum stopcocks; (N), inlet for nitrogen gas; (T), removable trap; (R), reflux condenser; (L), liquid-addition tube; (F), reaction flask; (M), open-end manometer; (U), U-tube trap.

for the trap could be a slush of solid-liquid chlorobenzene.   The system is evacuated through S_1, and the dimethyltin dibromide solution is added dropwise to the magnetically stirred sodium borohydride solution while the apparatus is continuously pumped.   The addition is complete in 10 minutes but the system is pumped for an additional 20 minutes.

Both diborane and dimethyltin dihydride are evolved as soon as the reactants are brought together.   During the reaction, the major portion of the diglyme is returned to the reaction flask by the reflux condenser, R.   A very small amount collects in T, but none is present in U.   The dimethyltin dihydride and diborane are trapped in U.   This mixture is separated by passing it through a $-123°C$. (chlorobutane) bath which traps the dimethyltin dihydride while allowing diborane to pass to a trap at $-196°C.$ (liquid nitrogen). Two such fractionations yielded tensiometrically pure dimethyltin dihydride having a vapor pressure of 60 torrs at $-23°C.$ (CCl_4 slush). A vapor-density molecular weight of 150 was determined (calcd. 151), and the infrared spectrum indicated no impurities. The yield of pure product was 93%. The entire preparation including purification can be carried out in less than 2 hours.

The diborane can be stored as a gas in conventional high-vacuum apparatus or, if such an apparatus is not used, it can be converted to the harmless pyridine borane by condensing it with pyridine at $-196°C.$ in a reaction vessel on the vacuum line and then allowing the system to warm up slowly. It can also be converted to sodium borohydride by passing it through a tube filled with powdered sodium trimethoxyborohydride (sodium hydridotrimethoxyborate) according to the equation:

$$2NaHB(OCH_3)_3 + B_2H_6 \rightarrow 2NaBH_4 + 2B(OCH_3)_3$$

The methyl borate produced in the diborane-disposal reaction can be removed from the solid by distillation.

In all preparations involving the use of diglyme it is essential that it be free of alcohols which might introduce trialkoxyboranes as impurities in the alkyltin hydrides. The diglyme can be purified best by distilling it from $LiAlH_4$. Because this *procedure can be hazardous unless properly done*, the procedure outlined below is recommended. It is suggested that quantities be kept small (50 ml.) as a safety precaution.

Small portions of lithium aluminum hydride are added to a 50-ml. aliquot of vigorously stirred diglyme. The diglyme is contained in a flask which will later be attached to a condenser for distillation; a magnetic stirring bar is recommended to stir the liquid. If the diglyme contains peroxides or more than small amounts of water, the reaction with lithium aluminum hydride may be so vigorous that the hydrogen evolved may ignite. For this reason, we recommend starting with spectroquality diglyme and using a pretest of the vigor of the reaction by adding a small amount of LiAlH$_4$ to about 1 cc. of diglyme on a watch glass. We have found that spectroquality diglyme and LiAlH$_4$ may react with evolution of small amounts of hydrogen and heat. The LiAlH$_4$ should be added slowly enough so that the temperature of the diglyme does not exceed 50°C. When there is no further indication of reaction upon addition of the last portion of LiAlH$_4$, about 0.3 g. of additional LiAlH$_4$ is added to the flask, which is then attached to the distilling condenser. The mixture is magnetically stirred while the temperature is gradually raised from room temperature with an electric heater. *Shielding of the flask during heating is recommended.* At temperatures above 100°C., hydrogen evolution is observed before refluxing of the diglyme begins. The evolution of hydrogen ceases before the boiling point of diglyme is reached. The final distillation is, therefore, carried out in the absence of LiAlH$_4$, which has been decomposed at the lower temperatures. The fraction of the distillate having a boiling point of 162 ± 1°C. is collected as purified diglyme.

References

1. C. A. Kraus and W. N. Greer, *J. Am. Chem. Soc.*, **44**, 2629 (1922).
2. A. E. Finholt, A. C. Bond, Jr., K. E. Wilzbach, and H. I. Schlesinger, *ibid.*, **69**, 2692 (1947).
3. G. J. M. van der Kerk and J. G. Noltes, *Chem. Ind. (London)*, 1290 (1958).
4. M. Ohara and R. Okawara, *J. Organometallic Chem.*, **3**, 484 (1965).
5. W. J. Considine and J. J. Ventura, *Chem. Ind. (London)*, 1683 (1962).
6. C. Tamborski, F. Ford, and E. Soloski, *J. Org. Chem.*, **28**, 237 (1963).
7. E. Amberger and M. R. Kula, *Chem. Ber.*, **96**, 2560 (1963).
8. W. P. Neumann and H. Niermann, *Liebigs Ann. Chem.*, **653**, 164 (1962).
9. G. J. M. van der Kerk, J. G. Noltes, and J. G. A. Luijten, *J. Appl. Chem.*, **7**, 366 (1957).
10. E. R. Birnbaum and P. H. Javora, *J. Organometallic Chem.*, **9**, 379 (1967).
11. R. A. Cummins and P. Dunn, "The Infrared Spectra of Organotin Compounds," p. 6.9-3, Australian Defence Scientific Service, Defence Standards Laboratories, Maribyrnong, Victoria, 1963.
12. W. P. Neumann, G. Burkhardt, E. Heymann, F. Kleiner, K. Konig, H. Niermann, J. Pedain, R. Schick, R. Sommer, and H. Weller, *Angew. Chem. Intern. Ed. Engl.*, **2**, 165 (1963).
13. Reference 11, p. 6.9-4.
14. C. R. Dillard and L. May, *J. Mol. Spectry.*, **14**, 250 (1964).

10. TETRAMETHYLGERMANE

$$4CH_3MgBr + GeCl_4 \rightarrow (CH_3)_4Ge + 4MgBrCl$$

Submitted by E. H. BROOKS* and F. GLOCKLING†
Checked by JOHN WARD‡ and W. A. G. GRAHAM‡

The most convenient and least hazardous method for the preparation of tetramethylgermane on a reasonably large scale is by the action of germanium tetrachloride [germanium(IV) chloride] on methylmagnesium halide.[1,2] The use of a high-boiling ether solvent is necessary to allow isolation of the product, and earlier workers experienced difficulties using low-boiling ethers.[1,3,4] This led to the introduction of other methylating agents for the preparation, notably trimethylaluminum,[5] dimethylzinc,[3] and dimethylcadmium.[4]

Procedure

Magnesium turnings (120 g., 4.9 moles) are covered with sodium-dried di-*n*-butyl ether in a 5-l. three-necked flask equipped with dropping funnel, water-cooled reflux condenser, and a motor-driven PTFE paddle stirrer§ equipped with a nitrogen inlet. Methyl bromide (5 ml.), cooled with Dry Ice, is cautiously added to initiate the exothermic reaction. The stirrer motor is started, and the volume of di-*n*-butyl ether in the flask is made up to 2 l. Methyl bromide (380 g., 4.0 moles) cooled with Dry Ice is added, and the reaction mixture is cooled in an ice-water bath during the addition. The

* I.C.I., Petrochemical and Polymer Laboratory, P. O. Box 11, The Heath, Runcorn, Cheshire, England.
† Department of Inorganic Chemistry, The Queen's University, Belfast, Northern Ireland.
‡ Department of Chemistry, University of Alberta, Edmonton, Alberta, Canada.
§ A gas-tight stirrer with N_2 inlet such as a commercial Trubore stirrer with a Teflon paddle is satisfactory.

mixture is then refluxed for one hour. After the Grignard is allowed to cool to room temperature, germanium tetrachloride (150 g., 0.70 mole) in 200 ml. of di-*n*-butyl ether is added dropwise in a static nitrogen atmosphere, and the mixture is maintained at 60°C. for an additional 3 hours, care being taken to ensure that vapor does not escape from the ground-glass joints during this time.

After cooling, the reflux condenser is replaced by the normal distillation arrangement, and the fraction containing tetramethylgermane is distilled off without hydrolysis. The mixture is heated progressively up to the boiling point of the solvent (142°C. for pure di-*n*-butyl ether) and about a 300-ml. aliquot of distillate is collected to ensure that no tetramethylgermane remains in the large bulk of the solvent. The product is separated from di-*n*-butyl ether by fractional distillation. The distillate which boils in the range 41.5–44°C. is collected. Yield is 66 g. (71%).

Properties

Tetramethylgermane is a colorless liquid, stable to air, water, aqueous base, and concentrated sulfuric acid. Physical constants are as follows:[2,4,6,7] b.p. 43.4°C./760 mm.; m.p. −88°C.; density, 1.006 g./ml.; chemical shift in CCl_4, τ(C—H) 9.863 relative to TMS, τ, 10.

References

1. L. M. Dennis and F. E. Hance, *J. Phys. Chem.*, **30**, 1055 (1926).
2. H. Siebert, *Z. Anorg. Allgem. Chem.*, **263**, 82 (1950).
3. J. H. Lengel and V. H. Dibeler, *J. Am. Chem. Soc.*, **74**, 2683 (1952).
4. L. H. Long and C. I. Pulford, *J. Inorg. Nucl. Chem.*, **30**(8), 2071 (1968).
5. F. Glockling and J. R. C. Light, *J. Chem. Soc.*, A, 623 (1967).
6. C. W. Young, J. S. Koehler, and D. S. McKinney, *J. Am. Chem. Soc.*, **69**, 1410 (1947).
7. D. F. Van de Vondel, *J. Organometallic Chem.*, **3**, 400 (1965).

11. ORGANOMETALLIC METAL-METAL BONDED DERIVATIVES

Submitted by R. E. J. BICHLER,* M. R. BOOTH,† H. C. CLARK,* and B. K. HUNTER*
Checked by J. E. H. WARD‡ and W. A. G. GRAHAM‡

The synthesis of numerous organometallic compounds containing covalent metal-metal bonds has been achieved in a variety of ways. The majority of such compounds contain silicon, germanium, or tin bonded to a transition metal; the latter frequently contains carbon monoxide among its ligands. The six principal methods of synthesis, with an example of each, are as follows:

a. Metathetical reactions with metal carbonyl anions,[1] e.g.:

$$(C_6H_5)_3SnCl + NaMn(CO)_5 \rightarrow (C_6H_5)_3Sn\!-\!Mn(CO)_5 + NaCl$$

b. Insertion of metal halides into metal-metal bonds,[2] e.g.:

$$GeI_2 + Co_2(CO)_8 \rightarrow GeI_2[Co(CO)_4]_2$$

c. Elimination reactions,[3] e.g.:

$$(\pi\text{-}C_5H_5)(CO)_3MoH + (CH_3)_3SiN(CH_3)_2 \rightarrow$$
$$(\pi\text{-}C_5H_5)(CO)_3Mo\!-\!Si(CH_3)_3 + HN(CH_3)_2$$

d. Reactions involving oxidation of the transition metal,[4] e.g.:

$$(C_6H_5)_3SnCl + [\{(C_6H_5)_3P\}_4Pt] \rightarrow$$
$$[(C_6H_5)_3Sn\!-\!Pt(Cl)\{(C_6H_5)_3P\}_2] + 2(C_6H_5)_3P$$

e. Reactions of transition-metal halides with lithium derivatives of group IV elements,[5] e.g.:

$$[\{(C_2H_5)_3P\}_2PtCl_2] + 2LiGe(C_6H_5)_3 \rightarrow$$
$$[\{(C_2H_5)_3P\}_2Pt\{Ge(C_6H_5)_3\}_2] + 2LiCl$$

* Department of Chemistry, University of Western Ontario, London, Ontario, Canada.
† Department of Chemistry, University College, London, England.
‡ Department of Chemistry, University of Alberta, Edmonton, Alberta, Canada.

f. Cleavage of metal-metal bonds by group IV hydrides,[6] e.g.:

$$2(C_6H_5)_3SiH + Mn_2(CO)_{10} \rightarrow 2(C_6H_5)_3Si—Mn(CO)_5 + H_2$$

The considerable current interest in these compounds stems from the relatively novel covalent metal-metal bonds and the need to understand fully their nature and behavior[7,8] and also from the potential applications of these compounds, particularly in the field of catalysis.[6]

We describe here typical procedures for the synthesis of compounds containing Sn—Mn, Sn—Mo, and Ge—Fe bonds. Such procedures can often be extended to the preparation of a wide variety of related compounds.

A. TRIMETHYLTIN-MANGANESE PENTACARBONYL

[Trimethylstannyl(pentacarbonyl)manganese]

$$Mn_2(CO)_{10} + 2Na \rightarrow 2NaMn(CO)_5{}^{13}$$
$$(CH_3)_3SnCl + NaMn(CO)_5 \rightarrow (CH_3)_3SnMn(CO)_5 + NaCl$$

Procedure

This procedure is basically that described by Clark and Tsai.[7] The reaction is best carried out in a nitrogen-filled dry-box or glove bag.[6]* ■ *Caution. Because of the toxicity and volatility of trimethyltin halides and metal carbonyl compounds, care should be taken to avoid inhalation of these materials.* Tetrahydrofuran must be dried before use. Refluxing the tetrahydrofuran under nitrogen with potassium and benzophenone until the solution is dark blue and then distilling give tetrahydrofuran which is both water- and oxygen-free (see also Appendix I).

In a nitrogen-filled glove bag, to 200 g. of 1% sodium amalgam is added 5.8 g. of decacarbonyldimanganese(0)† in 75 ml. of dry tetrahydrofuran in a 250-ml. round-bottomed flask. The mixture is stirred for 1½ hours, after which the solution is dark

---

* The checkers did not use a dry-box or glove bag but maintained a slow flow of N_2 through the apparatus.

† Available from Alfa Inorganics, Inc., P. O. Box 159, Beverly, Mass. 01915.

green.   The supernatant solution is decanted off the amalgam into another 250-ml. flask, and 6 g. of solid trimethyltin chloride (Alfa Inorganics) is added to the solution.   The mixture is stirred for an hour at room temperature and left to stand overnight.   The flask is then connected to a vacuum system, and the bulk of the tetrahydrofuran is pumped off.   Aqueous potassium fluoride solution (25 ml. of 10% solution) and pentane (50 ml.) are then added, and the mixture is stirred or shaken thoroughly for 10 minutes.   The aqueous layer which now contains the sodium chloride and the last traces of tetrahydrofuran as well as precipitated trimethyltin fluoride is removed from the pentane solution by means of a separating funnel.   The pentane solution is filtered (medium-porosity sintered-glass filter) to remove trimethyltin fluoride and is then cooled to $-78°C$. in a Dry Ice–acetone bath.   Trimethyltin-manganese pentacarbonyl (8.5 g., 80% yield) is precipitated and is quickly filtered on a medium-porosity sintered filter.   If a purer product is needed, it may be obtained by vacuum distillation (50°C. at 10^{-3} mm. Hg) or by vacuum sublimation at room temperature onto a $-78°C$. cold finger.   *Anal.* Calcd. for $C_8H_9SnMnO_5$: C, 26.8; H, 2.5; mol. wt., 358.8.   Found: C, 27.0; H, 2.8; mol. wt., 361.5.

Properties

Trimethyltin-manganese   pentacarbonyl   [i.e.,   trimethyl-stannyl(pentacarbonyl)manganese] is a white crystalline solid, m.p. 29.5°C. (uncorrected), which can be identified from its infrared and proton n.m.r. spectra.   In the infrared spectrum, characteristic carbonyl stretching bands[6] are observed at 2089, 1998, and 1991 cm.$^{-1}$; the methyltin protons cause absorption[6,7] at $\tau = 9.54 \pm 0.01$ p.p.m. with $J_{117SnCH} = 46.3 \pm 0.2$ Hz. and $J_{119SnCH} = 48.8 \pm 0.2$ Hz.   The crystal and molecular structures have been determined,[9] and reactions with multiple-bonded reagents such as ethylene, fluoro olefins, and sulfur dioxide have been investigated.[7,10,11]

B. TRIMETHYLTIN(π-CYCLOPENTADIENYL)MOLYBDENUM TRICARBONYL

[Trimethylstannyl-tricarbonyl(π-cyclopentadienyl)molybdenum]

$$NaC_5H_5 + Mo(CO)_6 \rightarrow Na[Mo(CO)_3(\pi\text{-}C_5H_5)] + 3CO^{14}$$
$$(CH_3)_3SnCl + Na[Mo(CO)_3(\pi\text{-}C_5H_5)] \rightarrow$$
$$(CH_3)_3SnMo(CO)_3(\pi\text{-}C_5H_5) + NaCl$$

Procedure

A 1.25-g. quantity of finely cut sodium and a 6.6-g. quantity of freshly distilled cyclopentadiene (from Eastman dicyclopenta-diene) are warmed to about 35°C. under dry nitrogen in 30 ml. of dry tetrahydrofuran in a 100-ml., two-necked, round-bottomed flask equipped with a water condenser and a magnetic stirrer. When all the sodium has reacted (about 30 minutes), a stream of nitrogen is passed through the apparatus to remove excess cyclopentadiene (10 minutes). At this stage the reaction mixture should be pale pink but may be dark red without significantly affecting the yield. A 10-g. quantity of molyb-denum hexacarbonyl (Alfa Inorganics) is added, and the mix-ture is refluxed under nitrogen with stirring overnight. The reaction mixture should now be yellow. A 7.7-g. portion of solid trimethyltin chloride (Alfa Inorganics) is added, and the reaction is stirred under nitrogen at room temperature for 3 hours. The reaction mixture is centrifuged and filtered through a medium-porosity sintered-glass filter.* The residue is washed with dichloromethane (2 × 25 ml.) and the washings com-bined with the tetrahydrofuran filtrate. The solution is evaporated to dryness on a flash evaporator, and the solid is crystallized from hexane to give 11 g. (71% yield) of pale tan trimethyltin(π-cyclopentadienyl)molybdenum tricarbonyl (m.p. 96–98°C.; corrected) [trimethylstannyl-tricarbonyl(π-cyclopen-tadienyl)molybdenum]. If a purer product is desired, the compound can be readily sublimed at 80°C. (10^{-3} mm. Hg) onto

* The checkers reported that the filtrate turned gelatinous at this point with no ill effect.

a cold finger to give pale yellow crystals (m.p. 99–100°C.; corrected). *Anal.* Calcd. for $C_{11}H_{14}SnMoO_3$: C, 32.3; H, 3.5; mol. wt., 408.8. Found: C, 32.6; H, 3.2; mol. wt., 405.

Properties

Trimethyltin(π-cyclopentadienyl)molybdenum tricarbonyl [i.e., trimethylstannyl-tricarbonyl(π-cyclopentadienyl)molybdenum] has been characterized spectroscopically.[12] Its infrared spectrum shows carbonyl stretching modes at 1997, 1922, and 1895 cm.$^{-1}$. Its proton n.m.r. spectrum shows absorption for C_5H_5 at $\tau = 4.59$ and for CH_3—Sn at $\tau = 9.52$ with $J_{117SnCH} = 47$ Hz. and $J_{119SnCH} = 49$ Hz. If the compound is pure (i.e., sublimed), it is air-stable; otherwise it decomposes in air over a period of weeks.

C. TRIMETHYLGERMANIUM(π-CYCLOPENTADIENYL)IRON DICARBONYL

[(Trimethylgermanyl)dicarbonyl(π-cyclopentadienyl)iron]

1. Trimethylbromogermane

$$(CH_3)_4Ge + Br_2 \rightarrow (CH_3)_3GeBr + CH_3Br$$

Trimethylgermanium halides are not readily available, but $(CH_3)_3GeBr$ can be made in good yield from the bromination of tetramethylgermane.*

Procedure

A 9-g. sample of tetramethylgermane and a 12-g. quantity of bromine are condensed into a Pyrex Carius tube (80-ml.) which is sealed and then warmed to about 45°C. for 2 days.† The volatiles are then condensed onto 5 g. of mercury in a similar tube which is then shaken to remove the excess bromine. The

* Tetramethylgermane can be prepared as in synthesis 10. It is also available from Alfa Inorganics, Inc., P.O. Box 159, Beverly, Mass. 01915.

† ■ *Caution. Explosion of a sealed Carius tube is always a potential hazard. Be sure a good tube with a proper seal is used. Warming should be done in a closed fume hood, behind a shield.*

reaction mixture is then decanted into a distillation apparatus, and the trimethylbromogermane is collected (100–104°C.) (yield 8–14 g.[6]).

2. Trimethylgermanium(π-cyclopentadienyl)iron Dicarbonyl

[(Trimethylgermanyl)dicarbonyl(π-cyclopentadienyl)iron]

$$[(\pi\text{-}C_5H_5)Fe(CO)_2]_2 + 2Na \rightarrow 2Na(CO)_2Fe(\pi\text{-}C_5H_5)^{15}$$
$$Na(CO)_2Fe(\pi\text{-}C_5H_5) + (CH_3)_3GeBr \rightarrow (CH_3)_3Ge\text{-}Fe(CO)_2(\pi C_5H_5)$$

Procedure

In a 500-ml., three-necked, round-bottomed flask fitted with dry nitrogen inlet and outlet and a magnetic stirrer, a 1% sodium amalgam is prepared from 2.55 g. of sodium in 255 g. of mercury. After the amalgam has cooled, 200 ml. of dry, freshly distilled tetrahydrofuran (see Appendix I and Sec. A of this synthesis) is added, followed by 9.0 g. of bis(π-cyclopentadienyl iron dicarbonyl) $[(\pi\text{-}C_5H_5)Fe(CO)_2]_2$ (Alfa Inorganics). The mixture is stirred for 15 hours under a slight positive pressure of nitrogen.

The flask is then stoppered and transferred to a nitrogen-filled glove bag, and the tetrahydrofuran solution is decanted into a similar flask. Ten grams of trimethylbromogermane is added with a dropper, and the solution is stirred with a magnetic stirrer for 2 hours. The flask is then attached to a trap (250-ml. capacity) which is cooled in Dry Ice–acetone and, under vacuum (3–5 cm.), the bulk of the tetrahydrofuran is removed. The remaining material is extracted with benzene (six 15-ml. portions); the benzene solution is centrifuged and then filtered on a medium-porosity sintered-glass filter. Most of the benzene is removed by passing dry nitrogen through the filtration apparatus. The benzene may also be removed on a flash evaporator as long as exposure to air is kept to a minimum. The remaining red oil is transferred to the still pot (25-ml.) of a one-piece microdistillation apparatus which is attached to a vacuum system through a $-196°$C. trap; the vacuum is applied very carefully (the remaining benzene foams badly); and the still pot is heated to about 100°C., at which point a yellow material begins to sublime into the still head.   All the remaining material is now transferred into a sublimation apparatus fitted with a water-cooled cold finger.   At 60°C., 7–8 g.[7] of $(CH_3)_3Ge—Fe(CO)_2(\pi\text{-}C_5H_5)$ sublimes. $[\pi\text{-}C_5H_5Fe(CO)_2]_2$ and other unknown material remain in the sublimer.   *Anal.* Calcd. for $C_{10}H_{14}GeFeO_2$: C, 40.80; H, 4.76.   Found: C, 40.64; H, 4.91.

Properties

$(CH_3)_3Ge—Fe(CO)_2(\pi\text{-}C_5H_5)$ is an orange waxy solid (m.p. 58–61°C.).   The proton n.m.r. spectrum in $CHCl_3$ shows C_5H_5 resonance at $\tau = 5.3$ and methylgermanium resonance at 9.5. As a liquid film, the compound shows carbonyl stretching absorptions at 1995 and 1940 cm.$^{-1}$.   After a period of months the solid appears wet and eventually decomposes.

Metathetical routes to organometallic compounds containing metal-metal bonds have been applied to many such syntheses. The method has been used to form metal-metal bonds between Ge, Sn, and Pb with transition metals for which anionic species are known (Cr, Mo, W, Mn, Re, Fe, and Co).   The method has also been used to form polynuclear species involving two or more metals, for example:

$$ClSn[Mn(CO)_5]_3 \quad \text{or} \quad (CO)_5Mn—Sn(CH_3)_2\text{-}Mo(CO)_3(\pi\text{-}C_5H_5)$$

There are no major problems to extension of this method except that the stabilities of the product may become sufficiently reduced that yields are low, and handling becomes difficult. An example is $(CH_3)_3Sn—Cr(CO)_3(\pi\text{-}C_5H_5)$ for which the yield may be less than 20%, and air decomposition is very rapid unless the product has been very carefully purified by sublimation.

One group of compounds for which the metathetical route cannot be used is that in which silicon is bonded to a transition metal, for example, $(CH_3)_3Si\text{-}Mo(CO)_3\pi\text{-}C_5H_5$.   For these compounds, the elimination reaction used by Lappert[3] is probably the most generally useful one:

$$(CH_3)_3SiN(CH_3)_2 + HMo(CO)_3(\pi\text{-}C_5H_5) \rightarrow$$
$$(CH_3)_3Si\text{---}Mo(CO)_3(\pi\text{-}C_5H_5) + HN(CH_3)_2$$

References

1. R. D. Gorsich, *J. Am. Chem. Soc.*, **84**, 2486 (1962).
2. D. J. Patmore and W. A. G. Graham, *Inorg. Chem.*, **5**, 1405 (1966).
3. D. J. Cardin, S. A. Keppie, B. M. Kingston, and M. F. Lappert, *Chem. Commun.*, 1035 (1967).
4. A. J. Layton, R. S. Nyholm, G. A. Pneuvraticakis, and M. L. Tobe, *Chem. Ind. (London)*, 465 (1967).
5. R. J. Cross and F. Glocking, *J. Chem. Soc.*, 5422 (1965).
6. W. Jetz, P. B. Simons, T. A. J. Thompson, and W. A. G. Graham, *Inorg. Chem.*, **5**, 2217 (1966).
7. H. C. Clark and J. H. Tsai, *ibid.*, **5**, 1407 (1966).
8. N. A. D. Carey and H. C. Clark, *ibid.*, in press.
9. R. F. Bryan, *Chem. Commun.*, 355 (1967).
10. M. R. Booth and H. C. Clark, *Inorg. Chem.*, submitted for publication.
11. N. A. D. Carey and H. C. Clark, *Can. J. Chem.*, in press.
12. H. R. H. Patil and W. A. G. Graham, *Inorg. Chem.*, **5**, 1401 (1966).
13. R. B. King and F. G. A. Stone, *Inorganic Syntheses*, **7**, 198 (1963).
14. R. B. King and F. G. A. Stone, *ibid.*, **7**, 108 (1963).
15. *Ibid.*, **7**, 112 (1963).

12. DIMETHYLGALLIUM(III) HYDROXIDE
(*Hydroxodimethylgallium*)

$$GaCl_3 + 3LiCH_3 \xrightarrow[\substack{(C_2H_5)_2O \\ -78°C. \text{ to } 25°C.}]{\text{pet. ether}} \begin{bmatrix} \text{solution;} \\ Ga(CH_3)_3 \\ \text{not} \\ \text{isolated} \end{bmatrix} \xrightarrow{H_2O} [Ga(CH_3)_2OH]_4$$

Submitted by M. J. SPRAGUE,* G. E. GLASS,* and R. S. TOBIAS*
Checked by JOHN RIE† and JOHN P. OLIVER†

The only published synthesis[1] of dimethylgallium(III) hydroxide, $(CH_3)_2GaOH$, which exists in the solid state as the tetrameric molecule, $[(CH_3)_2GaOH]_4$,[2] involves prior preparation of

* Department of Chemistry, Purdue University, Lafayette, Ind. 47907.
† Department of Chemistry, Wayne State University, Detroit, Mich. 48202.

the very air-reactive trimethylgallium, conversion to the ether addition compound, and subsequent careful hydrolysis at low temperature on the vacuum line. The compound also is formed when either $[(CH_3)_2GaCN]_4$[3] or $[(C_6H_5)_2PGa(CH_3)_2]_2$[4] is hydrolyzed by atmospheric moisture. A simple procedure requiring no special apparatus has now been devised, whereby dimethylgallium(III) hydroxide can be obtained readily from gallium(III) chloride in about 80% yield.

Procedure

Two weighed glass ampules of gallium(III) chloride, prepared by burning gallium metal in a stream of dry chlorine,[5] are opened in a nitrogen-filled glove bag and placed in a 500-ml. conical flask. The ampules are covered with approximately 50 ml. of petroleum ether (b.p. 30–50°C.), and then 75 ml. of dry diethyl ether is added, dropwise at first, to moderate the reaction of the ether with the halide. After complete reaction, the pink ethereal solution is transferred to a 500-ml., 24/40 S.T., round-bottomed flask containing a magnetic stirring bar; the empty ampules are washed with ether; the washings are added to the main solution; and the flask is stoppered. The broken ampules are removed from the glove bag, dried, and reweighed to give the weight of gallium(III) chloride taken (18.1 g., 0.103 mole, as $GaCl_3$). Inside the nitrogen-filled glove bag, a measured volume of standard methyllithium solution in ether—in this case 220 ml. of 1.49 M (0.328 mole, a 6% excess)—is placed in a pressure-equalized dropping funnel bearing 24/40 S.T., ground-glass joints, which is then stoppered. After its removal from the glove bag, the dropping funnel is fitted with a T-piece tube through which passes a slow stream of dry nitrogen, and it is then attached to the flask containing the gallium chloride etherate.

The halide solution is then cooled to −78°C., and the methyllithium solution is added slowly over a one-hour period with continuous stirring. After all the alkyl has been added, the liquid (which contains precipitated lithium chloride) is warmed to room temperature to ensure complete reaction, and after 30 minutes it is recooled to $-78°C$. To the cold, stirred liquid is then added dropwise 10 ml. of water over a 5-minute period. The mixture is allowed to warm slowly (2 hours), with continuous stirring, to room temperature, when a further 25 ml. of water is added to dissolve all the lithium chloride. The ethereal solution is then decanted and retained, and the aqueous phase is adjusted to an approximate pH value of 7 with dilute hydrochloric acid or sodium hydroxide, as necessary. The aqueous solution is extracted twice with 50-ml. portions of diethyl ether, which are added to the main organic solution. The ethereal solution is evaporated in a stream of dry nitrogen at room temperature to yield a brown oil. Treatment of this oil with 25 ml. of petroleum ether (b.p. 30–60°C.), followed by evaporation in a nitrogen stream—repeated several times if necessary—causes solidification of the oil to a buff-colored solid. Recrystallization of this solid from petroleum ether (b.p. 30–60°C.) at least three times* gives 8.6 g. (71% of theory) of pure dimethylgallium(III) hydroxide as a white crystalline solid melting with decomposition at 86.5–88.5°C. (literature,[1] 87–88.5°C.). *Anal.* Calcd. for $(CH_3)_2GaOH$: C, 20.57; H, 6.04; Ga, 59.69. Found: C, 20.27; H, 6.11; Ga, 59.86. A further 0.8 g. of pure compound can be isolated from the mother liquors to bring the total yield to 79%.

Properties

The pure hydroxide, which has an unpleasant, sweet odor, is reasonably air-stable and undergoes only very slow decomposition if stored in a stoppered vial in the freezing compartment of

---

* *Note:* If the crude solid is not recrystallized immediately, it should be stored in a refrigerator since there is rather rapid decomposition of the moist, impure hydroxide at room temperature. The compound cannot be purified satisfactorily by sublimation without considerable loss due to decomposition.

a refrigerator. Although the crystals are insoluble in water, they are readily soluble (without rupture of the gallium-carbon bonds) in organic solvents, dilute aqueous acids, and aqueous sodium hydroxide. The compound reacts rapidly and quantitatively with acetylacetone to give the air- and water-stable derivative, $(CH_3)_2Ga(C_5H_7O_2)$, m.p. 22°C.[6]

References

1. M. E. Kenney and A. W. Laubengayer, *J. Am. Chem. Soc.*, **76**, 4839 (1954).
2. G. S. Smith and J. L. Hoard, *ibid.*, **81**, 3907 (1959).
3. G. E. Coates and R. N. Mukherjee, *J. Chem. Soc.*, 229 (1963).
4. G. E. Coates and J. Graham, *ibid.*, 233 (1963).
5. N. N. Greenwood and K. Wade, *ibid.*, 1527 (1956).
6. G. E. Coates and R. G. Hayter, *ibid.*, 2519 (1953).

II. METAL β-KETO-ENOLATE SYNTHESES

Some General Observations on the Syntheses of Metal β-keto-enolate Compounds

Submitted by JOHN P. FACKLER*

(This section on metal β-keto-enolates was assembled in cooperation with Prof. John P. Fackler.)

The syntheses of metal complexes of enolizable β-diketones continue to enjoy considerable popularity. Some important procedures which have appeared since the 1957 article[1] by Fernelius, Terada, and Bryant have been reviewed[2] recently by this author. Among them, the preparation of β-diketone complexes through the reaction of metal carbonyls with the β-diketones[3] seems to be reasonably general. However, this reaction may lead to mixed ligand complexes such as the manganese(I) complexes of Wojcicki et al.[4] and the rhodium(I) and iridium(I) dicarbonyls of Bonati et al.[5]

* Department of Chemistry, Case Western Reserve University, Cleveland, Ohio 44106.

The syntheses of very volatile fluorine-containing metal β-keto-enolates led Sievers and co-workers[6] to use the volatility in the development of chromatographic techniques for metal-ion microanalyses.[7]

High-coordination-number complexes of β-keto-enolates continue to be obtained with the metals such as zirconium(IV),[8] hafnium(IV),[8] cerium(IV),[9] and the lanthanons(III),[10] the last being tetrakis anionic species. At least one example of a volatile tetrakis β-keto-enolate salt has been reported,[11] Cs[Y(CF$_3$-COCHCOCF$_3$)$_4$]. The ionic charge on the β-keto-enolate complex has been shown to produce[12] a high field nuclear magnetic resonance for anions and low field shifts for cations, relative to the positions observed for the neutral species.

Collman[13] described numerous ligand reaction methods for preparing methylene-substituted β-diketone complexes. Often these materials cannot be prepared by other techniques because of the lack of stability either of the ligand or its alkali metal salt. A particularly useful procedure leading to the introduction of a functionally active substituent has been reported.[14]

A summary of recent synthetic work in the field of β-keto-enolate complexes would be grossly incomplete without mention of the successful syntheses of both mono- and dithio-β-diketone metal derivatives. A wide variety of both the ligands and the complexes of the monothio derivatives has been reported by Livingstone and co-workers.[15] Martin and Stewart[16] succeeded in preparing cobalt(II), nickel(II), palladium(II), and platinum-(II) derivatives of 2,4-pentanedithione by the reaction of hydrogen sulfide with the β-diketone in ethanolic hydrogen chloride containing the appropriate metal ion. This ligand is itself unstable, and a dimeric species is formed if the metal ion is absent during the reaction.

References

1. W. C. Fernelius, K. Terada, and B. E. Bryant, *Inorganic Syntheses*, **6**, 147 (1957).

2. J. P. Fackler, Jr., *Progr. Inorg. Chem.*, **7**, 361 (1966).
3. J. C. Goan, C. H. Huether, and H. E. Podall, *Inorg. Chem.*, **2**, 1078 (1963).
4. M. Kilner and A. Wojcicki, *ibid.*, **4**, 591 (1965); F. A. Hartman, M. Kilner, and A. Wojcicki, *ibid.*, **6**, 34 (1967); see synthesis 15.
5. F. Bonati and G. Wilkinson, *J. Chem. Soc.*, 3156 (1964); F. Bonati and R. Ugo, *J. Organometallic Chem.*, **7**, 167 (1967).
6. See synthesis 13.
7. R. W. Moshier and R. E. Sievers, "Gas Chromatography of Metal Chelates," Pergamon Press, New York, 1965; D. W. Meek and R. E. Sievers, *Inorg. Chem.*, **6**, 1105 (1967).
8. J. W. Silverton and J. L. Hoard, *Inorg. Chem.*, **2**, 243 (1963).
9. See synthesis 14.
10. H. Bauer, J. Blanc, and D. L. Ross, *J. Am. Chem. Soc.*, **86**, 5125 (1964).
11. S. J. Lippard, *ibid.*, **88**, 4300 (1966); S. J. Lippard, F. A. Cotton, and P. Legzdins, *ibid.*, **88**, 5930 (1966).
12. R. C. Fay and N. Serpone, *ibid.*, **90**, 5701 (1968); see synthesis 21.
13. J. P. Collman, Reactions of Coordinated Ligands, *Advan. Chem. Ser. 37*, **78** (1963); *Angew. Chem. Intern. Ed. Engl.*, **4**, 132 (1965).
14. R. H. Barker, J. P. Collman, and R. L. Marshall, *J. Org. Chem.*, **29**, 3216 (1964); see synthesis 16.
15. R. K. Y. Ho, S. E. Livingstone, and T. N. Lockyer, *Australian J. Chem.*, **19**, 1179 (1966) and references therein.
16. R. L. Martin and I. M. Stewart, *Nature*, **210**, 522 (1966).

13. TRIS(1,1,1,2,2,3,3-HEPTAFLUORO-7,7-DIMETHYL-4,6-OCTANEDIONATO)IRON(III) AND RELATED COMPLEXES

$$a.\ \text{Fe} + 3\text{H(fod)} \rightarrow \text{Fe(fod)}_3 + \tfrac{3}{2}\text{H}_2$$
$$b.\ \text{FeCl}_3{\cdot}\text{XH}_2\text{O} + 3\text{H(fod)} \rightarrow \text{Fe(fod)}_3 + 3\text{HCl} + \text{XH}_2\text{O}$$

Submitted by ROBERT E. SIEVERS* and JOSEPH W. CONNOLLY*
Checked by ROBERT H. BARKER† and JAMES E. BOSTIC, JR.†

The iron(III) chelate of 1,1,1,2,2,3,3-heptafluoro-7,7-dimethyl-4,6-octanedione, $\text{CF}_3\text{CF}_2\text{CF}_2\text{C(OH)}{:}\text{CHCOC(CH}_3)_3$, hereafter called H(fod),[1] can be made by direct combination of elemental iron with H(fod) or by the reaction of iron(III) chloride[2] and

* Aerospace Research Laboratories, ARC, Wright-Patterson Air Force Base, Ohio.
† Department of Chemistry, Clemson University, Clemson, S.C. 29631.

H(fod). The former reaction occurs when a mixture of powdered iron and H(fod) is refluxed under a nitrogen atmosphere. This type of reaction, catalyzed by trace amounts of water, is the basis of a new simple technique for the microanalysis of metals by gas chromatography.[3]

Procedure (I)

A solution of 5.87 g. (19.81 mmoles) of H(fod)* in 60 ml. of ether is heated at reflux, with stirring, under a nitrogen atmosphere with 0.36 g. (6.45 mmoles) of powdered iron and 20 drops of water. A deep red solution, obtained after 8 hours of refluxing, is filtered hot to remove remnants of unreacted iron. The solvent and excess liquid are removed under a nitrogen stream on a steam bath. The resulting red crystals (m.p. 68–72°C.) are sufficiently pure to be used for most purposes without additional purification. However, to remove trace amounts of H(fod) and other possible contaminants, the crystals can be fractionally sublimed at 95–110°C. at a pressure of 0.5 mm. Hg. The first pale red fraction on the cold finger is discarded. The main fraction is a bright red aggregate (m.p. 76–78°C.). Yield is 92% (checkers reported yield of 87%). *Anal.* Calcd. for $Fe(C_{10}H_{10}F_7O_2)_3$: C, 38.27; H, 3.05; F, 42.38; Fe, 5.93; mol. wt., 941.4. Found: sample melting at 68–72°C.:† C, 38.10; H, 3.03; F, 42.06; Fe, 6.16; mol. wt. in CCl_4, 930. Found: sample melting at 76–78°C.:† C, 38.11; H, 3.24; F, 42.30; Fe, 6.09; mol. wt. in CCl_4, 935.

Procedure (II)

A 3-g. (10.0-mmole) sample of H(fod) is added with stirring to a solution containing an excess of initially anhydrous iron(III)

---

* The ligand may be synthesized by the method described in reference 1 or is available from Pierce Chemical Company, Rockford, Ill. All operations should be conducted in a fume hood.

† Melting points are uncorrected. Microanalyses are by Galbraith Laboratories, Inc., Knoxville, Tenn.

chloride (0.65 g., 4.0 mmoles), dissolved in 150 ml. of 95% ethanol. Water is added until two liquid phases are formed. This mixture is extracted with three 50-ml. portions of hexane; the combined hexane extract is washed well with water. Hexane is removed *in vacuo*, yielding 3.10 g. of bright red crystals. The crude product is fractionally sublimed (see Procedure I). In a typical run, 2.90 g. of bright red crystals (m.p. 73–75°C.) were collected. Yield is 93.5%. *Anal.* Calcd. for $Fe(C_{10}H_{10}F_7O_2)_3$: C, 38.27; H, 3.21; F, 42.38; Fe, 5.93; mol. wt., 941.4. Found: C, 38.22; H, 3.05; F, 42.36; Fe, 6.04; mol. wt. in CCl_4, 944.

Properties

The iron(III) complex is sufficiently volatile and thermally stable to permit chromatography in the gas phase. The retention volume (relative to *n*-hexadecane) was determined with a column containing 10% poly(dimethylsiloxane) on 60–80 mesh silanized calcined diatomaceous earth.* The value obtained at a column temperature of 170°C. is 1.29.

The compound freely dissolves in aliphatic and aromatic hydrocarbon solvents, chloroform, dichloromethane, carbon tetrachloride, ether, acetone, and the lower alcohols. It is insoluble in water but remarkably soluble in the above solvents.

Further Compounds

It is to be noted that further compounds in this series can be readily prepared by adaptations of this procedure. These include the fod complexes of In(III), Ga(III), Cr(III), Mn(III), Pb(II), V(III), and Sc(III). This synthetic approach is applicable to various other β-diketones as well.

The complexes of In(III), Ga(III), and Mn(III) are prepared by stirring a mixture of approximately 10.0 mmoles of powdered

---

* SE-30 silicone gum rubber (General Electric Co.) on Gas-Chrom Z (Applied Science Laboratories, Inc.).

TABLE I Elemental Analyses

Complex	Percent theoretical				Mol. wt.	Percent experimental				Mol. wt.	M.P., °C.*
	C	H	F	Metal		C	H	F	Metal		
$In(fod)_3$	36.03	3.02	39.90	11.40	1000.3	35.79	3.01	39.74	11.70	980($CHCl_3$)	<25
$Ga(fod)_3$	37.72	3.17	41.77	7.30	955.3	37.68	3.29	41.32	7.70	970($CHCl_3$)	<25
$Cr(fod)_3$	38.43	3.23	42.56	5.55	937.5	38.46	3.34	42.52	5.79	984(C_6H_6)	101–106
$Mn(fod)_3$	38.31	3.21	42.43	5.84	940.5	38.47	3.32	41.60	5.92	903($CHCl_3$)	71–73
$V(fod)_3$	38.47	3.23	42.61	5.44	936.5	38.67	3.24	42.42	5.21	935($CHCl_3$)	69–73
$Sc(fod)_3$	38.78	3.25	42.88	4.83	930.5	38.72	3.12	41.22	4.51	820($CHCl_3$)	<25
$Pb(fod)_2$	30.12	2.53	33.35	25.98	797.5	30.09	2.53	33.09	25.71	806($CHCl_3$)	74–75.1
$Sc(fod)_3 \cdot DMF$	39.49	3.72	39.76	4.48	.....	39.70	3.90	40.01	4.71	....	49–51

* The large melting-point ranges observed for some complexes may be caused by the presence of both cis and trans isomers arising from the asymmetry of the ligand.

metal, a 50% excess of H(fod), and 4 drops of water at reflux until dissolution is essentially complete. Complete dissolution of the metal requires 16 hours for gallium, 45 hours for indium, and 10 hours for manganese. Twenty-five milliliters of chloroform is added, and the resulting mixture is filtered to remove remnants of unreacted metal. The residue is washed with an additional 25-ml. portion and the washing combined with the filtrate. The solvent and excess ligand are removed *in vacuo* at 50°C. The resulting crude product is fractionally sublimed at atmospheric pressure in a thermal-gradient sublimation apparatus[4] for 12 hours at a temperature gradient of 50–150°C. with a helium flow rate of 20 ml./minute.

The fod complexes of Cr(III), V(III), and Pb(II) are prepared in a similar manner. Complete dissolution of the metal occurs in 26 hours for lead, in 42 hours for vanadium, and in 8 hours for chromium. After the excess ligand is removed *in vacuo* at 50°C., the complex is isolated from the reaction product in a conventional sublimation apparatus[5] at 95–105°C./0.5 mm. Hg.

In the instance of scandium, where dissolution is complete in 12 hours, the complex is converted to its dimethylformamide (DMF) adduct,[6] $Sc(fod)_3 \cdot DMF$, by heating the reaction product in excess DMF. Three recrystallizations of the crude adduct from DMF at 10°C. yield a pure DMF adduct. The scandium complex, $Sc(fod)_3$, is regenerated from the adduct by heating at 60°C. *in vacuo* for 12 hours.

References

1. R. E. Sievers, K. J. Eisentraut, D. W. Meek, and C. S. Springer, Jr., "Proceedings of 9th International Conference on Coordination Chemistry," W. Schneider (ed.), p. 479, Verlag Helvetica Chimica Acta, Basle, Switzerland, 1966; C. S. Springer, Jr., D. W. Meek, and R. E. Sievers, *Inorg. Chem.*, **6**, 1105 (1967).
2. R. N. Hazeldine, W. K. R. Musgrave, F. Smith, and L. M. Turton, *J. Chem. Soc.*, 609 (1951).
3. R. E. Sievers, J. W. Connolly, and W. D. Ross, *J. Gas Chromatography*, **5**, 241 (1967); W. D. Ross and R. E. Sievers, *Anal. Chem.*, **41**, 1109 (1969).

4. C. S. Springer, Jr., Ph. D. thesis, The Ohio State University, 1967.
5. A. I. Vogel, "Practical Organic Chemistry," p. 156, Longmans, Green & Co., London, 1956.
6. C. S. Springer, Jr., unpublished results.

14. TETRAKIS(2,4-PENTANEDIONATO)CERIUM(IV) AND TETRAKIS(1,1,1-TRIFLUORO-2,4-PENTANEDIONATO)CERIUM(IV)

[*Cerium(IV) Acetylacetonate and Cerium(IV) Trifluoroacetylacetonate*]

Submitted by THOMAS J. PINNAVAIA* and ROBERT C. FAY†
Checked by J. P. FACKLER,‡ J. FETCHIN,‡ and WILLIAM SEIDEL‡

Tetrakis(2,4-pentanedionato)cerium(IV) has been prepared by air oxidation of tris(2,4-pentanedionato)cerium(III) in hot benzene solution[1] and by the reaction of hydrous cerium(IV) oxide with excess 2,4-pentanedione (acetylacetone) in water.[2] The procedures presented here are based on the former approach.

A. TETRAKIS(2,4-PENTANEDIONATO)CERIUM(IV)

[Cerium(IV) Acetylacetonate]

$$Ce(NO_3)_3 \cdot 6H_2O + 3C_5H_8O_2 + 3NH_3 \rightarrow$$
$$Ce(C_5H_7O_2)_3 + 3NH_4NO_3 + 6H_2O$$
$$4Ce(C_5H_7O_2)_3 \xrightarrow{O_2} 3Ce(C_5H_7O_2)_4 + \cdots$$

Procedure

To a solution containing 10.8 g. (0.0249 mole) of cerium(III) nitrate 6-hydrate (Fisher Scientific Co.) in 26 ml. of 2.8 M nitric acid is added, with stirring, a second solution containing 12.7 ml. (0.124 mole) of freshly distilled 2,4-pentanedione (Matheson Coleman and Bell; b.p. 136–140°C.) in 36 ml. of

* Michigan State University, East Lansing, Mich. 48823.
† Cornell University, Ithaca, N.Y. 14850. Financial support by the National Science Foundation is gratefully acknowledged.
‡ Case Western Reserve University, Cleveland, Ohio 44106.

2.0 M ammonia.   The pH of the mixture is increased to 6.0
by dropwise addition of 2.0 M ammonia; the resulting yellow
precipitate of hydrous tris(2,4-pentanedionato)cerium(III) is
filtered,* washed with three 25-ml. portions of water, and dried
in air at room temperature.   The precipitate is then slurried
in 250 ml. of benzene in an open beaker at a temperature slightly
below the boiling point.   (■ *Hood!*)

After a 1½-hour heating time, the dark red-brown solution
of tetrakis(2,4-pentanedionato)cerium(IV) is transferred to a
glass-stoppered 500-ml. Erlenmeyer flask, boiled down to a
volume of 40 ml., diluted to 300 ml. with hot hexane, and
finally cooled in a Dry Ice–acetone bath.   The fine, needle-
shaped, red-black crystals are quickly filtered and washed with
two 25-ml. portions of hexane.   A small amount (*ca.* 0.35 g.) of
the 6.55 g. of product is a powdery, insoluble, yellow impurity.
The yield based on cerium(III) nitrate 6-hydrate is 62%.†

Two recrystallizations from benzene–hexane are required for
purification of the product.   The recrystallizations are carried
out by dissolving the crystals in 50 ml. of benzene, filtering,
boiling off enough benzene to obtain 20–30 ml. of hot solution,
diluting to 125 ml. with hot hexane, and cooling in an ice–salt
water bath (−5°C.).   After the second recrystallization, the
crystals are washed with two 20-ml. portions of hexane and
dried *in vacuo* at 60°C. for 20 minutes.   *Anal.* Calcd. for
$Ce(C_5H_7O_2)_4$: C, 44.61; H, 5.36; Ce, 26.04.   Found: C, 44.77;
H, 5.26; Ce, 26.09.

---

* All filtrations are carried out under suction, using a glass-frit Büchner funnel
of medium (10–20 μ) porosity.

† Addition of 0.0249 mole of 2,4-pentanedione before the air-oxidation step
increased the yield to 10.4 g., of which *ca.* 0.7 g. was the insoluble, yellow impurity.
Under these conditions the reaction presumably proceeds according to the equation:

$$2Ce(C_5H_7O_2)_3 + 2C_5H_8O_2 + \tfrac{1}{2}O_2 \rightarrow 2Ce(C_5H_7O_2)_4 + H_2O$$

This modification is not of general utility, however.   Addition of a stoichiometric
amount of 1,1,1-trifluoro-2,4-pentanedione prior to air oxidation of $Ce(C_5H_4F_3O_2)_3$
to $Ce(C_5H_4F_3O_2)_4$ in the procedure of Sec. B produced extensive decomposition
and resulted in low yields of the tetrakis chelate.

B. TETRAKIS(1,1,1-TRIFLUORO-2,4-PENTANEDIONATO)-CERIUM(IV)

[Cerium(IV) Trifluoroacetylacetonate]

$$Ce(NO_3)_3 \cdot 6H_2O + 3C_5H_5F_3O_2 + 3NH_3 \rightarrow$$
$$Ce(C_5H_4F_3O_2)_3 + 3NH_4NO_3 + 6H_2O$$
$$4Ce(C_5H_4F_3O_2)_3 \xrightarrow{O_2} 3Ce(C_5H_4F_3O_2)_4 + \cdots$$

Procedure

An 8-ml. (0.064-mole) quantity of freshly distilled 1,1,1-trifluoro-2,4-pentanedione(trifluoroacetylacetone, b.p. 106.5–107°C.) (Columbia Organic Chemicals Co., Inc.) is added to 23 ml. of 2.0 M ammonia, and 45 ml. of water is added to dissolve the white precipitate. The resulting solution is added dropwise with rapid stirring to a solution containing 5.80 g. (0.0134 mole) of cerium(III) nitrate 6-hydrate (Fisher Scientific Co.) in 14 ml. of 1.4 M nitric acid. After the pH of the mixture has been adjusted to 6.0 by dropwise addition of 2.0 M ammonia, the yellow precipitate of hydrous tris(1,1,1-trifluoro-2,4-pentanedionato)cerium(III) is filtered, washed with three 25-ml. portions of water, and dried in air at room temperature. The dry precipitate (9.7 g.) is slurried in 250 ml. of chlorobenzene at 120–125°C. in a 500-ml. round-bottomed flask, while a gentle stream of air is simultaneously passed through the mixture by means of a gas dispersion tube. (■ *Hood!*) After a 1½-hour heating time, the airflow is increased, and the red-brown solution of tetrakis(1,1,1-trifluoro-2,4-pentanedionato)cerium(IV) is allowed to evaporate down to a volume of 50 ml. Then 75 ml. of hot hexane is carefully added, and the solution is cooled in a Dry Ice–acetone bath. The fine, needle-shaped, red-brown crystals are quickly filtered and washed with two 20-ml. portions of hexane. The product weighs 6.85 g. but contains about 0.5 g. of a powdery, yellow impurity which is only slightly soluble in benzene; the yield based on cerium(III) nitrate 6-hydrate is 84%. The product is purified by performing three recrystallizations from benzene–hexane as described in the procedure of Sec. A. The crystals are finally dried *in vacuo* at 80°C. for 20 minutes. *Anal.* Calcd. for $Ce(C_5H_4F_3O_2)_4$: C, 31.92; H, 2.14; Ce, 18.62; F, 30.30. Found: C, 32.06; H, 2.33; Ce, 18.82; F, 30.11.

Properties

Tetrakis(2,4-pentanedionato)cerium(IV) is an eight-coordinate metal complex which crystallizes from organic solvents as dark red-black, needle-shaped, monoclinic crystals ($a = 11.70$, $b = 12.64$, $c = 16.93$ A., $\beta = 112°15'$, four molecules per unit cell). The coordination polyhedron is a slightly distorted square antiprism.[3] The compound exhibits a melting point which varies with heating time; rather sharp melting points can be observed in the range 140–165°C. Tetrakis(2,4-pentanedionato)cerium(IV) is soluble in benzene, chloroform, ethanol, and acetone but only sparingly soluble in hexane and diethyl ether. It undergoes destructive oxidation in diphenyl ether solution under 1 atmosphere of pure oxygen at 100°C.[1] The n.m.r. spectrum in chloroform solution (concentration, 10 g./100 ml. of solvent) shows a single ring-proton resonance at $\tau = 4.69$ and a single methyl resonance at $\tau = 8.09$.

Tetrakis(1,1,1-trifluoro-2,4-pentanedionato)cerium(IV) is a dark red-brown crystalline substance which melts at 132–132.5°C. Its solubility properties are similar to those of tetrakis(2,4-pentanedionato)cerium(IV). In chloroform solution (concentration, 10 g./100 ml. of solvent) the compound exhibits n.m.r. lines at $\tau = 4.19$ (ring protons) and $\tau = 7.84$ (methyl protons).

References

1. M. Mendelsohn, E. M. Arnett, and H. Freiser, *J. Phys. Chem.*, **64**, 660 (1960).
2. A. Job and P. Goissedet, *Compt. Rend.*, **157**, 50 (1913).
3. B. Matković and D. Grdenić, *Acta Cryst.*, **16**, 456 (1963).

15. TETRACARBONYL(1,1,1,5,5,5-HEXAFLUORO-2,4-PENTANEDIONATO)MANGANESE(I)

[*Manganese(I) Tetracarbonyl Hexafluoroacetylacetonate*]

$$Tl_2CO_3 + 3C_5H_2F_6O_2 \rightarrow 2Tl(C_5HF_6O_2) + C_5H_2F_6O_2{\cdot}H_2O + CO_2$$
$$Mn(CO)_5Br + Tl(C_5HF_6O_2) \rightarrow Mn(CO)_4(C_5HF_6O_2) + TlBr + CO$$

Submitted by FREDERICK A. HARTMAN,* PAULA L. JACOBY,* and
ANDREW WOJCICKI*
Checked by KENNETH V. TAKVORIAN† and ROBERT H. BARKER†

The reaction of metal carbonyls with 1,3-diketones generally results in a complete displacement of carbon monoxide accompanied by oxidation of the metal to yield 1,3-diketonato complexes. For example, iron pentacarbonyl, chromium hexacarbonyl, and molybdenum hexacarbonyl afford $Fe(C_5H_7O_2)_3$,[1] $Cr(C_5H_7O_2)_3$,[2] and $Mo(C_5H_7O_2)_3$,[2,3] respectively, when allowed to react with 2,4-pentanedione.

Recently Bonati and Wilkinson[4] reported that tetracarbonyl-di-μ-chloro-dirhodium(I) reacts with various 1,3-diketones in the presence of barium carbonate to give compounds such as $Rh(CO)_2(C_5H_7O_2)$. Similarly, chloro(pentacarbonyl)manganese(I) and 1,1,1,5,5,5-hexafluoro-2,4-pentanedione afford the complex $Mn(CO)_4(C_5HF_6O_2)$ in low yields (14%).[5]

The synthetic procedure reported here, the reaction of bromo(pentacarbonyl)manganese(I) and thallium(I) 1,1,1,5,5,5-hexafluoro-2,4-pentanedionate, is simple and gives good yields of $Mn(CO)_4(C_5HF_6O_2)$. It can be readily adapted to the corresponding trifluoro derivative, $Mn(CO)_4(C_5H_4F_3O_2)$, a number of phosphine-substituted manganese(I) 1,3-diketonatocarbonyls, and various 1,3-diketonatodicarbonylrhodium(I) complexes by using $Rh_2(CO)_4Cl_2$.

* Department of Chemistry, The Ohio State University, Columbus, Ohio 43210.
† Department of Chemistry, Clemson University, Clemson, S.C. 29631.

A. THALLIUM(I) 1,1,1,5,5,5-HEXAFLUORO-
2,4-PENTANEDIONATE

[Thallium(I) Hexafluoroacetylacetonate]

Procedure

■ *Caution. Metal carbonyls and thallium compounds, espe-cially the volatile 1,3-diketonates, are toxic. All experiments with these chemicals should be conducted carefully in a well-ventilated fume hood.*

1,1,1,5,5,5-Hexafluoro-2,4-pentanedione (Columbia Organic Chemicals Co., Columbia, S.C.) is washed with twice its volume of concentrated sulfuric acid and distilled immediately before use, b.p. 63–65°C. Thallium(I) carbonate may be purchased from Alfa Inorganics, Inc., Beverly, Mass. Chloroform and pentane are reagent- and technical-grade solvents, respectively.

A 100-ml. round-bottomed flask equipped with a magnetic stirring bar and containing a slurry of 10 g. (0.021 mole) of finely powdered thallium(I) carbonate in 10 ml. of chloroform is immersed in an oil bath placed on a stirrer–hot plate. A solution of 13.4 g. (0.064 mole) of 1,1,1,5,5,5-hexafluoro-2,4-pentanedione in 15 ml. of chloroform is introduced in one portion to the flask, which is then connected to a gas bubbler by means of a ground-glass adapter and Tygon tubing. Evolution of carbon dioxide commences immediately as the contents of the flask are stirred. When effervescence subsides, the mixture is heated, with stirring, for 15 minutes at 50°C. The bath is then removed, and the contents of the flask are allowed to cool to room temperature.* The mixture is filtered, and 25 ml. of pentane is added, with stirring, to the filtrate. Cooling this solution to *ca.* −75°C. in a Dry Ice–ethanol bath causes deposi-tion of pale yellow needles and platelets. Further purification

---

* The checkers noted that the mixture became only slightly cloudy on cooling to room temperature. However, the cooling due to evaporation during filtration gave premature precipitation of 1.2 g. of the product.

can be effected by sublimation at 45°C./0.1 mm. to give 7.4 g. (86% yield) of Tl($C_5HF_6O_2$). *Anal.* Calcd. for Tl($C_5HF_6O_2$): C, 14.60; H, 0.24; F, 27.71. Found: C, 14.71; H, 0.25; F, 27.68.

B. TETRACARBONYL(1,1,1,5,5,5-HEXAFLUORO-2,4-PENTANEDIONATO)MANGANESE(I)

[Manganese(I) Tetracarbonyl Hexafluoroacetylacetonate]

The preparation of tetracarbonyl(1,1,1,5,5,5-hexafluoro-2,4-pentanedionato)manganese(I) is carried out in a 100-ml., three-necked, round-bottomed flask equipped with a magnetic stirring bar, a dropping funnel with a pressure-equalizing sidearm, a ground-glass adapter connected by means of Tygon tubing to a gas bubbler, and a stopper. Bromo(pentacarbonyl)-manganese(I) (1.29 g., 0.0047 mole), prepared as described by Abel and Wilkinson,[6] dissolved or suspended in 30 ml. of chloroform is introduced into the flask, which is then flushed thoroughly with nitrogen. Thallium(I) 1,1,1,5,5,5-hexafluoro-2,4-pentanedionate (1.90 g., 0.0047 mole) dissolved in 30 ml. of chloroform is added dropwise from the funnel into the flask. Immediately the resulting mixture becomes cloudy as thallium(I) bromide begins to precipitate. ■ *Caution. Carbon monoxide is evolved.* The contents are stirred for 4 hours at room temperature and then for 15 minutes at 50°C. After the mixture cools to room temperature, it is poured through a 2.5 × 5 cm. column packed with powdered cellulose (Whatman, 200 mesh) in order to remove finely dispersed thallium(I) bromide. The collected orange solution is concentrated to 15–20 ml. in a stream of nitrogen and then introduced onto a 2.5 × 20 cm. column packed with Florisil (Fisher, 60–100 mesh) in chloroform. A single orange band which forms at the top of the column is eluted with a 1:1 mixture by volume of pentane and chloroform.* The solvent is removed from the

---

* Although the band spreads out extensively instead of being cleanly eluted from the column, this procedure ensures removal of volatile contaminants from the product.

eluate in a water-aspirator vacuum at room temperature. The resultant residue is then sublimed at 55°C./0.1 mm. to give 1.1 g. (62%) of the product. *Anal.* Calcd. for $Mn(CO)_4(C_5HF_6O_2)$: C, 28.9; H, 0.3; F, 30.5; Mn, 14.7. Found: C, 28.9; H, 1.0; F, 30.1; Mn, 14.8.

Properties

The bright yellow $Mn(CO)_4(C_5HF_6O_2)$ melts with decomposition at 99–100°C. It is stable to air in the solid; however, its solutions decompose slowly on storage. The compound is readily soluble in polar organic solvents, sparingly soluble in nonpolar solvents, and insoluble in water. It is diamagnetic and monomeric in solution. The proton magnetic resonance spectrum of $Mn(CO)_4(C_5HF_6O_2)$ in deuterochloroform consists of a single signal at $\tau = 3.84$ and the infrared metal carbonyl stretching bands of its pentane solutions occur at 2123 (weak), 2049 (strong), 1956 (strong), and 1947 (strong) cm.$^{-1}$.

The tetracarbonyl undergoes facile substitution reactions in solution.[7] Tertiary phosphines such as triphenylphosphine, tri-*n*-butylphosphine, and methyldiphenylphosphine as well as the phosphites—triphenyl phosphite and trimethyl phosphite—replace two carbonyl groups to give $Mn(CO)_2(C_5HF_6O_2)L_2$, where L is the entering ligand. It has been inferred from infrared and proton magnetic resonance spectral data that the ligands L are trans and the carbonyl groups are cis. Pyridine and triphenylarsine replace only one carbonyl group from $Mn(CO)_4(C_5HF_6O_2)$ to yield $Mn(CO)_3(C_5HF_6O_2)L$, in which L is trans to carbon monoxide. All these dicarbonyl and tricarbonyl derivatives are orange to purple crystalline solids, stable to air, and soluble in a wide range of organic solvents.

References

1. W. Hieber, *Sitzber. Heidelberg. Akad. Wiss.*, **3**, 3 (1929).
2. T. G. Dunne and F. A. Cotton, *Inorg. Chem.*, **2**, 263 (1963).

3. M. L. Larson and F. W. Moore, *ibid.*, **1**, 856 (1962).
4. F. Bonati and G. Wilkinson, *J. Chem. Soc.*, 3156 (1964).
5. M. Kilner and A. Wojcicki, *Inorg. Chem.*, **4**, 591 (1965).
6. E. W. Abel and G. Wilkinson, *J. Chem. Soc.*, 1501 (1959).
7. F. A. Hartman, M. Kilner, and A. Wojcicki, *Inorg. Chem.*, **6**, 34 (1967).

16. TRIS(3-CYANOMETHYL-2,4-PENTANEDIONATO)-CHROMIUM(III)

[*Chromium(III) 3-Cyanomethylacetylacetonate*]

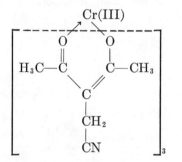

a. $Cr[H_3CC(O)CHC(O)CH_3]_3 + 3CH_2[N(CH_3)_2]_2 \xrightarrow[\text{HOOCCH}_3]{\text{HCHO}}$
$Cr\{H_3CC(O)C[CH_2N(CH_3)_2]C(O)CH_3\}_3 + 3HN(CH_3)_2$
b. $Cr\{H_3CC(O)C[CH_2N(CH_3)_2]C(O)CH_3\}_3 + 3CH_3I \rightarrow$
$(Cr\{H_3CC(O)C[CH_2N(CH_3)_3]C(O)CH_3\}_3)^{3+} + 3I^-$
c. $(Cr\{H_3CC(O)C[CH_2N(CH_3)_3]C(O)CH_3\}_3)^{3+} + 3I^- + 3KCN \rightarrow$
$Cr[H_3CC(O)C(CH_2CN)C(O)CH_3]_3 + 3(CH_3)_3N + 3KI$

Submitted by KENNETH B. TAKVORIAN* and ROBERT H. BARKER*
Checked by ROBERT B. VON DREELE† and ROBERT C. FAY†

Tris(3-cyanomethyl-2,4-pentanedionato)chromium(III), [Cr{H$_3$CC(O)C(CH$_2$CN)C(O)CH$_3$}$_3$], has recently been prepared by the aminomethylation of tris(2,4-pentanedionato)-chromium(III) followed by quaternization of the amine and subsequent displacement of trimethylamine with alkali metal cyanide.[1] The procedure described below is a modification of

* Clemson University, Clemson, S.C. 29631.
† Department of Chemistry, Cornell University, Ithaca, N.Y. 14850.

this approach, giving higher yields. Preliminary experiments in which laboratory animals were injected with solutions of the crude salt indicated a high degree of toxicity.

Procedure

To a mixture of 11.0 g. (0.031 mole) of tris(2,4-pentanedionato)chromium(III)[2] and 4.0 g. (0.12 mole) of powdered paraformaldehyde in 150 ml. of glacial acetic acid is added 12.5 ml. (0.12 mole) of $N,N,N'N'$-tetramethyldiaminomethane, prepared according to Lindsay and Hauser.[3] After stirring for 12 hours at room temperature, the acid is neutralized by slow addition to 700 ml. of saturated aqueous potassium carbonate. Sufficient quantities of ice are added periodically to keep the neutralization solution below room temperature. An excess of ammonium hydroxide is then added to make the solution basic. The triamine is extracted into chloroform from which it is recovered as a red-purple tar by evaporation (end of step *a*).

The crude triamine is dissolved in 200 ml. of 50% ethanol–water solution (v/v), and 30 ml. (68.4 g., 0.48 mole) of methyl iodide is added. After stirring half an hour, 20 ml. of ethanol is added. Stirring is continued for an additional 2 hours, after which the mixture is diluted with 50 ml. of water (end of step *b*). A solution containing 14 g. (0.21 mole) of potassium cyanide dissolved in 75 ml. of water is then added. After stirring for $2\frac{1}{2}$ hours,* the solution is extracted with chloroform to give a dark organic solution which is evaporated to a purple tar. Chromatography on Merck alumina,† with 2:1 benzene–chloroform as the eluent, effects the separation of pure tris(3-cyanomethyl-2,4-pentanedionato)chromium(III) (end of step *c*). The pure chelate

---

* A portion of the trinitrile may precipitate at this point, in which case it may be purified by filtration and recrystallization from benzene–hexane.

† Checkers found Merck (71707) reagent aluminum oxide to be satisfactory, after encountering difficulties that may have been due to use of Merck acid-washed alumina.

trinitrile is recrystallized from benzene–hexane to give 10.0 g. (68.0%) of dichroic crystals, m.p. 226°C.

Properties

The tris(3-N,N-dimethylaminomethyl-2,4-pentanedionato)-chromium(III) can be isolated as a dark, low-melting glass by drying under vacuum. Its infrared spectrum exhibits a strong absorption maximum at 1565 cm.$^{-1}$ but no other absorption in the 1500–1600 cm.$^{-1}$ region, indicating the complete substitution of all three rings. As further confirmation of complete substitution, a small peak which occurs in the 1200 cm.$^{-1}$ region of the spectra of the unsubstituted chelates and is usually associated with a wagging mode of the ring hydrogen, is absent in the spectrum of the aminomethylated product.

The triamine is extremely hygroscopic and soluble in both water and n-heptane. Because of this amazing solubility range, it cannot be crystallized easily. The crude triamine can be partially purified by chromatography on a column of alumina, with 50:50 (v/v) benzene–chloroform as the eluent, but is extremely difficult to isolate in a pure state.

The trisquaternary salt (product of equation b) is a pale purple powder which can be isolated by addition of dioxane to the ethanol–water mixture followed by filtration. In cases where the isolation of the salt is desired, better yields can be obtained by carrying out the quaternization reaction in dioxane solution. Isolation is then accomplished by addition of ethyl ether and filtration. Preliminary experiments in which laboratory animals were injected with solutions of the crude salt indicated a high degree of toxicity.

The pure trinitrile exists as dichroic crystals (m.p. 226°C.) having infrared absorption maxima at 2240 and 1570 cm.$^{-1}$ but no other absorption in the 1500–1600 cm.$^{-1}$ region. Ultraviolet absorption maxima occur at 274 mμ ($\epsilon \simeq 12,400$), 339 mμ ($\epsilon \simeq 9,500$), and 564 mμ ($\epsilon \simeq 80$).

References

1. R. H. Barker, J. P. Collman, and R. L. Marshall, *J. Org. Chem.*, **29**, 3216 (1964).
2. W. C. Fernelius and Julian E. Blanch, *Inorganic Syntheses*, **5**, 130 (1957).
3. J. K. Lindsay and C. R. Hauser, *J. Org. Chem.*, **22**, 355 (1957).

17. HALO(2,4-PENTANEDIONATO)ZIRCONIUM(IV) COMPLEXES

$$\text{ZrX}_4 + n\text{C}_5\text{H}_8\text{O}_2 \xrightarrow{\text{solvent}} \text{Zr}(\text{C}_5\text{H}_7\text{O}_2)_n\text{X}_{4-n} + n\text{HX}$$
$$(\text{X} = \text{Cl, Br}; n = 2 \text{ and } 3)$$
$$\text{ZrI}_4 + 3\text{Zr}(\text{C}_5\text{H}_7\text{O}_2)_4 \xrightarrow{\text{solvent}} 4\text{Zr}(\text{C}_5\text{H}_7\text{O}_2)_3\text{I}$$

Submitted by THOMAS J. PINNAVAIA*† and ROBERT C. FAY*
Checked by JAMES MOYER‡ and EDWIN M. LARSEN‡

The behavior of zirconium and hafnium in the formation of 2,4-pentanedione (acetylacetone) derivatives is similar to that of titanium in that all three metals form six-coordinate dihalo metal complexes of the type $M(\text{C}_5\text{H}_7\text{O}_2)_2\text{X}_2{}^{1-8}$ (where X is Cl or Br).§ On the other hand, zirconium and hafnium resemble cerium and thorium in that they give eight-coordinate complexes of the type $M(\text{C}_5\text{H}_7\text{O}_2)_4$.[9-11] Unlike the other elements of the titanium subgroup, zirconium and hafnium are also known to form $M(\text{C}_5\text{H}_7\text{O}_2)_3\text{X}$ complexes (where X = Cl, Br, and M = Zr, Hf) in which the metal has a coordination number of seven.[4,7,8] Procedures are described here for preparation of dichlorobis(2,4-pentanedionato)-, chlorotris(2,4-pentanedionato)-, bromotris(2,4-pentanedionato)-, and iodotris(2,4-pentanedionato)zirconium(IV).

* Cornell University, Ithaca, N.Y. 14850. Financial support by the National Science Foundation is gratefully acknowledged.
† Present address: Michigan State University, East Lansing, Mich. 48823.
‡ University of Wisconsin, Madison, Wis. 53706.
§ Although $\text{Ti}(\text{C}_5\text{H}_7\text{O}_2)_2\text{F}_2$ and $\text{Ti}(\text{C}_5\text{H}_7\text{O}_2)_2\text{I}_2$ are known,[5,6] the zirconium and hafnium analogs have not yet been prepared.

Dichlorobis(2,4-pentanedionato)zirconium(IV) was first prepared by Jantsch[12] by reaction of anhydrous zirconium(IV) chloride with an excess of 2,4-pentanedione in refluxing diethyl ether. An attempt to prepare the hafnium analog by the method of Jantsch afforded instead chlorotris(2,4-pentanedionato)hafnium(IV).[4] The procedure presented here is similar to that of Jantsch, but a limited amount of solvent is used so that the dihalo complex precipitates from solution as it is formed. This minimizes the danger of converting the product to the chlorotris(2,4-pentanedionato) complex by homogeneous reaction with excess 2,4-pentanedione. The same procedure is equally successful for preparation of dichlorobis(2,4-pentanedionato)hafnium(IV) although a somewhat longer reaction time (24 hours) is required. The corresponding dibromo complexes can also be prepared by this procedure; however, these syntheses are best carried out at room temperature. The recommended reaction times are 24 hours for $Zr(C_5H_7O_2)_2Br_2$ and 32 hours for $Hf(C_5H_7O_2)_2Br_2$.

Chlorotris(2,4-pentanedionato)zirconium(IV) has been prepared (1) by reaction of excess 2,4-pentanedione with zirconium(IV) chloride in refluxing chloroform[13-15] or benzene;[4] (2) by reaction of zirconium(IV) chloride with bis(2,4-pentanedionato)copper(II) in benzene,[13] tetrakis(2,4-pentanedionato)zirconium(IV) in tetrahydrofuran,[16] or tris(2,4-pentanedionato)iron(III) in ether;[4] and (3) by reaction of acetyl chloride with tetrakis(2,4-pentanedionato)zirconium(IV) in benzene.[14] Yields, when reported, have generally ranged from 50 to 70%. The identity of some of the products, however, is in doubt since the four groups of workers who have studied this compound reported four different melting points,[4,13-16] the melting points apparently depending on the research group rather than on the method of preparation. Although the possibility of several crystalline forms cannot be ruled out, it is more likely that the melting-point discrepancies are due to partial hydrolysis. The procedure presented here involves reaction of zirconium(IV) chloride with excess 2,4-pentanedione in benzene solution *under anhydrous conditions.* The synthesis affords a reproducible, high-purity, well-characterized[7,8] product in nearly quantitative yield. The procedure given for preparation of bromotris(2,4-pentanedionato)zirconium(IV) is similar, but carbon tetrachloride is used as the solvent.* The procedures described for the chloro and bromo zirconium complexes may be employed as well for synthesis of the corresponding hafnium analogs except that longer reaction times are required [Hf$(C_5H_7O_2)_3$Cl, 10 hours; Hf$(C_5H_7O_2)_3$Br, 15 hours].

Iodotris(2,4-pentanedionato)zirconium(IV) has been prepared by reaction of zirconium(IV) iodide with 2,4-pentanedione in isopropyl ether and by the ligand-exchange reaction between zirconium(IV) iodide and tetrakis(2,4-pentanedionato)zirconium(IV) in tetrahydrofuran.[7] The latter approach, which yields a higher-purity product, is described here.

General Techniques

The anhydrous zirconium(IV) halides† are freshly sublimed *in vacuo* at 250–300°C.; 2,4-pentanedione (Matheson Coleman and Bell; b.p. 136–140°C.) is freshly distilled through a small fractionating column before use. The anhydrous tetrahalides and the products are readily hydrolyzed and must be handled in a dry-box or in a plastic bag filled with dry nitrogen.‡ All glassware is dried at 180°C. and is cooled, whenever possible,

---

* Attempts to prepare the bromo derivative in benzene yielded an intractable red oil.

† Zirconium(IV) chloride (99.8 + %) was obtained from Chemicals Procurement Laboratories, Inc., 18-17 130th St., College Point, N.Y. 13356. Zirconium(IV) bromide was synthesized from the elements, using apparatus described by Young and Fletcher.[17] Zirconium(IV) iodide was prepared from the elements by the method of Lowry and Fay.[18]

‡ Commercial prepurified-grade nitrogen is satisfactory and may be used without further purification.

in a desiccator containing calcium sulfate. All solvents are reagent grade. Diethyl ether and tetrahydrofuran are dried by distillation under dry nitrogen from lithium aluminum hydride and are subsequently handled in a nitrogen-filled bag. Benzene, hexane, and carbon tetrachloride are refluxed over calcium hydride for at least 12 hours and then distilled.

All reactions are conducted under dry nitrogen or argon. With the exception of iodotris(2,4-pentanedionato)zirconium(IV), the compounds are prepared in a three-necked reaction flask equipped with 24/40 S.T. joints and fitted with a purge gas inlet, a dropping funnel, and a condenser topped with a P_2O_5 drying tube. Throughout the synthesis, a slow stream of dry nitrogen is passed into the reaction mixture through the gas inlet tube, which extends below the surface of the solvent. A magnetic stirrer is used for stirring.

Filtration and washing of the products are conveniently carried out under anhydrous conditions, using the apparatus illustrated in Fig. 5. The 24/40 S.T. joints are held together with spring clamps, and the stopcocks are fitted with retainer clips. The joints are not greased except when a vacuum is required. To filter, the medium-frit filter stick A and the sidearm Erlenmeyer flask B are attached to the reaction flask (or recrystallizing flask), and the apparatus is rotated into the position shown. To wash a precipitate on the frit, the nitrogen flow is reversed, the three-necked flask is removed, and the 125-ml. dropping funnel C, which is filled with freshly distilled washing solvent, is attached to A, using adapter D. The desired amount of wash solvent is delivered to the frit, and the nitrogen flow is allowed to pass up through the frit to agitate the precipitate. The wash solvent is then flushed from the frit by applying nitrogen pressure through the three-way stopcock of the dropping funnel. The sidearm flask and funnel are replaced with adapters E. The filter stick is evacuated, and the product is subsequently handled in a dry atmosphere.

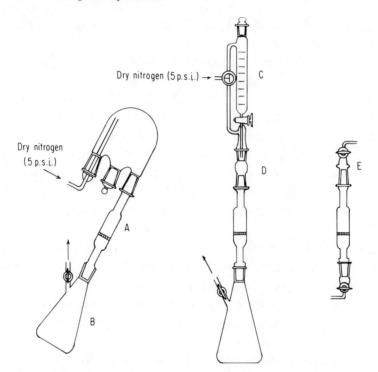

Dry nitrogen (5 p.s.i.) →

Dry nitrogen (5 p.s.i.) →

Fig. 5. Apparatus for the synthesis and purification of halo(2, 4-pentanedionato)zirconium(IV) complexes.

During all manipulations, it is important that nitrogen pressure be continuously applied to the filter stick except for the brief moments when it is necessary to reverse the direction of nitrogen flow.

Recrystallizations are performed in a glass-stoppered Erlenmeyer flask fitted with a sidearm and stopcock (similar to *B* in Fig. 5). During recrystallizations, solutions are kept out of contact with moist air, while being heated, by passing dry nitrogen into the sidearm and over the surface of the solution. The recrystallized product is filtered, washed, and collected as described above except that the recrystallizing flask is substituted for the reaction flask in the filtration step.

A. DICHLOROBIS(2,4-PENTANEDIONATO)ZIRCONIUM(IV)

[Dichlorobis(acetylacetonato)zirconium(IV)]

$$ZrCl_4 + 2C_5H_8O_2 \rightarrow Zr(C_5H_7O_2)_2Cl_2 + 2HCl$$

■ *Note. It is important that the methods described in the General Techniques subsection for excluding moisture be followed carefully.*

Procedure

To a suspension of 9.35 g. (0.0401 mole) of anhydrous zirconium(IV) chloride in 400 ml. of diethyl ether is added dropwise, with rapid stirring, 23.3 ml. (0.227 mole) of 2,4-pentanedione. After the reaction mixture is allowed to reflux for 12 hours, the white precipitate which forms is filtered, washed with two 50-ml. portions of ether, and then with two 50-ml. portions of hexane. The yield of crude product is 8.70 g. (60%). The product is recrystallized twice from benzene–hexane. Recrystallizations are carried out by dissolving the product in about 150 ml. of benzene, filtering, boiling off some solvent to obtain about 100 ml. of hot solution, adding about 50 ml. of hot hexane to obtain turbidity, and cooling in an ice bath. The colorless, prismatic crystals are dried *in vacuo* at 80°C. for half an hour. A yield of 5.65 g. of purified product was reported by the checkers. *Anal.* Calcd. for $Zr(C_5H_7O_2)_2Cl_2$: C, 33.33; H, 3.92; Cl, 19.68; Zr, 25.11. Found: C, 33.12; H, 3.88; Cl, 19.64; Zr, 25.54.

B. CHLOROTRIS(2,4-PENTANEDIONATO)ZIRCONIUM(IV)

[Chlorotris(acetylacetonato)zirconium(IV)]

$$ZrCl_4 + 3C_5H_8O_2 \rightarrow Zr(C_5H_7O_2)_3Cl + 3HCl$$

■ *Note. It is important that the methods described in the General Techniques subsection for excluding moisture be followed carefully.*

Procedure

Thirteen milliliters (0.127 mole) of 2,4-pentanedione is added dropwise to a slurry of 9.00 g. (0.0386 mole) of anhydrous zirconium(IV) chloride in 100 ml. of refluxing benzene. After a 4-hour reaction time, the condenser is removed to allow about half of the solvent to boil away, and then hot hexane is added to the clear, pale yellow solution until turbidity appears. The nitrogen flow is maintained throughout the entire operation. The flask is then stoppered and cooled in an ice bath. The yield of colorless, prismatic crystals is 15.6 g. (95%). The product is recrystallized twice from benzene–hexane as described in Sec. A, washed with two 50-ml. portions of hexane, and finally dried *in vacuo* at 80°C. for half an hour. (The checkers obtained about 7 g. of purified product.) *Anal.* Calcd. for $Zr(C_5H_7O_2)_3Cl$: C, 42.49; H, 4.99; Cl, 8.36; Zr, 21.51. Found: C, 42.26; H, 5.00; Cl, 8.36; Zr, 21.82.

C. BROMOTRIS(2,4-PENTANEDIONATO)ZIRCONIUM(IV)

[Bromotris(acetylacetonato)zirconium(IV)]

$$ZrBr_4 + 3C_5H_8O_2 \rightarrow Zr(C_5H_7O_2)_3Br + 3HBr$$

■ *Note. It is important that the methods described in the General Techniques subsection for excluding moisture be followed carefully.*

Procedure

To a slurry of 9.60 g. (0.0234 mole) of anhydrous zirconium(IV) bromide in 300 ml. of refluxing carbon tetrachloride is added dropwise 11.9 ml. (0.116 mole) of 2,4-pentanedione. After a $6\frac{1}{2}$-hour reaction time, the clear, yellow solution is cooled to room temperature and then decolorized by treatment with about 40 g. of Norit A activated carbon* which has been

---

* The checkers used Darco G-60 charcoal from Matheson Coleman and Bell. It appeared to be very fine and plugged the filter during filtration. Norit A is recommended.

previously dried in a tube furnace for an hour at 450°C. under a stream of dry nitrogen. The solution is allowed to stand over the carbon for half an hour with intermittent stirring. It is emphasized that all operations are conducted without interruption of the nitrogen flow through the flask. The carbon is filtered off and the product recovered from the filtrate by boiling off about 100 ml. of solvent, adding about 200 ml. of hot hexane, and cooling to room temperature. The yield of white dendritic crystals is 6.84 g. (62%). The product is recrystallized from carbon tetrachloride–hexane, as described in Sec. A, washed with 25 ml. of hexane, and finally dried *in vacuo* at 80°C. for half an hour. (The checkers obtained about 3 g. of purified product.) *Anal.* Calcd. for $Zr(C_5H_7O_2)_3Br$: C, 38.46; H, 4.52; Br, 17.06; Zr, 19.47. Found: C, 38.13; H, 4.21; Br, 17.23; Zr, 19.62.

D. IODOTRIS(2,4-PENTANEDIONATO)ZIRCONIUM(IV)

[Iodotris(acetylacetonato)zirconium(IV)]

$$ZrI_4 + 3Zr(C_5H_7O_2)_4 + 4C_4H_8O \rightarrow 4Zr(C_5H_7O_2)_3I \cdot C_4H_8O$$
$$Zr(C_5H_7O_2)_3I \cdot C_4H_8O \xrightarrow[\text{vacuum}]{80°C.} Zr(C_5H_7O_2)_3I + C_4H_8O$$

■ *Note. It is important that the methods described in the General Techniques subsection for excluding moisture be followed carefully.*

Procedure

A solution of 3.81 g. (0.00636 mole) of anhydrous zirconium(IV) iodide in 250 ml. of tetrahydrofuran is prepared in a glass-stoppered Erlenmeyer flask equipped with a sidearm and stopcock (similar to flask *B* in Fig. 5). While a stream of argon is passed through the stopcock and over the surface of the solution, 9.31 g. (0.0191 mole) of tetrakis(2,4-pentanedionato)zirconium(IV) [zirconium(IV) acetylacetonate][19] is added. The solution is stirred for an hour at room temperature, and then the volume is reduced to about 80 ml. by vacuum distillation at room temperature. The resulting white to pale yellow crystals of $Zr(C_5H_7O_2)_3I\cdot C_4H_8O$ are filtered and washed with four 50-ml. portions of tetrahydrofuran. The tetrahydrofuran is removed from the solvate by heating *in vacuo* at 80°C. for at least 4 hours but not longer than 5 hours. A shorter heating time fails to remove all the tetrahydrofuran, whereas excessive heating results in partial decomposition of the compound. The yield is 10.75 g. (82%). *Anal.* Calcd. for $Zr(C_5H_7O_2)_3I$: C, 34.95; H, 4.12; I, 24.62; Zr, 17.70. Found: C, 33.73; H, 4.17; I, 24.66; Zr, 17.66.

Properties

Dichlorobis(2,4-pentanedionato)zirconium(IV), chlorotris-(2,4-pentanedionato)zirconium(IV), and bromotris(2,4-pentanedionato)zirconium(IV) are white crystalline solids. When crystallized from benzene–hexane solution, these compounds melt, respectively, with slight decomposition in sealed capillaries at 180.5–182°C., 159.5–161°C., and 162.5–164°C. The same melting point is observed for the monobromo complex when the compound is crystallized from carbon tetrachloride–hexane. Iodotris(2,4-pentanedionato)zirconium(IV) is a pale yellow crystalline solid which gradually decomposes on heating but melts rather sharply to a yellow-orange liquid in the range 178–183°C., depending on the heating rate. All the compounds are readily hydrolyzed by atmospheric moisture; when treated with liquid water, chlorotris(2,4-pentanedionato)zirconium(IV) is partially converted to tetrakis(2,4-pentanedionato)zirconium(IV).[20] Nujol mulls of the anhydrous compounds show no infrared bands in the region 3000 to 3500 cm^{-1}. These compounds are soluble in benzene, dichloromethane, and chloroform but are nearly insoluble in diethyl ether and hexane. Nuclear magnetic resonance spectra in dichloromethane solution (concentration 10 g./100 ml. of solvent) show one ring-proton resonance and one methyl resonance:

$Zr(C_5H_7O_2)_2Cl_2$, $\tau = 4.01$, 7.87; $Zr(C_5H_7O_2)_3Cl$, $\tau = 4.22$, 7.96; $Zr(C_5H_7O_2)_3Br$, $\tau = 4.20$, 7.96; $Zr(C_5H_7O_2)_3I$, $\tau = 4.11$, 7.94.

Dichloro-bis(2,4-pentanedionato)zirconium(IV) is monomeric and a weak electrolyte in nitrobenzene solution; n.m.r. chemical shifts,[7] infrared and Raman spectra,[8] and dipole-moment studies[21] indicate that this compound exists in solution as the octahedral *cis* geometrical isomer. Chloro- and bromotris(2,4-pentanedionato)zirconium(IV) are seven-coordinate complexes which are monomeric in benzene and only slightly dissociated in nitrobenzene and 1,2-dichloroethane.* Iodotris(2,4-pentanedionato)zirconium(IV), however, is appreciably dissociated both in nitrobenzene and in 1,2-dichloroethane.[7]

References

1. K. C. Pande and R. C. Mehrotra, *Chem. Ind. (London)*, 1198 (1958).
2. D. M. Puri and R. C. Mehrotra, *J. Less-Common Metals*, **3**, 247 (1961).
3. D. M. Puri, K. C. Pande, and R. C. Mehrotra, *ibid.*, **4**, 481 (1962).
4. M. Cox, J. Lewis, and R. S. Nyholm, *J. Chem. Soc.*, 6113 (1964).
5. R. C. Fay and R. N. Lowry, *Inorg. Chem.*, **6**, 1512 (1967).
6. R. C. Fay, N. Serpone, and R. N. Lowry, abstracts of papers, 153d National Meeting, American Chemical Society, Miami Beach, Fla., 1967, p. L-71.
7. T. J. Pinnavaia and R. C. Fay, *Inorg. Chem.*, **7**, 502 (1968).
8. R. C. Fay and T. J. Pinnavaia, *ibid.*, **7**, 508 (1968).
9. D. Grdenić and B. Matković, *Nature*, **182**, 465 (1958).
10. B. Matković and D. Grdenić, *Acta Cryst.*, **16**, 456 (1963).
11. J. V. Silverton and J. L. Hoard, *Inorg. Chem.*, **2**, 243 (1963).
12. G. Jantsch, *J. Prakt. Chem.*, **115**, 7 (1927).
13. G. T. Morgan and A. R. Bowen, *J. Chem. Soc.*, **125**, 1252 (1924).
14. R. Kh. Freidlina, E. M. Brainina, and A. N. Nesmeyanov, *Bull. Acad. Sci. USSR, Div. Chem. Sci.*, 45 (1957).
15. V. Doron, R. K. Belitz, and S. Kirschner, *Inorganic Syntheses*, **8**, 38 (1966).
16. E. M. Brainina and R. Kh. Freidlina, *Bull. Acad. Sci. USSR, Div. Chem. Sci.*, 1331 (1964).
17. R. C. Young and H. G. Fletcher, *Inorganic Syntheses*, **1**, 49 (1939).
18. R. N. Lowry and R. C. Fay, *ibid.*, **10**, 1 (1967).
19. R. C. Young and A. Arch, *ibid.*, **2**, 121 (1946).
20. E. M. Brainina and R. Kh. Freidlina, *Bull. Acad. Sci. USSR, Div. Chem. Sci.*, 1489 (1961).
21. N. Serpone and R. C. Fay, *Inorg. Chem.*, **8**, 2379 (1969).

* Our results[7,8] appear to contradict an earlier report in Volume VIII of this series[15] which implies that $Zr(C_5H_7O_2)_3Cl$ is an ionic compound containing six-coordinated zirconium.

Chapter Three

SOME SIGNIFICANT COMPOUNDS
OF BORON

I. POLY(1-PYRAZOLYL)BORATES AND THEIR COMPLEXES, BORANE ADDUCTS OF BASES, AND HALO-BORANE ADDUCTS OF BASES

18. POLY(1-PYRAZOLYL)BORATES, THEIR TRANSITION-METAL COMPLEXES, AND PYRAZABOLES

Submitted by S. TROFIMENKO*
Checked by JOHN R. LONG,† THOMAS NAPPIER,† and SHELDON G. SHORE†

The recently discovered[1] poly(1-pyrazolyl)borate ions represent a novel class of chelating ligands which give rise to a host of remarkably stable transition-metal complexes.[2] Ligands containing two pyrazolyl groups are bidentate, forming square planar or tetrahedral chelates, and those with three or four pyrazolyl groups are the first example of a symmetrical uninega-

* Central Research Department, Experimental Station, E. I. du Pont de Nemours & Company, Wilmington, Del. 19898.
† Department of Chemistry, The Ohio State University, Columbus, Ohio 43210.

tive tridentate ligand and form octahedral coordination compounds. Alkali metal polypyrazolylborates containing substituents on carbon or on boron have also been prepared;[3] such substitution gives rise to some unexpected changes in the properties of transition-metal chelates derived therefrom.[4,5] Single-crystal e.p.r.[6] and Mossbauer[7,8] studies on some transition-metal poly(1-pyrazolyl)borates have also been published. Finally, it has been shown that half-sandwiches, analogous to those derived from $C_5H_5^-$ and containing in addition to the tris(1-pyrazolyl)borate ligand, carbonyl, π-allyl, or alkyl groups, may also be easily prepared.[9] Nuclear magnetic resonance studies showing stereochemical nonrigidity in some of these compounds have been reported.[10]

A. POTASSIUM DIHYDROBIS(1-PYRAZOLYL)BORATE— ALKALI METAL POLY(1-PYRAZOLYL)BORATES

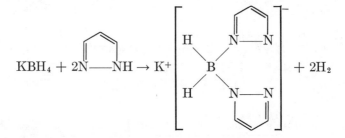

Syntheses of the potassium salts of the parent bidentate and the tridentate ligands are described. Other alkali metal salts can be prepared in similar fashion. Quaternary ammonium salts are made by neutralization of the free acids.[2] C- and B-substituted ligands have been prepared by similar procedures.[3]

Procedure

■ *Caution. Since large amounts of hydrogen are evolved, this reaction should be run in an efficient hood, and open flames or sparks are to be avoided.*

A mixture of 54 g.* (1 mole) of finely divided potassium boro-hydride and 272 g. (4 moles) of pyrazole† is placed, together with a magnetic stirring bar, into a 1-l. single-necked flask equipped with a thermometer sidearm. The flask is connected through a 50-cm.-long air condenser to a wet-test meter, or some other volumetric device, and is placed into an oil bath resting on a heating and stirring plate. The oil bath is heated to 90°C., and the mixture is allowed to melt with stirring. Hydrogen evolution starts at this point and continues briskly as potassium borohydride dissolves. The melt temperature is raised gradually to maintain a steady yet controlled evolution of hydrogen but is kept below 125°C.‡ When about 50 l. (2 moles) of hydrogen has been evolved, the melt should be clear or contain just a few undissolved potassium borohydride particles. These are removed mechanically, and the still-hot melt is poured into 600 ml. of stirred toluene. The resulting mixture is stirred for about 5 minutes and filtered hot. The solid is washed on the funnel with two 150-ml. portions of hot (80°C.) toluene and then with two 200-ml. portions of warm hexane; after air-drying it is obtained in 130–140-g. (70–75%) yield.§

Properties

Potassium dihydrobis(1-pyrazolyl)borate, m.p. 171–172°C., is a white crystalline solid, highly soluble in water and polar solvents. It may be recrystallized from anisole, but with large solubility losses. The infrared spectrum has a complicated BH_2 stretch multiplet in the 2200–2500-cm.$^{-1}$ range. It is very soluble in water and alcohols, but the ligand undergoes slow solvolysis, and solutions should be made up just before use.

* The checker carried out these procedures on $\frac{1}{4}$ scale.
† Available from Aldrich Chemical Co., Inc., 2371 North 30th St. Milwaukee, Wis. 53210, or J. T. Baker Chemical Co., 40 Avenue A, Bayonne, N.J. 07002.
‡ At higher temperatures trisubstitution becomes appreciable.
§ The filtrates may be saved for subsequent recovery of pyrazole and poly-(1-pyrazolyl)borates.

B. POTASSIUM HYDROTRIS(1-PYRAZOLYL)BORATE— ALKALI METAL POLY(1-PYRAZOLYL)BORATES

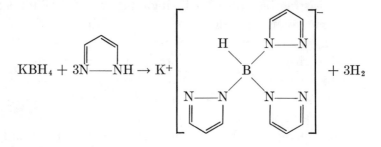

$$KBH_4 + 3N\text{---}NH \rightarrow K^+ [\ldots] + 3H_2$$

Procedure

The reaction is started as in Sec. A. When about 35 l. of hydrogen has been evolved, the oil bath is replaced with a heating mantle, and the temperature is raised gradually to about 190°C. until a total of 75 l. (3 moles) of hydrogen has been evolved.* The melt is cooled to about 150°C. and is then poured into 600 ml. of stirred toluene; the resulting mixture is stirred for 5 minutes and then filtered hot. The product is washed with hot toluene and then with hexane as in Sec. A. After air-drying it is obtained in about 190-g. (80%) yield.†

Properties

Potassium hydrotris(1-pyrazolyl)borate is a white crystalline solid. As prepared, m.p. 185–190°C., it is suitable for the synthesis of transition-metal chelates. Its melting point may be raised to 188–189°C. by recrystallization from anisole but this entails large solubility losses. Its infrared spectrum has a sharp BH peak at 2500 cm.$^{-1}$. This salt is very soluble in water, alcohols, and polar organic solvents. It is stored best as a solid, and its reactions should be carried out on freshly prepared solutions.

* Alternatively one can use potassium borohydride pellets (available from Metal Hydrides, Inc.), which react more slowly, and use a heating mantle from the start.

† The filtrates may be saved for subsequent recovery of pyrazole and poly-(1-pyrazolyl)borates.

C. POTASSIUM TETRAKIS(1-PYRAZOLYL)BORATE—
ALKALI METAL POLY(1-PYRAZOLYL)BORATES

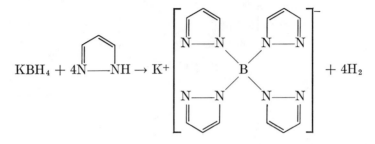

$$KBH_4 + 4N\text{—}NH \rightarrow K^+ \left[\begin{array}{c} \end{array} \right]^- + 4H_2$$

Procedure

A mixture of 54 g. (1 mole) of potassium borohydride pellets and 340 g. (5 moles) of pyrazole is stirred and heated, as in Sec. B, until about a 100-l. (4-mole) volume of hydrogen has been evolved. The fourth mole of hydrogen is evolved slowly, and it takes over 24 hours for completion of the reaction.* The melt, the temperature of which should be 220–230°C., is cooled to about 180°C. and poured carefully into 1200 ml. of stirred xylene. The solution is cooled to room temperature, and the solid is filtered. It is then stirred and boiled for a few minutes in 1 l. of heptane and filtered hot. The product is washed a few times with hot hexane to remove traces of pyrazole and is obtained, after air-drying, in about 270-g. (85%) yield. It is purified by recrystallizing from boiling water and drying *in vacuo* at 130–150°C.

Properties

Potassium tetrakis(1-pyrazolyl)borate is a white solid, m.p. 253–254°C. Its infrared spectrum is devoid of BH bands. This salt is quite soluble in dimethylformamide (DMF) and dimethyl sulfoxide but less so in water and alcohols. Stable 0.1 M aqueous solutions may be prepared and used as needed.

---

* Since this reaction, unlike the reactions in Secs. A and B, cannot be overrun, it is often convenient to use the evaporated filtrate residues from A and B and reflux them in excess pyrazole for 2 to 3 days, thus converting all incompletely pyrazolylated species to tetrakis(1-pyrazolyl)borate ion.

D. BIS[DIHYDROBIS(1-PYRAZOLYL)BORATO]NICKEL(II)— TRANSITION-METAL POLY(1-PYRAZOLYL)BORATES

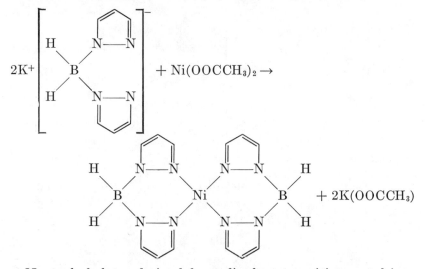

Neutral chelates derived from divalent transition-metal ions are prepared readily by metathetical reactions in water or, when the solubility of the ligand in water is low, in DMF–water mixtures. With the exception of some bidentate chelates of Mn(II) and Fe(II), which are air-sensitive, the transition-metal compounds are stable to air, and an inert atmosphere is not required during their synthesis or storage. The following procedures are applicable to the preparation of chelates derived from other transition-metal ions.

Procedure

Potassium dihydrobis(1-pyrazolyl)borate (18.6 g., 0.1 mole) of the solid as prepared in Sec. A is stirred in 300 ml. of distilled water, and any insolubles are removed by filtration. To the resulting clear solution is added, with stirring, 100 ml. of 0.5 M solution of nickel(II) acetate [or any other Ni(II) salt]. A yellowish precipitate forms immediately. The mixture is stirred at room temperature for 10 minutes and is then filtered. The solid is washed with copious amounts of water, then with methanol, and is obtained, after air-drying, in 16–17-g. (90–96%) yield.   It is purified by recrystallization from toluene.

Properties

Bis[dihydrobis(1-pyrazolyl)borato]nickel(II),   m.p.   181–182°C., forms well-shaped orange crystals, readily soluble in methylene chloride, chloroform, and aromatic hydrocarbons. The latter are preferred since polyhalogenated hydrocarbons react slowly with this compound at room temperature and rapidly in hot solutions.   Its infrared spectrum exhibits a complex BH_2 multiplet in the 2200–2500-cm.$^{-1}$ region.

E. BIS[HYDROTRIS(1-PYRAZOLYL)BORATO]COBALT(II)— TRANSITION-METAL POLY(1-PYRAZOLYL)BORATES

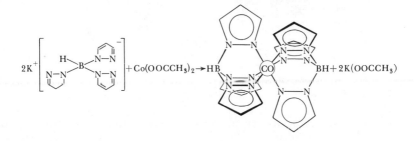

Procedure

Potassium hydrotris(1-pyrazolyl)borate, 25.2 g. (0.1 mole), of the solid, as prepared in Sec. B, is stirred in 300 ml. of distilled water, and any insolubles are removed by filtration.   To this solution is added, with stirring, 100 ml. of 0.5 M cobalt(II) acetate [or any other Co(II) salt] solution.   A yellowish precipitate forms promptly.   The mixture is stirred at room temperature for 10 minutes and is filtered; after washing with copious amounts of water and then with methanol, the solid is air-dried.   It is obtained in 22–23-g. (91–95%) yield and may be purified either by recrystallization from toluene or by chromatography (on acid-washed aluminum oxide packing and eluting with dichloromethane).

Properties

Bis{hydrotris(1-pyrazolyl)borato}cobalt(II) forms yellow crystals, m.p. 277–278°C. It is readily soluble in dichloromethane, chloroform, and aromatic hydrocarbons. It sublimes at 200°C./1 mm. and has a simple infrared spectrum displaying a sharp BH spike at 2470 cm.$^{-1}$.

F. BIS[TETRAKIS(1-PYRAZOLYL)BORATO]MANGANESE(II)— TRANSITION-METAL POLY(1-PYRAZOLYL)BORATES

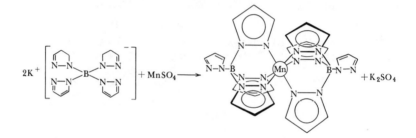

Procedure

To a stirred solution of 31.8 g. (0.1 mole) of potassium tetrakis(1-pyrazolyl)borate in 300 ml. of dimethylformamide is added rapidly 100 ml. of 0.5 M manganese(II) sulfate solution. A precipitate forms immediately. The mixture is stirred for 10 minutes at room temperature and is then diluted with 300 ml. of water and filtered. The white solid is washed with copious amounts of water, then with methanol, and, finally, with a small amount of ether. After air-drying, there is obtained 29–30 g. (94–98%) of white solid.

Properties

Bis[tetrakis(1-pyrazolyl)borato]manganese(II) is a white crystalline solid, melting, after recrystallization from toluene at 342–343°C. and subliming at 280°C./1 mm. It is soluble in dichloromethane, hot aromatic hydrocarbons, and dilute aqueous mineral acids. From the acid solutions it may be recovered by treatment with base. The infrared spectrum is devoid of BH bands.

G. PYRAZABOLE

Submitted by S. TROFIMENKO*
Checked by CHARLES SCHULTZ†

If the chelating termini of a bis(1-pyrazolyl)borate ligand are bridged by a BR_2 group instead of a metal ion, a pyrazabole is formed. Pyrazaboles are boron-nitrogen heterocycles exhibiting remarkable oxidative and hydrolytic stability and a wealth of derivative chemistry. They are prepared in good yields by the reaction of boranes or amine borane complexes with pyrazole or substituted pyrazoles. Substituents such as alkyl, aryl, halo, nitro, cyano, carboxy, amino, and formyl have been reported for this system. Standard operations on various functional groups have been carried out under diverse conditions without destruction of the pyrazabole ring.[1,11,12]

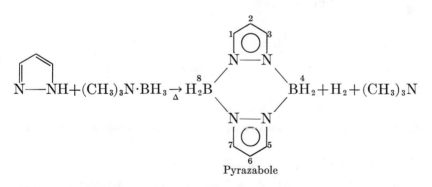

Pyrazabole

* Central Research Department, Experimental Station, E. I. du Pont de Nemours & Company, Wilmington, Del. 19898.
† Department of Chemistry, University of Michigan, Ann Arbor, Mich. 48104.

Procedure

A mixture of 68 g. of pyrazole and 73 g. of trimethylamine-borane* (pyridine-borane* or diborane dissolved in tetrahydrofuran may be used as well) (both 1.0 mole) is stirred and refluxed in 1 l. of toluene. The emanating gases are conducted through a $-80°C$. trap to a wet-test meter or another suitable volumetric device. When about 25 l. of hydrogen has been evolved (about 4–6 hours, or overnight reflux), the solvent is removed under 30-mm. pressure through a small Vigreux column. The residual oil solidifies on cooling, and the product, which is obtained in 72-g. (90%) yield, is purified further by recrystallization from hexane.

Properties

Pyrazabole is a white crystalline solid, m.p. 80–81°C., with a pungent, camphoraceous odor. It is readily soluble in most organic solvents, is less so in alcohols, and is insoluble in water. Its H^1 n.m.r. spectrum consists of a doublet and triplet ($J = 2.0$) at τ values of 2.49 and 3.83, respectively, in 2:1 ratio, whereas the B^{11} n.m.r. spectrum displays a triplet ($J = 108$) at $+27.1$ p.p.m. from trimethyl borate.

Substituted Pyrazaboles

Pyrazaboles substituted on either the ring or the boron may be obtained by reaction between the appropriately substituted pyrazole and the appropriately substituted borane.[11] For example the 2,6-dibromo-4,4,8,8-tetraethylpyrazabole can be made by the reaction between 4-bromopyrazole and pyrophoric triethylborane in xylene at reflux temperature. Another route to such compounds involves the dihydrobis(1-pyrazolyl)borate

---

*Trimethylamine-borane and pyridine-borane are available from Callery Chemical Co., Callery, Pa.

anion as shown in the equation below:[13]

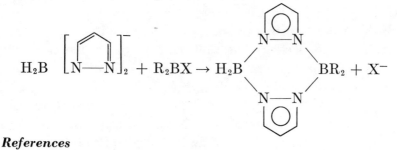

$$H_2B \left[\begin{array}{c} \boxed{} \\ N\!-\!N \end{array} \right]_2 + R_2BX \rightarrow H_2B \diagup\!\!\!\!\diagdown BR_2 + X^-$$

References

1. S. Trofimenko, *J. Am. Chem. Soc.*, **88**, 1842 (1966).
2. S. Trofimenko, *ibid.*, **89**, 3170 (1967).
3. S. Trofimenko, *ibid.*, **89**, 6288 (1967).
4. J. P. Jesson, S. Trofimenko, and D. R. Eaton, *ibid.*, **89**, 3148 (1967).
5. J. P. Jesson, S. Trofimenko, and D. R. Eaton, *ibid.*, **89**, 3158 (1967).
6. J. P. Jesson, *J. Chem. Phys.*, **45**, 1049 (1966).
7. J. P. Jesson, *ibid.*, **46**, 1995 (1967).
8. J. P. Jesson, J. F. Weiher, and S. Trofimenko, *ibid.*, **48**, 2058 (1968).
9. S. Trofimenko, *J. Am. Chem. Soc.*, **89**, 3904 (1967); **91**, 588 (1969).
10. S. Trofimenko, *ibid.*, **91**, 3183 (1969).
11. S. Trofimenko, *ibid.*, **89**, 3165, 4948 (1967).
12. C. W. Heitsch, Abstracts, 153d National Meeting, American Chemical Society, Miami Beach, Florida, 1967, p. L-109.
13. S. Trofimenko, *Inorg. Chem.*, **8**, 1714 (1969).

19. BORANE ADDITION COMPOUNDS OF BASES

Submitted by ROBERT C. MOORE,* SIDNEY S. WHITE, JR.,* and HENRY C. KELLY*
Checked by DOUGLAS L. DENTON† and SHELDON G. SHORE†

A. ETHYLENEDIAMINE-BISBORANE

$$3NaBH_4 + 4(C_2H_5)_2OBF_3 + 4C_4H_8O \rightarrow$$
$$4C_4H_8OBH_3 + 3NaBF_4 + 4(C_2H_5)_2O$$
$$4C_4H_8OBH_3 + 2H_2NCH_2CH_2NH_2 \rightarrow$$
$$2H_3BH_2NCH_2CH_2NH_2BH_3 + 4C_4H_8O$$

Ethylenediamine-bisborane (1,2-ethanediamine-bisborane) has been prepared by the reaction of ethylenediamine with diborane

* Texas Christian University, Fort Worth, Texas 76129. Financial support of the Robert A. Welch Foundation (Grant P-162) is gratefully acknowledged.
† Department of Chemistry, The Ohio State University, Columbus, Ohio 43210.

in vacuo,[1] or with solvated borane,[1,2] and by the reaction of ethylenediamine dihydrochloride with alkali metal borohydrides (alkali metal tetrahydroborates) by a modification[3,4] of the method of Schaeffer and Anderson.[5] These methods are general for the preparation of compounds in which the BH_3 group is coordinated to amine nitrogen, but, since properties of amine-boranes, e.g., kinetic stabilities, depend greatly on the nature of the amine group, some preparative methods are preferred for certain adducts, and methods of purification are selective. The present syntheses involve the reaction of the electron-donor group (Lewis base) with tetrahydrofuran-borane, which is prepared by treating sodium borohydride with the boron trifluoride–diethyl ether complex in tetrahydrofuran.[6,7] The authors have found this to be the most advantageous method from the standpoint of the time required, convenience, and yield.

As a result of the hydridic nature of the hydrogen attached to boron, amine-boranes are interesting and useful reducing agents and have been employed in the reduction of numerous organic carbonyl compounds. They have been utilized in studying the kinetics and mechanism of hydride reactions and are precursors for the synthesis of substituted boranes, borazines, boronium ions, higher boron hydrides, and carboranes.

Procedure

■ *Caution. The reaction must be carried out in a well-ventilated hood.*

A 1-l. four-necked flask is equipped with a sealed stirrer, pressure-equalizing dropping funnel, gas inlet tube, and cold-water condenser. The end of the condenser is equipped with a gas outlet tube to which is attached, by means of Tygon tubing, a drying tube filled with anhydrous calcium chloride. The apparatus is purged with nitrogen gas (dried by passing through calcium chloride) for about 15 minutes. This flow is interrupted as the flask is quickly charged with 12.8 g. (0.32 mole) of sodium borohydride* and 200 ml. of freshly distilled tetrahydrofuran.†

The flow of nitrogen is resumed, and the suspension is stirred and cooled in an ice-water bath. The flow of gas is stopped while a solution containing 55 g. (0.39 mole) of boron trifluoride–diethyl ether complex‡ in 100 ml. of tetrahydrofuran is added dropwise to the stirred slurry.

The dropping funnel is washed with small portions of tetrahydrofuran to remove all traces of boron trifluoride etherate and then is charged with a solution containing 10.57 g. (0.176 mole) of dry ethylenediamine in 100 ml. of tetrahydrofuran. Commercial ethylenediamine (Eastman 98%) is dried by careful distillation from sodium behind a safety shield in a hood or by the procedure described in reference 11. This solution is added dropwise (about 2–2½ hours) and the reaction mixture warmed with stirring to room temperature.

The suspension is allowed to settle. Solids are collected by filtration through a coarse-fritted-glass funnel under slight nitrogen pressure and are washed with tetrahydrofuran.§ The combined filtrate and washings are collected in a 1-l. round-

---

* This must be dry and of high purity. The product obtained from Ventron Corporation analyzed 95% purity by iodate analysis[8] and was warmed *in vacuo* at 60–75°C. for several hours before use.

† ■ *Warning.* See *Appendix I.* Tetrahydrofuran, obtained from Baker, was heated with potassium hydroxide pellets for several hours, cooled, decanted onto lithium tetrahydroaluminate, and heated under reflux for over 12 hours before being collected by fractional distillation (b.p. 65°C.) from this reagent. ■ *Warning. Other investigators have reported serious explosions on treatment of impure tetrahydrofuran with solid or aqueous potassium hydroxide.[9] This is believed to be due to the presence of peroxides in the solvent. It has been recommended that, prior to any purification, tests for peroxides in the tetrahydrofuran be carried out, using acidic aqueous iodide solution. Traces of peroxide can be removed by using copper(I) chloride.[10] Solvent containing more than trace quantities of peroxide should be discarded. It has been suggested that tetrahydrofuran which is moist but free from peroxides may be suitable for predrying over potassium hydroxide. In all cases, considerable precaution should be taken, and the authors suggest that purification be carried out in a hood behind a suitable safety shield.* See also page 317.

‡ The colored commercial product (Eastman or Matheson Coleman and Bell) should be distilled *in vacuo* to obtain a colorless distillate.

§ Procedures for filtration under nitrogen are described in Chapter One, page 9.

bottomed flask which is protected from the atmosphere with a drying tube.   This flask is attached to a small vacuum manifold which is connected through two cold traps (cooled, respectively, with a Dry Ice slush bath and a liquid-nitrogen bath) to a mechanical vacuum pump.   The product is recovered by evaporation of solvent *in vacuo* (overnight).   ■ *Caution. The solution contains tetrahydrofuran-borane which is toxic and reacts vigorously with hydroxylic substances. Care must be exercised, therefore, in disposing of the solvent which is collected in the cold traps. It is recommended that a cold solution of ethanolamine* in tetrahydrofuran be poured into the cold solvent. The mixture is then discarded by carefully pouring it down a sink in a well-ventilated hood.   The container should be rinsed thoroughly with water.*

The product is obtained in nearly quantitative yield. Although it may be useful as a reagent in the form in which it is recovered from solution, this material is usually impure and decomposes more rapidly than the purified compound.

Purification and Analysis

Various procedures for the purification of specific amine-boranes have been reported.[12]   Ethylenediamine-bisborane can be purified conveniently by aqueous extraction to remove water-soluble impurities.   The crude product obtained from the tetrahydrofuran solution is ground to a fine powder in a mortar and added in small portions, with stirring, to about 75 ml. of water which has been cooled in an ice-water bath.   Foaming may occur during this treatment.   The white crystalline solid which does not dissolve is collected by filtration through a coarse-fritted-glass funnel and washed several times with small portions of cold water (about 5°C.).   The final wash filtrate should be neutral toward pH paper.   The product is dried by

---

* Obtainable from a number of commercial sources, for example, Pfaltz and Bauer, Inc., 126-02 Northern Blvd., Flushing, N.Y. 11368.

evaporation of water *in vacuo*. The yield is 11.1 g. (71.9%). *Anal.* Calcd. for $(CH_2NH_2BH_3)_2$: H (hydridic), 6.89; B, 24.65; N, 31.91. Found: H (hydridic), 6.67; B, 23.9; N, 31.6.

A 1.362-g. sample of the compound is redissolved in a minimal amount (about 30 ml.) of warm (40°C.) tetrahydrofuran. Traces of insoluble material are removed by filtration, following which about 100 ml. of benzene* is added to the solution. The resulting precipitate is collected by filtration and washed twice with 25-ml. portions of benzene; the benzene is removed by evaporation in a stream of dry nitrogen. Yield is 1.173 g. (86.1% recovery). *Anal.* Found: H (hydridic), 6.83; B, 24.31; N, 31.91.

Properties

Ethylenediamine-bisborane is a white crystalline solid which is quite stable on standing and can be stored at room temperature in a desiccator for several months. It evolves hydrogen slowly on warming and decomposes rapidly at about 90°C. The use of this compound for the reduction of acetone, acrolein, cinnamaldehyde, and acetyl chloride to the corresponding alcohols (isopropyl alcohol, allyl alcohol, cinnamyl alcohol, and ethyl alcohol) has been described.[1] Its solubility characteristics and its thermal and hydrolytic stability have been reported.[1-4]

B. TRIPHENYLPHOSPHINE-BORANE

$$(C_6H_5)_3P + C_4H_8OBH_3 \rightarrow (C_6H_5)_3PBH_3 + C_4H_8O$$

The procedure described for the preparation of ethylenediamine-bisborane is general for the synthesis of amine-boranes and of borane adducts of other electron donors in which elements such as phosphorus and sulfur serve as the donor atom. In addition to the procedure used here for triphenylphosphine-borane, borane adducts of phosphines have been prepared by

---

* Analytical reagent grade (Mallinckrodt).

(1) the direct interaction of the phosphine with diborane *in vacuo*,[13,14] (2) the reaction of a phosphine-trihaloborane adduct with diborane,[15] (3) the reaction of the appropriate phosphine with an amine-borane,[16] and (4) the pyrolysis of a quaternary phosphonium borohydride salt.[17,18]

Procedure

■ *Note. The precautions prescribed in Sec. A should be observed in this procedure.*

A suspension prepared from 4.25 g. of 95% sodium borohydride (0.11 mole) and 150 ml. of tetrahydrofuran is cooled in an ice-water bath and stirred while a solution containing 17 ml. (0.13 mole) of boron trifluoride–diethylether in 60 ml. of tetrahydrofuran is added dropwise over a period of 45 minutes. Thirty grams (0.11 mole) of triphenylphosphine* is dissolved in 130 ml. of tetrahydrofuran, and this solution is added through a dropping funnel to the cooled suspension (1 hour). The mixture is allowed to warm to room temperature. Solids are collected by filtration, and the crude product is recovered by evaporation of the solvent from the filtrate *in vacuo*. Yield is 32 g. (100% theory).

A 29.3-g. sample of crude product is washed with 150 ml. of cool distilled water. The white crystalline product is collected by filtration, washed with water, and dried *in vacuo*. Yield is 27.6 g. (94.2% theory). *Anal.* Calcd. for $C_{18}H_{18}PB$ (276.13): C, 78.29; H, 6.57. Found: C, 78.27; H, 6.56; m.p. 184.5–187°C.

Properties

A characteristic feature of this compound is the high resistance to hydrolysis of boron-bonded hydrogens shown, in part, by the fact that the compound can be recovered unchanged from

* Obtained from Eastman Organic Chemicals, 343 State St., Rochester, N.Y. 14650.

boiling concentrated hydrochloric acid. Studies of bonding, reactivity, and spectral characteristics of this compound have been described.[17, 19, 20]

C. OTHER BASE-BORANES

Other base-borane adducts which can be obtained by base displacement of tetrahydrofuran from tetrahydrofuran-borane include triethylamine-borane (65% yield obtained), morpholine-borane (72% yield obtained), triethylphosphine*-borane (91% yield), and di(*n*-propyl) sulfide-borane (61% yield). Triethylamine-borane and di(*n*-propyl)sulfide-borane are best purified by vacuum sublimation or distillation from the crude product.

References

1. H. C. Kelly and J. O. Edwards, *J. Am. Chem. Soc.*, **82**, 4842 (1960).
2. H. C. Kelly, *Inorg. Chem.*, **5**, 2173 (1966).
3. J. Goubeau and H. Schneider, *Ber.*, **94**, 816 (1961).
4. H. C. Kelly and J. O. Edwards, *Inorg. Chem.*, **2**, 226 (1963).
5. G. W. Schaeffer and E. R. Anderson, *J. Am. Chem. Soc.*, **71**, 2143 (1949).
6. H. C. Brown and P. A. Tierney, *ibid.*, **80**, 1552 (1958).
7. H. C. Brown, K. J. Murray, L. J. Murray, J. A. Snover, and G. Zweifel, *ibid.*, **82**, 4233 (1960).
8. D. A. Lyttle, E. H. Jensen, and W. A. Struck, *Anal. Chem.*, **24**, 1843 (1952).
9. *Organic Syntheses*, **46**, 105 (1966).
10. K. Heusler, P. Wieland, and C. Meystre, *ibid.*, **45**, 57 (1965).
11. C. L. Rollinson and J. C. Bailar, Jr., *Inorganic Syntheses*, **2**, 197 (1946).
12. K. Niedenzu and J. W. Dawson, "Boron-Nitrogen Compounds," Academic Press, Inc., New York, 1965, and references therein.
13. F. Hewitt and A. K. Holliday, *J. Chem. Soc.*, 530 (1953).
14. A. B. Burg and R. I. Wagner, *J. Am. Chem. Soc.*, **75**, 3872 (1953).
15. W. A. G. Graham and F. G. A. Stone, *J. Inorg. Nucl. Chem.*, **3**, 164 (1957).
16. R. A. Baldwin and R. M. Washburn, *J. Org. Chem.*, **26**, 3549 (1961).
17. H. G. Heal, *J. Inorg. Nucl. Chem.*, **16**, 208 (1961).
18. P. A. Chopard and R. F. Hudson, *ibid.*, **25**, 801 (1963).
19. M. A. Frisch, H. G. Heal, H. Mackle, and I. O. Madden, *J. Chem. Soc.*, 899 (1965).
20. M. Becke-Goehring and H. Thielemann, *Z. Anorg. Allgem. Chem.*, **308**, 33 (1961).

* Triethylphosphine can be obtained from Strem Chemicals, Inc., Danvers, Mass. 01923.

20. TRIMETHYLAMINE-HALOBORANES

$$(CH_3)_3NBH_3 \xrightarrow[X_2]{HX \text{ or}} (CH_3)_3NBH_2X \text{ or } (CH_3)_3NBHX_2$$

$$(HX = HCl, HBr, \text{ or } HI; X_2 = Br_2 \text{ or } I_2)$$

Submitted by G. E. RYSCHKEWITSCH* and J. W. WIGGINS*
Checked by R. L. SNEATH† and L. J. TODD†

Addition compounds of borane, BH_3, and various amines can be readily formed by the symmetrical cleavage of diborane with amines,[1,2] by the reaction of ammonium salts with alkali metal borohydrides,[3] or by displacement from an ether borane (synthesis 19). These methods cannot be extended easily to the synthesis of monohaloboranes, since the required halodiboranes readily disproportionate and since monohaloborohydride salts have not been isolated. Amine-boranes can, however, be monohalogenated with hydrogen halides,[4] halogen,[4] boron trihalides,[4] mercury(II) halides,[5] or a variety of halides or oxyhalides of elements in a high oxidation state.[5] Some of these reactions eventually produce dihalo- or trihaloborane addition compounds, depending on the nature of the halogen, stoichiometry, or reaction temperature. We have selected the most convenient laboratory procedures in which polyhalogenation is easily avoided while giving high yields of the desired product. The procedures can be extended, in general, to addition compounds of other amines.

Trimethylamine-borane of good purity is commercially available‡ or may be prepared on a large scale from $NaBH_4$.[3b] (See Sec. 19C.) It may be easily purified by vacuum sublimation. Solvents may be dried over calcium hydride or molecular sieves. ■ *Caution. All syntheses should be carried out in a*

* University of Florida, Gainesville, Fla. 32601.

† Department of Chemistry, University of Indiana, Bloomington, Ind. 47401.

‡ Trimethylamine-borane is commercially available from Callery Chemical Company, Callery, Pa.

hood and away from flames or spark sources since the evolved hydrogen may cause an explosion or fire hazard and the hydrogen halides, used in large excess, create a corrosion and health hazard.

A. TRIMETHYLAMINE-MONOCHLOROBORANE

$$(CH_3)_3NBH_3 + HCl \rightarrow (CH_3)_3NBH_2Cl + H_2$$

The addition compound of trimethylamine and monochloroborane was first synthesized by the reaction of gaseous hydrogen chloride with trimethylamine-borane at low temperature in a vacuum line. Although boron trichloride[4] or mercury(II) chloride[5] can be used, the first method can easily be modified for large-scale runs by using aqueous concentrated hydrochloric acid.[6] No special precautions for the exclusion of air or water are necessary.

Procedure

Into a 1-l. Erlenmeyer flask containing a magnetic stirring bar are placed 20.06 g. (0.276 mole) of trimethylamine-borane and 500 ml. of reagent-grade benzene. The flask is closed with a stopper containing a gas inlet tube reaching below the liquid level and an outlet tube connected to a mercury bubbler. Gaseous hydrogen chloride from a storage cylinder is passed in slowly for 6 hours while the solution is vigorously stirred. The solution is then quickly decanted into a tared 1-l., standard-taper, round-bottomed flask, and dissolved hydrogen chloride is largely removed by applying vacuum from a water aspirator through a calcium chloride tube. The solvent is then completely evaporated, using a mechanical pump and a rotary evaporator while the flask is dipping in an oil bath at 60°C. Yield is 27.9 g., or 94%; m.p. 84–85°C. (Checker obtained 88% yield, m.p. 83–84°C.)

Synthesis can be accomplished more rapidly, but in lower yield, with aqueous HCl. In an Erlenmeyer flask, 2.78 g.

(0.038 mole) of trimethylamine-borane and 200 ml. of carbon tetrachloride are rapidly stirred. Twenty milliliters of concentrated hydrochloric acid (36.5–38% HCl) is then added, and stirring is continued for 10 minutes. The carbon tetrachloride phase is then removed in a separatory funnel, and the solvent is stripped under vacuum in a rotary evaporator. The remaining solid is pure trimethylamine-monochloroborane, 2.45-g. (59%) yield. Prolonged stirring decreases the yield of pure compound, whereas substantially smaller quantities of HCl may leave unreacted $(CH_3)_3NBH_3$ in the product. *Anal.* by checker: Calcd. for $C_3H_{11}NBCl$: C, 33.45; H, 10.21; N, 13.02. Found: C, 33.53; H, 10.28; N, 13.13.

Properties

$(CH_3)_3NBH_2Cl$ forms white crystals, m.p. 84–85°C., easily soluble in benzene and hot carbon tetrachloride and somewhat less soluble in cold carbon tetrachloride. The compound may be precipitated from carbon tetrachloride solution with cyclohexane to obtain good recovery of product and may be sublimed *in vacuo* at 60°C. Heating to 150°C. in an evacuated tube causes decomposition to $(CH_3)_3NBHCl_2$ and other products. $(CH_3)_3NBH_2Cl$ is only difficultly soluble in water and hydrolyzes in cold aqueous methanol only slowly, even if the solution is made basic or acidic. The solid can be handled in the atmosphere without difficulty for short periods of time (a day or less).

B. TRIMETHYLAMINE-MONOBROMOBORANE

$$(CH_3)_3NBH_3 + HBr \rightarrow (CH_3)_3NBH_2Br + H_2$$

The addition compound of trimethylamine and monobromoborane has been prepared by reaction of trimethylamine-borane with either bromine or boron tribromide,[4] but an excess of either reagent gives further bromination. Hydrogen bromide, however, gives only the monobromo derivative, even when added in excess. The synthesis is similar to that of trimethylamine-monochloroborane and does not require extreme precautions for the exclusion of moisture.

Procedure

A 250-ml. Erlenmeyer flask is fitted through a two-hole stopper with a gas inlet tube and an outlet protected with a calcium chloride tube. Trimethylamine-borane (7.28 g., 0.100 mole) is partially dissolved in 70 ml. of carbon tetrachloride contained in the flask. A brisk stream of hydrogen bromide is passed through a mercury bubbler and then through the mixture till gas evolution ceases (about 15 minutes). The amine-borane dissolves completely during the reaction; any solid impurities are quickly filtered off under aspirator suction. To the filtrate is now added 120 ml. of petroleum ether* to precipitate $(CH_3)_3NBH_2Br$. After cooling and settling for about 30 minutes, the product is filtered off, washed several times with petroleum ether or cyclohexane, and then dried under vacuum. The procedure gives an 85% yield of quite pure product whose melting point does not change on recrystallization from CCl_4-cyclohexane. (Checker obtained a yield of 88%.) *Anal.* Calcd. for $C_3H_{11}NBBr$: C, 23.68; H, 7.23; N, 9.20. Found: C, 23.98; H, 7.20; N, 9.25.

Properties

$(CH_3)_3NBH_2Br$ is a white crystalline solid which melts at 67–68°C.[4] It is very soluble in carbon tetrachloride and benzene but only sparingly soluble in saturated hydrocarbons, such as hexane, or in water. It dissolves readily in methanol and evolves hydrogen when aqueous NaOH or HCl is added to such a solution. In this respect it is much more reactive than the analogous chloro compound, but it nevertheless can be

---

* The solution turns light pink at this point.

transferred in the atmosphere without apparent reaction. (Checker observed a $1:2:1$ triplet in the ^{11}B n.m.r. spectrum.)

C. TRIMETHYLAMINE-MONOIODOBORANE

$$2(CH_3)_3NBH_3 + I_2 \rightarrow 2(CH_3)_3NBH_2I + H_2$$

$(CH_3)_3NBH_2I$ has been prepared by Nöth through the action of hydrogen iodide or iodine on trimethylamine-borane.[4] Unlike the other haloboranes, the compound reacts very readily with water and must therefore be synthesized with due precautions for the exclusion of atmospheric moisture. For some purposes it may be unnecessary to isolate the pure material. In this case the solution prepared in the first step of the synthesis below may be used directly.

Procedure

The apparatus suitable for a bench preparation in 40-g. quantities consists of two 300-ml. three-necked distilling flasks, one serving as a reaction vessel and the other as a filtrate receiver, which are connected by a filtering tube with a glass frit of medium porosity (Fig. 6). The connection should be secured with strong springs in order to avoid separation during pressure filtration. The reaction flask is also fitted with a solvent take-off condenser; the third neck of the reaction flask and the top of the condenser are protected with nitrogen-flushed addition tubes *A* and *B*. The components of the apparatus are dried in an oven and assembled hot according to the diagram while flushing with dry nitrogen which escapes through the· neck *D* of the receiver flask. The nitrogen pressure through the two supply lines is controlled with screw clamps and limited to about 25 mm. Hg by means of a mercury bubbler. Before starting the reaction it is wise to check that all connections will withstand this gage pressure.

The receiver flask is stoppered, the condenser delivery tube *C* is opened, and a magnetic stirring bar is added through addition

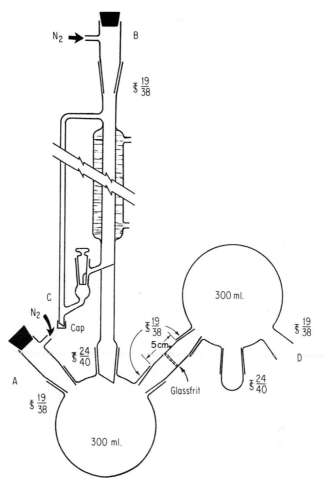

Fig. 6. Apparatus for the preparation of trimethyl-amine-monoiodoborane.

tube *A* of the reaction flask. The gas delivery arm is turned in an upward direction, and the flow is adjusted to a minimum before 14.56 g. (0.200 mole) of trimethylamine-borane and 100 ml. of dry benzene are added. The amine-borane is nearly dissolved with stirring while a 25.38-g. (0.100-mole) sample of iodine is weighed out. The solution is then cooled with an ice bath and vigorously stirred while the iodine is added with a porcelain spoon through the addition tube. The exothermic reaction, which is accompanied by rapid gas evolution, is easily controlled by adding the iodine in 0.2–0.3-g. portions over a 15-minute period. Outlet C is now closed, and the addition tube is removed and washed with benzene. All washings (25 ml.) are added to the reaction mixture.

Outlet A is now stoppered with a ground joint, the ice bath is replaced with a heating mantle, and the reaction mixture is brought slowly to incipient reflux over a 10-minute period in order to avoid excessive foaming. If at this time the solution is not transparently amber, or if the reflux is pale pink, an additional 0.15–0.30 g. of trimethylamine-borane is added in small portions through the top of the condenser, and reflux is continued for another 10 minutes. Thereafter 70 ml. of solvent is stripped off over a 10-minute period.

The heating mantle is then removed, and the nitrogen flushing tube is again firmly attached at A while the stopper at D is replaced with a short tube drawn out to a capillary tip. The condenser is now removed and replaced with a tight stopper. Nitrogen should pass from A to D at a low flow rate. After the receiver flask is turned 180°, the flasks are slowly tipped to allow decantation of the hot reaction mixture through the frit.* The capillary plug at D is then replaced by the flushing tube from A, and a coarse-porosity filter tube is substituted for the used one. A stoppered addition funnel with pressure-equalizing tube containing 150 ml. of dry petroleum ether (boiling range 30–60°C.) is now attached to the center neck. After the solution has cooled, $(CH_3)_3NBH_2I$ is precipitated by rapid addition of the petroleum ether and finally by cooling in an ice bath for 10 minutes. The product is isolated by filtration under nitrogen pressure through the attached filter plug. The last traces of solvent are removed by evacuation in a desiccator. After transfer in a dry-box, the yield is 36.00 g. (90%), m.p. 70.5–72°C.

* If filtration is too slow, the pressure may be momentarily increased by pinching off the relief valve. It is wise to keep a spare filtering tube in case of clogging.

The compound is sufficiently pure for many uses but may be recrystallized in the dry-box. A solution in 60 ml. of hot CCl_4 is filtered through a preheated medium-porosity frit, using suction from a rubber aspirator bulb. The washings of 25 ml. of hot CCl_4 are added, and the product is reprecipitated with a total of 170 ml. of cyclohexane, added in portions to the hot solution. Two liquid phases appear at first but, as the mixture cools with vigorous stirring, large crystals are formed and only one liquid phase remains. After standing for 12 hours, the product is filtered, washed with 50 ml. of cyclohexane, and dried *in vacuo*. The yield is 32.00 g., m.p. 72–73°C. (sealed capillary). Small samples of pure product may also be obtained by vacuum sublimation near the melting point.

Properties

The white crystalline solid melts at 73°C.[4] Like the other haloboranes, it is soluble in carbon tetrachloride and benzene, is insoluble in saturated hydrocarbons, but reacts very rapidly with methanol. It should be stored in tightly sealed, dark containers in a desiccator or a dry-box; otherwise its storage life is very short.

D. TRIMETHYLAMINE-DIBROMOBORANE

$$(CH_3)_3NBH_3 + HBr \rightarrow (CH_3)_3NBH_2Br + H_2$$
$$(CH_3)_3NBH_2Br + Br_2 \rightarrow (CH_3)_3NBHBr_2 + HBr$$

Submitted by M. A. MATHUR* and G. E. RYSCHKEWITSCH*
Checked by Br. F. SWICKER, F.S.C.,† and J. C. CARTER†

Trimethylamine-dibromoborane has been reported in the literature only once.[7] It was prepared by the reaction of boron tribromide with trimethylamine-borane, in carefully controlled

* Department of Chemistry, University of Florida, Gainesville, Fla. 32601.
† Department of Chemistry, University of Pittsburgh, Pittsburgh, Pa. 15213.

stoichiometry. This synthesis is inconvenient because of the hydrolytic instability of BBr_3 and the necessity to dispose of the diborane formed in the reaction. An alternative route, direct bromination of trimethylamine-borane, presents difficulties in predicting the reaction stoichiometry, since the hydrogen bromide evolved in the successive brominations reacts competitively with the starting material, and the end point of dibromination is not readily detected. Excess bromine will produce the boron tribromide adduct, which complicates the purification procedure.

A two-step process, using HBr for the first replacement, and then a stoichiometric amount of Br_2 for the second, avoids the difficulties.

Procedure

A 1-l. three-necked flask is charged with a solution of 7.28 g. (100 mmoles) of trimethylamine-borane (Callery Chemical Co.) in 150 ml. of dry benzene. One neck is fitted with a gas inlet dipping below the liquid level, which can be connected through a mercury bubbler to a tank of hydrogen bromide; the center neck is fitted through a ground joint to a buret containing a 1.23 M solution of bromine in benzene; the remaining neck is vented through a drying tube containing calcium chloride and calcium sulfate to a hood or absorbent for HBr.* A stirring bar is added, and the apparatus is flushed with nitrogen for 10 minutes before hydrogen bromide is passed through the stirred solution for 30 minutes† at a rate of two to three bubbles per minute. The gas inlet is now connected to the nitrogen supply, and the bromine solution is added drop by drop to the stirred solution while the hydrogen bromide is purged by a stream of nitrogen. The bromine color is discharged rapidly until an

---

* Dry conditions are maintained mainly to facilitate final removal of HBr. Reactant and product boranes are rather stable toward water.

† After this time the effluent gas should be free of hydrogen and be completely soluble in aqueous NaOH.

easily detected end point is reached and the solution attains a persistent yellow color (90.5 ml.).* The nitrogen flow is maintained for an additional 40 minutes; the flask then is transferred to a dry-box for filtration. Excess bromine is removed with a drop of cyclohexene. After removal of benzene by evacuation on a rotary evaporator, the product is collected and evacuated in a desiccator for an additional 3 hours. The yield is 21.9 g. (95% yield), m.p. 128–130°C. (literature, 126–127°C.). The melting point is unchanged if the product is redissolved in benzene and precipitated with cyclohexane. Bromine analysis by the Volhard method on a sample solvolyzed in NaOH/1-propanol gives 69.26% (calculated 69.28%).

Properties

The compound is soluble in acetone, chloroform, methylene chloride, and nitromethane. It is insoluble in alkanes and cold water and stable toward water in the cold but gives an acidic solution when boiled with water. The infrared spectrum shows a single band at 2510 cm.$^{-1}$, about 50 cm.$^{-1}$ higher wave number than in the monobromo compound. The ^{11}B spectrum gives a doublet 18.9 p.p.m. upfield from trimethyl borate: $J_{B-H} = 155$ Hz. The proton spectrum in methylene chloride shows a sharp singlet at 2.91 p.p.m. downfield from external tetramethylsilane and 0.20 p.p.m. farther downfield than the monobromo compound. Boron-attached hydrogens are not detectable.

References

1. A. B. Burg and H. I. Schlesinger, *J. Am. Chem. Soc.*, **59**, 780 (1937).
2. H. C. Brown and L. Domash, *ibid.*, **78**, 5384 (1956).
3a. E. R. Anderson and G. W. Schaffer, *ibid.*, **71**, 2143 (1949).
3b. J. Bonham and R. S. Drago, *Inorganic Syntheses*, **9**, 8 (1967).
4. H. Nöth and H. Beyer, *Chem. Ber.*, **93**, 2251 (1960).

* The checker observed a color change from yellow to orange-brown at the stoichiometric end point.

5. J. W. Wiggins, doctoral dissertation, University of Florida, 1966.
6. K. Boer and J. Dewing, British patent 881,376 (Nov. 1, 1961).
7. H. Nöth and H. Beyer, *Chem. Ber.*, **93**, 2251 (1960).

II. COMPOUNDS CONTAINING COMPLEX BORON CATIONS

Although anions containing boron have been known for many years, boron cations are much less common. The first such boron cation to be reported was bis(2,4-pentanedionato)-boron(III) found in 1906 by Dilthey[1] in compounds of the types:

$$[B(C_5H_7O_2)_2]^+X^- \quad \text{and} \quad [B(C_5H_7O_2)_2]_2^+X^{2-}$$

where $C_5H_7O_2$ represents the 2,4-pentanedionate group (i.e., acetylacetonate); X^- is $FeCl_4^-$, $AuCl_4^-$, I^-, I_3^-, or $ZnCl_3^-$; and X^{2-} is $PtCl_6^{2-}$ or $SnCl_6^{2-}$. More conventional coordination cations of boron were first identified by a group[2] at the University of Michigan as a structural component of the "diammoniate of diborane," $[H_2B(NH_3)_2]^+[BH_4]^-$. [See *Inorganic Syntheses*, **9**, 4 (1967).] Shortly thereafter, a group[3] in Germany demonstrated that other amines such as methylamine could be placed in the cation instead of ammonia. In 1963 Miller and Muetterties[4] systematically extended the replacement process. They showed that many amines, as well as phosphines and sulfides, can be used to replace the ammonia, and the hydrides can be replaced by other anions such as fluoride. Moews and Parry[5] reported the relatively simple *chelated* cation containing ethylenediamine in which methyl groups replace hydrogens of the original cation, $[H_2B(base)_2]^+$:

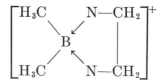

More recently Ryschkewitsch and associates[6] prepared chelated boron cations with charges ranging from 1 to 3. Cations of this and related types are reported here. Many show relatively high stability.

References

1. W. Dilthey, *Liebigs Ann. Chem.*, **344**, 300 (1906).
2. D. R. Schultz, S. G. Shore, G. Kodama, P. R. Girardot, R. C. Taylor, and R. W. Parry, *J. Am. Chem. Soc.*, **80**, 1 (1958); C. E. Nordman and C. R. Peters, *ibid.*, **81**, 3551 (1959).
3. H. Nöth and H. Beyer, *Angew. Chem.*, **71**, 383(1959); H. Nöth and H. J. Vetter, *Ber.*, **97**, 110 (1964); H. Nöth and H. Beyer, *ibid.*, **93**, 1078 (1960); H. Nöth, *Angew. Chem.*, **72**, 638 (1960).
4. N. E. Miller and E. L. Muetterties, *J. Am. Chem. Soc.*, **86**, 1033 (1964).
5. P. C. Moews, Jr., and R. W. Parry, *Inorg. Chem.*, **5**, 1552 (1966).
6. G. E. Ryschkewitsch, *J. Am. Chem. Soc.*, **89**, 3145 (1967); G. E. Ryschkewitsch and J. M. Garrett, *ibid.*, **89**, 4240 (1967); C. W. Makowsky, G. L. Galloway, and G. E. Ryschkewitsch, *Inorg. Chem.*, **6**, 1972 (1967); K. C. Nainan and G. E. Ryschkewitsch, *ibid.*, **7**, 1316 (1969); *J. Am. Chem. Soc.*, **91**, 330 (1969).

21. CATIONIC BIS(2,4-PENTANEDIONATO) BORON(III) COMPLEXES

Submitted by NICK SERPONE*† and ROBERT C. FAY*
Checked by PHILIP J. POLLICK‡ and ANDREW WOJCICKI‡

The cationic chelates which boron forms with β-diketones are of interest because relatively few elements give positively charged metal diketonates,[1] and the boron complexes are the only known cationic diketonates in which the central atom exhibits a coordination number of four. Dilthey[2] reported 2,4-pentane-dione (acetylacetone) complexes of the type $[B(C_5H_7O_2)_2]^+X^-$ and $[B(C_5H_7O_2)_2]_2^+X^{2-}$, where $X^- = FeCl_4^-$, $AuCl_4^-$, I^-, I_3^-, or $ZnCl_3^-$, and $X^{2-} = PtCl_6^{2-}$ or $SnCl_6^{2-}$. The halometallate salts were prepared in chloroform or glacial acetic acid solution

* Cornell University, Ithaca, N.Y. 14850. Financial support by the National Science Foundation is gratefully acknowledged.
† Present address: Sir George Williams University, Montreal, Quebec, Canada.
‡ Department of Chemistry, The Ohio State University, Columbus, Ohio 43210.

by reaction of the appropriate metal halide with the product of the reaction of 2,4-pentanedione with boron(III) chloride. The latter reaction product has recently been shown to be the hydrogen dichloride, $[B(C_5H_7O_2)_2][HCl_2]$.[3] Dilthey's approach is straightforward and rapid, and it appears to be quite generally applicable for preparation of bis(2,4-pentanedionato)boron(III) halometallate salts.* His method is applied here for preparation of the hexachloroantimonate salt;[3] a detailed procedure is also given for synthesis of the hydrogen dichloride intermediate.

A. BIS(2,4-PENTANEDIONATO)BORON(III) HYDROGEN DICHLORIDE

[Bis(acetylacetonato)boron(III) Hydrogen Dichloride]

$$BCl_3 + 2C_5H_8O_2 \rightarrow [B(C_5H_7O_2)_2][HCl_2] + HCl$$

Procedure

Approximately 125 ml. of anhydrous ether is placed in a dry,† 250-ml. Erlenmeyer flask which is equipped with a 24/40 S.T. joint and a sidearm gas inlet (see Fig. 7). The stopcock on the sidearm is fitted with a retainer clip, the flask is stoppered, and the stopper is held in place with a spring clamp. The flask is weighed and is then attached to a lecture bottle of boron(III) chloride,‡ as shown in Fig. 7, by means of a 1-foot length of Tygon tubing which is punctured with a needle at least four times to allow release of excess pressure of boron(III) chloride. ■ *This apparatus is set up in the hood.* The boron(III) chloride cylinder is opened cautiously,§ and BCl_3 is delivered into the flask for approximately 3 minutes while the flask is shaken vigorously to facilitate solution of the gas in the

---

* The same approach has been used for synthesis of bis(1-phenyl-1,3-butanedionato)boron(III) complexes.[2]

† All glassware is dried at 120°C. and is allowed to cool in a nitrogen atmosphere.

‡ Obtainable as 99.9% BCl_3 from J. T. Baker Chemical Co., 222 Red School Lane, Phillipsburg, N.J. 08865.

§ It is wise to demonstrate that the valve on the lecture bottle is free and working by opening it briefly in the hood before hooking up to the Tygon tubing.

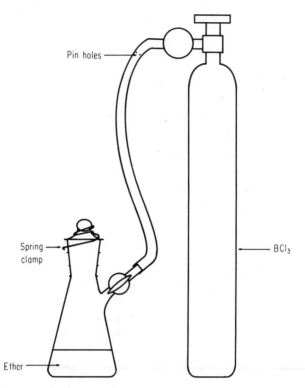

Fig. 7. *Apparatus for the preparation of an ether solution of boron(III) chloride.*

ether.   The flask is then reweighed.   A 3-minute delivery time results in solution of about 3.5 g. (0.030 mole) of boron(III) chloride.   The ether solution is then filtered under nitrogen* through a medium-frit filter stick into a second 250-ml. sidearm Erlenmeyer flask.

Filtration† is necessary in order to remove about 0.04 g. of white precipitate, presumably boric acid, which results from reaction of boron(III) chloride with traces of water in the ether. While nitrogen is passed into the sidearm and over the surface

---

* Commercial prepurified-grade nitrogen is satisfactory and may be used without further purification.

† All filtrations and the washing of precipitates are carried out under nitrogen. A detailed description of these operations is given in synthesis 17 of this volume.

of the ether solution, a 5% excess of 2,4-pentanedione (6.5 ml., 0.063 mole) is added with a hypodermic syringe. A magnetic stirring bar is added, the flask is stoppered, the stopper is clamped with a spring clamp, and the sidearm stopcock is closed. The reaction mixture is then stirred vigorously for about 10 minutes, the stopcock being opened occasionally for a moment to relieve excess pressure of hydrogen chloride. The resulting white-yellow solid is quickly filtered under nitrogen* from the red solution. The product is washed* with two 20-ml. portions of anhydrous ether and is dried *in vacuo* at room temperature for 10 minutes. The yield is 5.5 g. (65%). *Anal.* Calcd. for $B(C_5H_7O_2)_2HCl_2$: C, 42.75; H, 5.38; Cl, 25.24. Found: C, 42.95; H, 5.49; Cl,† 24.79.

B. BIS(2,4-PENTANEDIONATO)BORON(III) HEXACHLOROANTIMONATE(V)

[Bis(acetylacetonato)boron(III) Hexachloroantimonate(V)]

$$[B(C_5H_7O_2)_2][HCl_2] + SbCl_5 \rightarrow [B(C_5H_7O_2)_2][SbCl_6] + HCl$$

Procedure

In a nitrogen-filled glove bag, 0.90 g. (0.0032 mole) of freshly prepared bis(2,4-pentanedionato)boron(III) hydrogen dichloride is dissolved in 25 ml. of dry chloroform. It is convenient to prepare this solution in a 125-ml. Erlenmeyer flask equipped with a standard-taper joint; the chloroform is dried by refluxing for one day over calcium hydride and then distilling. To the bis(2,4-pentanedionato)boron(III) solution, 0.50 ml. (0.0039 mole) of antimony(V) chloride is added from a 1-ml. syringe. Twenty milliliters of anhydrous ether is then added slowly with vigorous swirling, which results in formation of an oily product. The flask is stoppered and transferred to a refrigerator for about 2 hours, whereupon the product crystallizes. The crystals

---

* All filtrations and the washing of precipitates are carried out under nitrogen. A detailed description of these operations is given in synthesis 17 of this volume.

† Determined gravimetrically as silver chloride.

are filtered under nitrogen,* washed with two 20-ml. portions of anhydrous ether, and dried *in vacuo* at room temperature for half an hour. The yield is 1.30 g. (75%). *Anal.* Calcd. for $B(C_5H_7O_2)_2SbCl_6$: C, 22.10; H, 2.60; B, 1.99; Sb, 22.40; Cl, 39.14. Found: C, 22.00; H, 2.57; B, 2.02; Sb, 22.41; Cl, 39.25.

Properties

Bis(2,4-pentanedionato)boron(III) hydrogen dichloride is a cream-colored, free-flowing solid, which melts with decomposition at 88–91°C. It is very unstable in air, turning within minutes to a sticky, red-brown solid with loss of hydrogen chloride. It also decomposes on standing *in vacuo* at room temperature and is best stored under nitrogen in a refrigerator. The compound is soluble in chloroform, dichloromethane, acetone, and glacial acetic acid but is nearly insoluble in ether and hexane. Its proton n.m.r. spectrum in deuteriochloroform solution (concentration, 10.0 g./100 ml. of solvent) shows resonance lines at $\tau = ca. -1.4$ (HCl_2^- proton),† $\tau = 3.16$ (ring protons), and $\tau = 7.44$ (methyl protons). It is interesting that the acetylacetonate proton resonances of bis(2,4-pentane-dionato)boron(III) complexes occur at unusually low field, owing primarily to the net positive charge on the complex;[3] the τ values for the hydrogen dichloride salt are lower than those reported for any other acetylacetonate complex.

Bis(2,4-pentanedionato)boron(III) hexachloroantimonate(V) crystallizes from chloroform–ether as long, flat, cream-colored needles which melt with decomposition at 155–157°C. Its solubility properties are similar to those of the hydrogen dichloride salt. In deuteriochloroform solution (concentration, 10 g./100 ml. of solvent) the compound exhibits proton n.m.r. lines at $\tau = 3.44$ (ring protons) and $\tau = 7.45$ (methyl protons).

---

* Checker reported that the [SbCl$_6$] compound can be filtered and dried in air with no change in melting point.

† The chemical shift of the HCl_2^- proton depends strongly on concentration.

References

1. J. P. Fackler, *Progr. Inorg. Chem.*, **7**, 361 (1966).
2. W. Dilthey, *Liebigs Ann. Chem.*, **344**, 300 (1906).
3. R. C. Fay and N. Serpone, *J. Am. Chem. Soc.*, **90**, 5701 (1968).

22. (4-METHYLPYRIDINE)(TRIMETHYLAMINE)- DIHYDROBORON(III) CATION SALTS

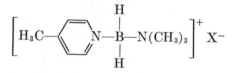

$$2(CH_3)_3NBH_3 + I_2 \rightarrow H_2 + 2(CH_3)_3NBH_2I$$
$$(CH_3)_3NBH_2I + CH_3C_5H_4N \rightarrow [H_3CC_5H_4NBH_2N(CH_3)_3]^+ + I^-$$

Submitted by JAMES M. GARRETT* and G. E. RYSCHKEWITSCH*
Checked by L. E. SENOR† and J. C. CARTER†

Boron cations with a single positive charge have been prepared by a number of methods. The synthetic routes can be placed in a number of different categories: (1) unsymmetrical cleavage of diborane;[1,2] (2) abstraction of hydride from a BH_3 group by a protonic acid in the presence of a donor molecule;[3] (3) hydride abstraction from coordinated BH_3 by other electrophiles, such as iodine[4,5] or mercury(II) ion[6] in the presence of a donor; and (4) displacement of halide ion from a haloborane adduct by a suitable neutral base.[6-8] Dihydroboron cations appear to be most stable when the two neutral ligands are tertiary amines; the last method offers the simplest and mildest reaction conditions and makes it possible to substitute cleanly two different amines on boron,[9] especially when an iodo derivative is used. One can avoid the somewhat tedious isolation of the iodoborane

* Department of Chemistry, University of Florida, Gainesville, Fla. 32601.
† Department of Chemistry, University of Pittsburgh, Pittsburgh, Pa. 15213.

adduct[10] by reaction *in situ.* The procedure outlined below can easily be extended to combinations of other amines.[11]

Procedure*

Trimethylamine-borane† (32.4 g., 0.444 mole; 10% excess) is dissolved in 500 ml. of dry, reagent-grade benzene in a 1-l. conical flask. Solid iodine (51.2 g., 0.404 g. atom) is then added in small portions over a 10-minute period while the solution is stirred magnetically. After each addition, the flask is lightly stoppered to prevent access of air and any loss of HI which results from the vigorous exothermic reaction of iodine with the borane. ■ *Caution. About 10 l. of hydrogen gas is evolved over a 10-minute period. Reaction should be carried out in a hood and away from open flames and spark sources.*

The mixture is stirred for an additional 30 minutes after completion of the I_2 addition, and a steel-gray solution with a small amount of white flocculent precipitate is obtained. The sides of the flask are now rinsed with 200 ml. of benzene while dry nitrogen is blown over the surface of the solution to assure complete removal of traces of HI gas and exclusion of moisture.

Ninety-four milliliters (90 g., 0.97 mole) of 4-picoline which has been freshly distilled from calcium hydride is then added at once. The color immediately changes from gray to yellow with the yellow color fading as a white solid precipitates. The reaction vessel warms up and is quickly placed in an ice bath and then in the refrigerator for an hour.

The white solid is filtered and washed, in turn, with two 100-ml. portions of benzene and two 100-ml. portions of anhydrous ether. The solid is dried in a vacuum desiccator by pumping overnight. Yield is 115 g. (97%). (Checker obtained 84% on $\frac{1}{10}$ scale.) The compound melts around 150°C., but its melting point and range depend on the heating rate.

---

* The reaction was carried out by the checkers on $\frac{1}{10}$ scale.
† Available from Callery Chemical Co., Callery, Pa. 16024.

The salt prepared in the above manner is substantially pure, provided that moisture has been successfully excluded in the reagents and during reaction. In the preparation of this and similar salts, there was occasionally evidence for the presence of small impurities of ammonium salts, indicated by a weak infrared absorption around 2700 cm.$^{-1}$. Such impurities can be eliminated most conveniently by conversion to the hexafluorophosphate salt.

A solution of 8.70 g. (29.8 mmoles) of the iodide salt in 50 ml. of water is warmed to about 80°C., and a warm solution of NH_4PF_6 (11.8 g.; threefold excess) in 75 ml. of water is added slowly while the mixture is stirred with a glass rod. Precipitation of the sparingly soluble hexafluorophosphate starts immediately and is completed by cooling in an ice bath.

The fine white needles are filtered, washed with 50 ml. of cold water, and dried over $CaCl_2$ in a vacuum desiccator. Yield is 8.56 g. (93%) (checker obtained 94% on $\frac{1}{10}$ scale); m.p. 147°C.

Precipitation in more concentrated solutions tends to produce an oil, which absorbs impurities and later solidifies to a cake. Slow recrystallization of dilute solutions from boiling water yields needles 1 cm. or longer.

Properties

The iodide salt is insoluble in benzene, alkanes, carbon tetrachloride, or diethyl ether but is easily soluble in water, methanol, acetone, or methylene chloride. It can be recovered from methylene chloride in nearly quantitative yield by precipitation with ether. The hexafluorophosphate dissolves in the same solvents as does the iodide, with the exception of cold water, in which it is only slightly soluble (about 0.01 M). The cation is stable in aqueous HCl or NaOH but rapidly loses trimethylamine in pyridine solution. It can be chlorinated or brominated on boron by the elements in methylene chloride.[9] The hexafluorophosphate ^{11}B resonance occurs at 17.8 p.p.m. [triplet, $J_{B-H} = 90$ Hz., external $B(OCH_3)_3$]. The proton resonances in CH_2Cl_2 appear at δ(p.p.m.) 2.97 (9H), 2.63 (3H), 7.78 (2H, doublet, $J = 7$ Hz.), and 8.55 (2H, doublet, $J = 7$ Hz.) downfield from external tetramethylsilane.

References

1. D. R. Schultz, S. G. Shore, G. Kodama, P. R. Girardot, R. C. Taylor, and R. W. Parry, *J. Am. Chem. Soc.*, **80**, 1 (1958).
2. G. E. McAchran and S. G. Shore, *Inorg. Chem.*, **4**, 125 (1965).
3. N. E. Miller and E. L. Muetterties, *J. Am. Chem. Soc.*, **86**, 1033 (1964).
4. J. E. Douglass, *ibid.*, **86**, 5431 (1964).
5. K. C. Nainan and G. E. Ryschkewitsch, *Inorg. Chem.*, **7**, 1316 (1968).
6. G. E. Ryschkewitsch, *J. Am. Chem. Soc.*, **89**, 3145 (1967).
7. H. Nöth, P. Schweizer, and F. Ziegelgänsberger, *Chem. Ber.*, **99**, 1089 (1966).
8. C. W. Makosky, G. L. Galloway, and G. E. Ryschkewitsch, *Inorg. Chem.*, **6**, 1972 (1967).
9. G. E. Ryschkewitsch and J. M. Garrett, *J. Am. Chem. Soc.*, **89**, 4240 (1967); *ibid.*, **90**, 7234 (1968).
10. Synthesis 20, Sec. C, of this volume.
11. K. C. Nainan and G. E. Ryschkewitsch, *J. Am. Chem. Soc.*, **91**, 330 (1969).

23. BIS(TRIMETHYLPHOSPHINE)-DIHYDROBORON(III) IODIDE

$$(CH_3)_3P + \tfrac{1}{2}B_2H_6 \rightarrow (CH_3)_3PBH_3$$
$$2(CH_3)_3PBH_3 + I_2 \rightarrow 2(CH_3)_3PBH_2I + H_2$$
$$(CH_3)_3PBH_2I + (CH_3)_3P \rightarrow [H_2B\{P(CH_3)_3\}_2]^+I^-$$

Submitted by N. E. MILLER*
Checked by K. C. NAINAN† and G. E. RYSCHKEWITSCH†
Checked independently by GOJI KODAMA‡

Borane monovalent cations with tertiary phosphine bases were first prepared by the action of hydrogen iodide on a mixture of phosphine-borane and phosphine, by the displacement of sulfides from $[H_2B(SR_2)_2]^+$ cations, or directly from phosphine

* Chemistry Department, University of South Dakota, Vermillion, S.D.
† University of Florida, Gainesville, Fla. 32601.
‡ Department of Chemistry, University of Utah, Salt Lake City, Utah 84112.

and diborane.[1,2]   It is now possible to make them by using a more direct displacement of iodide from iodoborane adducts.[3,4,5] Iodine displacement is the method of choice here.

Procedure

■ *Caution. Both diborane and trimethylphosphine are toxic and very easily ignited. This synthesis is therefore properly carried out in a well-ventilated hood behind safety shields.* Diborane forms explosive mixtures over a wide range of compositions with air and therefore presents a special hazard. *Personnel should be protected by shields from any glass apparatus containing diborane at less than atmospheric pressure.* Accidental breakage of the glass will introduce air and almost assuredly lead to a glass-shattering explosion.

Trimethylphosphine* (1.05 g., 13.8 mmoles), measured as a gas in a known volume in a high-vacuum line, is condensed with liquid nitrogen into a 100-ml. flask equipped with a magnetic stirring bar and a sidearm capable of being fitted with a serum cap. See Fig. 8. About 5 ml. of dry, purified pentane (shaken with sulfuric acid to remove olefin and dried over calcium hydride) is also condensed into the flask, and the mixture is warmed to form a homogeneous liquid and then refrozen. Diborane† (0.19 g., 6.85 mmoles), measured as a gas, is condensed into the flask with liquid nitrogen. The flask is closed off from the vacuum line and allowed to warm slowly, whereupon trimethylphosphine-borane forms as a white solid. With these quantities of reactants, this method of making the adduct is safe, but with much larger quantities the reaction could easily become uncontrollable. [For an alternative preparation of $(CH_3)_3PBH_3$ which uses easily handled $NaBH_4$ instead of B_2H_6 as a starting

* Prepared and purified by standard procedures and stored over calcium hydride,[6] or available commercially from Pfaltz and Bauer, Inc., 126-02 Northern Blvd., Flushing, N.Y. 11368.

† Available commercially from Callery Chemical Co., Callery, Pa. The technique and precautions for storage and use from stainless-steel cylinders have been given by Miller and Muetterties.[7]

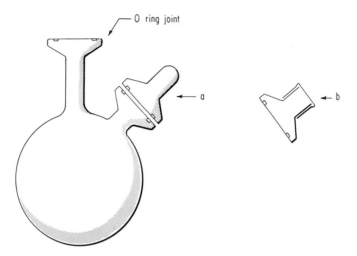

Fig. 8. Apparatus for the preparation of $[H_2B\{P(CH_3)_3\}_2]^+I^-$.

material, see Sec. 19C.] The heavy slurry* which is formed is stirred by an external driving magnet for about 30 minutes after there is no more visible evidence of reaction. The solvent (and any excess of either reactant due to measuring inaccuracies) is then removed under vacuum by *slowly* opening the stopcock attaching the vessel to the vacuum line. When most of the pentane is gone, as evidenced by the vapor pressure, the stopcock is closed and opened intermittently (every 5 minutes) to prevent excessive loss of the volatile adduct. Dry nitrogen is introduced to bring the flask to atmospheric pressure, and, as rapidly as possible, cap *a* is replaced with the serum-cap adapter *b*. The flask is connected through the vacuum line to a mercury bubbler to permit the escape of hydrogen formed in the next step.

A solution of 1.75 g. (6.9 mmoles) of I_2 in 60 ml. of dry chloroform (distilled from diphosphorus pentoxide) is prepared in a stoppered flask. Agitation of some kind must be provided, since this is nearly a saturated solution. About a 5-ml. sample of dry chloroform is added to the reaction flask with a syringe to produce a slurry† with the trimethylphosphine-borane. Then

---

* The checker obtained a clear solution at this point.
† The checker obtained a paste.

the iodine solution is added with a syringe to the stirred slurry over a 30-minute period during which time hydrogen is slowly evolved. Stirring is continued for an hour, and the solution, which was a dark red color at the last stages of the addition, is now a light yellow. Cap *a* is replaced, and the vessel is evacuated slowly while stirring continues. When a 10-ml. quantity has distilled, the flask can be considered sufficiently evacuated for the next step.

The reaction flask is cooled to −78°C., and a 1.05-g. (13.8-mmole) sample of trimethylphosphine, measured as a gas, is introduced. The colorless solution which forms is allowed to warm to room temperature and is stirred for an hour. Solvent is removed slowly under vacuum, and the white solid which remains is pumped for one hour under high vacuum. Dry nitrogen is admitted, and the flask is removed from the vacuum line. The solid product is broken up with a spatula and washed from the flask with ether. It is collected by filtration, washed with 50 ml. of ether in small portions, and allowed to air-dry in the hood for half an hour. Yield is 3.5 g. (87%). The white salt is pure $[H_2B\{P(CH_3)_3\}_2]^+I^-$ as judged by its infrared spectrum, but it has a strong trimethylphosphine odor.

Properties

The salt is a white solid, soluble in water, acetonitrile, methylene chloride, and alcohol. The cation is stable in neutral and acidic aqueous solution, but it is degraded in aqueous base. Because the cation is stable in aqueous solution, many of its salts can be readily prepared by metathesis or ion exchange. Principal infrared absorption bands of the iodide salt, run as a mineral-oil mull (exclusive of that in common with the mineral oil), occur at 2410(m), 2380(m), 1300(m), 1065*(w), 1037(w), 990(m), 950(s), 900(m), 855(w), 772(m), and 740(m) cm.$^{-1}$.

* This absorption was not observed by the checker.

References

1. N. E. Miller and E. L. Muetterties, *J. Am. Chem. Soc.*, **86**, 1033 (1964).
2. H. C. Miller, N. E. Miller, and E. L. Muetterties, *Inorg. Chem.*, **3**, 1456 (1964).
3. J. E. Douglass, *J. Am. Chem. Soc.*, **86**, 5431 (1964).
4. G. E. Ryschkewitsch, *ibid.*, **89**, 3145 (1967).
5. R. J. Rowatt and N. E. Miller, *ibid.*, **89**, 5509 (1967).
6. R. Thomas and K. Eriks, *Inorganic Syntheses*, **9**, 59 (1967).
7. H. C. Miller and E. L. Muetterties, *ibid.*, **10**, 81 (1967).

24. TRIS(PYRIDINE)HYDROBORON(III) CATION SALTS

$$(CH_3)_3NBHBr_2 + 3C_5H_5N \rightarrow [(C_5H_5N)_3BH]^{2+} + 2Br^- + (CH_3)_3N$$

Submitted by M. A. MATHUR* and G. E. RYSCHKEWITSCH*
Checked by T. C. LONG† and J. C. CARTER†

Divalent cations of boron have been synthesized by partial displacement of bromide from boron tribromide adducts by using substituted pyridines.[1] This reaction may lead to complete substitution of bromine with sufficiently basic amines or if the reaction conditions are not properly controlled. The present synthesis avoids this side reaction by blocking the fourth coordination position on boron with hydride which is not readily displaced by amines.[2] The starting material in this case is an adduct of dibromoborane, $(CH_3)_3NBHBr_2$, which is readily synthesized and is described in Sec. 20B.

Procedure

A solution of 2.31 g. (10.0 mmoles) of trimethylamine-dibromoborane in 40 ml. of dry pyridine is placed into a 125-ml.‡

* Department of Chemistry, University of Florida, Gainesville, Fla. 32601.
† Department of Chemistry, University of Pittsburgh, Pittsburgh, Pa. 15213.
‡ Checker suggested use of a 250-ml. flask to minimize difficulties from lumping.

flask equipped with reflux condenser and protected from the atmosphere by a drying tube filled with Drierite. The solution is brought to boiling and refluxed for 24 hours while the fine crystalline bromide salt slowly precipitates. After cooling, the solid is filtered through Whatman No. 42 paper,* washed with a total of 100 ml. of benzene, and finally with 30 ml. of diethyl ether. After drying under vacuum, the tan-colored dibromide salt weighs 3.95 g. A filtered solution of this salt (0.818 g. in 5 ml. of water and 10 ml. of wash) is added slowly to 240 ml. of acetone. The white precipitate is filtered and washed with acetone, yielding 0.700 g. of purified product, m.p. 222–224°C. If exposed to a humid atmosphere, the product is a dihydrate, as indicated by its bromine analysis (35.83%; calcd. 35.91%) and the fact that the (reversible) weight loss on pumping for 40 hours at room temperature corresponds closely to the calculated one.

The salt is easily converted to the water-insoluble hexafluorophosphate. A filtered solution of 0.766 g. of the crude salt in 15 ml. of water is precipitated by slowly adding excess saturated aqueous ammonium hexafluorophosphate while the solution is stirred. The filtered solid is washed (50 ml. of water) until free of bromide ion and then washed with 30 ml. of anhydrous diethyl ether. Yield of the dried product is 0.973 g. (96%), m.p. 248–250°C. Recrystallization by letting a warm solution (20 ml. of 1:1 water–acetone) evaporate slowly gives no change in melting point.

Properties

The PF_6^- salt is soluble in acetone, nitromethane, and hot 1-propanol and is stable in water. The proton resonance spectrum of the bromide in D_2O is complex: three closely spaced doublets at -8.79, -8.68, and -8.56 p.p.m. (vs. TMS) and

---

* More rapid filtration was obtained by the checker by using a frit with an aspirator.

three singlets with some fine structure at -8.25, -8.13, and -8.01 p.p.m. The relative intensity of the two sets of resonances is $3:2$, typical of coordinated pyridine. [The checker reported the ^{11}B n.m.r. as a doublet at 5.8 p.p.m. downfield from BF_3OEt_2 (measured at 32.1 MHz.).]

References

1. C. W. Makosky, G. L. Galloway, and G. E. Ryschkewitsch, *Inorg. Chem.*, **6**, 1972 (1967).
2. G. E. Ryschkewitsch, M. A. Mathur, and T. E. Sullivan, *Chem. Commun.*, **17**, 1970.

25. TETRAKIS(4-METHYLPYRIDINE)BORON(III) CATION SALTS

$$(CH_3)_3NBBr_3 + 4CH_3C_5H_4N \rightarrow [(CH_3C_5H_4N)_4B]^{3+} + 3Br^- + (CH_3)_3N$$

Submitted by G. E. RYSCHKEWITSCH* and G. L. GALLOWAY*
Checked by R. F. BRATTON† and J. C. CARTER†

Triply charged boron cations can be prepared by a reaction analogous to the synthesis of boron cations with lower charges,[1-3] that is, by displacement of halide ion from a boron trihalide adduct by a suitable amine. See syntheses 22 to 24. The reaction proceeds readily and leads to complete substitution of all halogens, provided that the amine is a sufficiently strong base and does not introduce steric strains in the tetra-substituted boron cation.[2] The 4-alkylpyridines satisfy these requirements in the displacement of bromide from boron.[4] It is convenient, but not necessary, to use the boron tribromide adduct of trimethylamine as starting material, since this compound is readily prepared and purified and has a good storage life.

* Department of Chemistry, University of Florida, Gainesville, Fla. 32601.
† Department of Chemistry, University of Pittsburgh, Pittsburgh, Pa. 15213.

Procedure

Trimethylamine–boron tribromide[5] is prepared by distilling boron tribromide from magnesium turnings directly into *n*-pentane in a three-necked flask and passing excess trimethylamine into the stirred solution.* The pure white solid is filtered off and recrystallized twice from ethanol;† m.p. 240–242°C. A 200-ml. boiling flask is fitted with a reflux condenser which has attached to its top a T-fitting connected to a supply of dry nitrogen. The apparatus is flushed with the gas, and the flask is charged with 13.5 g. of trimethylamine–boron tribromide (43.5 mmoles) and 75 ml. of freshly distilled‡ dry 4-methylpyridine (about fourfold excess). During subsequent reactions, a protective atmosphere is maintained by passing nitrogen through one arm of the T-connection. The solid adduct dissolves completely as the mixture is heated with magnetic stirring. As reflux is reached, solid product starts to precipitate while the odor of trimethylamine is noted in the effluent gas. Heating is now adjusted barely to maintain reflux, since the accumulation of solid tends to interfere with stirring and efficient heat transfer, and overheating may cause excessive bumping. After 2 hours, the mixture is cooled, diluted with 50 ml. of benzene which is added through the condenser, filtered with aspirator suction, and washed with 100 ml. of benzene and finally with 60 ml. of diethyl ether. The light yellow bromide salt, dried at about 10 μ, weighs 27.4 g. (98% yield, based on a monohydrate). Recrystallization of 3.00 g. from 20 ml. of boiling ethanol (20 ml. of 2:1 acetone-alcohol wash) yields 1.67 g. of dry product. A second crop of 0.25 g. is obtained after dilution of the filtrate with 170 ml. of acetone, for a total yield of 64%. The salt, when taken directly

* The following procedural details were recommended by the checker. At least ten times as much *n*-pentane should be used as boron tribromide. A reflux condenser is needed to return the *n*-pentane, and a bubbler is needed to monitor the addition of trimethylamine.

† Checker reported recrystallization from benzene gives a better melting point.

‡ The checker distilled the 4-methylpyridine from KOH.

from the desiccator, melts at 253–254°C., with decomposition. It reversibly absorbs water from the atmosphere to give a dihydrate, m.p. 236–237°C. (decomp.). An acidified aqueous solution of the dried salt quantitatively precipitates silver bromide with excess silver nitrate; gravimetric analysis gives 37.28% Br, if the analytical sample is transferred to the weighing bottle in a dry-box.

A hexafluorophosphate salt is prepared from the crude product by dissolving 3.68 g. in 20 ml. of water acidified with a few drops of concentrated HCl and mixing the filtered solution with 10 ml. of 4 M aqueous NH_4PF_6. The precipitate is washed with 20 ml. of water and recrystallized from a minimum amount of acidified boiling water (65 ml.) to yield dry hexafluorophosphate salt in 65% yield. This salt melts with decomposition at 220–225°C. *Anal.:* C, 35.1; H, 3.6; N, 6.7%. Calcd.: C, 35.2; H, 3.5; N, 6.8%.

Properties

The boron cation is stable in acidic aqueous solution, somewhat less stable in water, but decomposes in alkaline solution. The proton magnetic resonance spectrum in D_2O relative to external TMS shows three bands: −8.60 (broad), −8.19 (doublet, $J = 7$ Hz.), and −2.85 (singlet), with the expected intensities corresponding to the α- and β-ring hydrogen atoms and the methyl group, respectively. The bromide is very soluble in cold water and hot ethanol but insoluble in acetone. The hexafluorophosphate is slightly soluble in cold water but very insoluble in acetone.

References

1. G. E. Ryschkewitsch, *J. Am. Chem. Soc.*, **89**, 3145 (1967).
2. K. C. Nainan and G. E. Ryschkewitsch, *ibid.*, **91**, 330 (1969).
3. G. E. Ryschkewitsch, M. A. Mathur, and T. E. Sullivan, *Chem. Commun.*, 17, 1970.
4. C. W. Makosky, G. L. Galloway, and G. E. Ryschkewitsch, *Inorg. Chem.*, **6**, 1972 (1967).
5. R. C. Osthoff, C. A. Brown, and H. Clark, *J. Am. Chem. Soc.*, **73**, 4045 (1951).

Chapter Four

SOME SIGNIFICANT SOLIDS

I. ELEMENTAL BORON, TUNGSTEN BRONZES, AND METAL SULFIDES

26. ALPHA-RHOMBOHEDRAL BORON

$$2BBr_3 + 3H_2 \xrightarrow{700°C} 2B_{(s)} + 6HBr\uparrow$$

Submitted by ROGER NASLAIN*
Checked by RICHARD JONES† and GRANT URRY‡

The α-rhombohedral form of boron (α-boron) was first reported by L. V. McCarty, J. S. Kasper, F. H. Horn, B. F. Decker, and A. E. Newkirk.[1] Of the many allotropic forms of boron, it has the simplest structure.[2] It may be prepared by the pyrolysis of boron(III) iodide on a tantalum filament at 800–1000°C., but the product is usually contaminated by other allotropic varieties of boron.[3,4] Recently, Hagenmuller and Naslain showed that boron(III) bromide may be reduced by

* Department of Structural Inorganic Chemistry of The Sciences Faculty of The University of Bordeaux, associated with the C.N.R.S. 33 Talence, France.
† Department of Chemistry, Purdue University, West Lafayette, Ind.
‡ Department of Chemistry, Tufts University, Medford, Mass. 02155.

H_2 at 700°C. in a quartz tube to give finely divided α-rhombohedral boron. X-ray diffraction reveals only a single phase in the product.[5,6]   The reduction of BBr_3 with H_2 is described here.

A. BORON(III) BROMIDE*

$$2B + 3Br_2 \rightarrow 2BBr_3$$

Bromine vapor is allowed to react with a large excess of commercial boron at 700°C. The apparatus used is represented schematically in Fig. 9. The reaction vessel (R) is a silica tube (4.5 cm. i.d., 90 cm. high). In the center of the tube is a silica grating (g) supported on a silica rod or tube. The reactor is heated by means of an electric furnace equipped with a temperature controller. The evaporation chamber (V) for the bromine and the condenser (C) for the product are 1-l. round-bottomed flasks. All stopcock joints are lubricated with a fluorinated grease.

Procedure

A 120-g. sample† of commercial‡ boron (10 moles) is introduced as tablets (about 17-mm. diam., 10 mm. thick) onto the grating g of the reactor. With stopcock R_2 closed, the boron charge is heated to 700–750°C. and degassed under vacuum,

---

* *Editor's note:* The synthesis of boron(III) bromide using the reaction between BF_3 and Al_2B_6 is described in *Inorganic Syntheses*, **3**, 29 (1950). The pertinent equation is

$$Al_2Br_6 + 2BF_3 \rightarrow 2BBr_3 + \frac{2}{n}(AlF_3)_n$$

The product prepared in this way without the use of elemental bromine should be satisfactory for the starting material in Sec. B, but both the original preparation and the checking procedure used BBr_3 made by the direct reaction of boron and bromine as described herein.

† In his Ph.D. dissertation, Cornell University, Ithaca, N.Y. (1943), Dr. A. E. Newkirk described the preparation of BBr_3 using 6 g. of boron instead of 120 g. The procedure can be scaled down if desired.

‡ 90–92% amphorous boron is available from the U.S. Borax Co., 412 Crescent Way, Anaheim, Calif. The checkers used boron from Matheson Coleman and Bell, P.O. Box 85, East Rutherford, N.J. 07073.

Editor's note: Very recently U.S. Borax announced the availability of boron ranging in purity from 98.0 to 99.7% with undefined crystal properties.

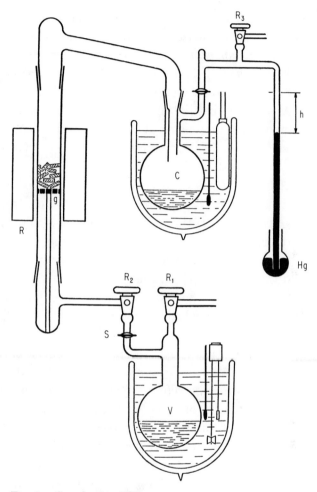

Fig. 9. **Synthesis of BBr₃.**

through stopcock R_3, with the aid of a good vacuum pump. A liquid-nitrogen trap is placed between R_3 and the pump. Degassing for 12 hours is recommended to eliminate traces of air and water vapor from the reactor. A 1280-g. sample of bromine (8 moles) is distilled under reduced pressure into the evaporator (V, Fig. 9) which has been disconnected from the reaction tube at the ground-glass ball joint (S). ■ *Warning. Always use a hood for handling and distilling bromine. The handling of large amounts of bromine is hazardous. A bottle of 5% sodium thiosulfate solution should be kept at hand and used to reduce any bromine that is spilled or that comes into contact with the body. Always wear goggles. Severe burns and eye damage may result from the spattering of liquid bromine.*

After the distillation, the flask containing the bromine is attached to the reactor at the ball joint S. Stopcock R_2 is kept closed. The bromine is cooled to $-40°C$. (alcohol bath*), and the system is evacuated through stopcock R_1. After evacuation, R_1 is closed, the condenser C is cooled to $-40°C$., and flask V is warmed to 25°C. Bromine vapor is introduced into the reactor through stopcock R_2. The rate of bromine addition should be adjusted so that the product condensing in C is colorless. R_3 should be closed until the rate of bromine addition is properly adjusted. (Br_2 attacks Hg rapidly.) If the pressure in the reactor system rises much above 100 mm. Hg because of gas leakage, gases can be pumped out periodically through R_3.

The apparatus described permits the preparation of 1300 g. of BBr_3 in about 40 hours. The unreacted boron is washed, dried, and may be reused. The yield is about 98%.

The crude boron(III) bromide may be light yellow in color due to free bromine. The color can be removed by adding a few milliliters of liquid mercury to flask C after the completion of the reaction. After standing 10 minutes at room temperature above mercury, the BBr_3 is transferred to the previously dried distillation apparatus shown in Fig. 10. The transfer takes place through a fine-porosity frit (p) under a small positive pressure of argon. Following transfer, the distillation unit is sealed off at XX, and the liquid BBr_3 is distilled at 60°C. under 280 mm. Hg pressure. The column B is a tube 2×60 cm. filled with glass beads. The purified BBr_3 is collected into ampules (diam., 3.6 cm.; height, 20 cm.) which are connected to a collecting adapter (D). The ampules are sealed off at the constriction as soon as they are filled.

* The alcohol bath used here was maintained at $-40°C$. by a small cryostat.

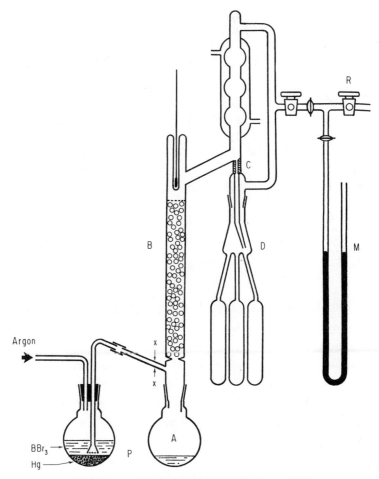

Fig. 10.   Purification by fractional distillation of **BBr₃**.

Boron(III) bromide is obtained as a perfectly colorless liquid. It is rapidly hydrolyzed by atmospheric moisture.[7, 8]

B. α-BORON

$$2BBr_3 + 3H_2 \xrightarrow{700°C} 2B + 6HBr\uparrow$$

Since the yield from the reduction of BBr₃ with hydrogen is low, the apparatus shown in Fig. 11 is used to permit recycling of unconverted BBr₃. The reaction zone is shifted along the

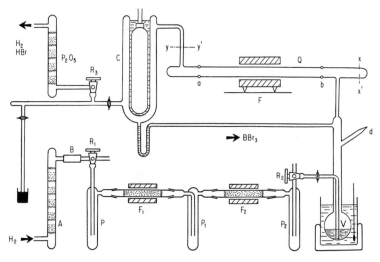

Fig. 11.   Preparation of α-boron.

axis of the tube during reaction, since prolonged heating results in the deposition of some β-rhombohedral form.[6]   Commercial hydrogen (99%) is dried over CaH_2 (section *A* of Fig. 11); oxygen, which is a particularly undesirable impurity, is reduced on a Deoxo catalyst (*B*) at room temperature.   The water formed is retained in a liquid-nitrogen trap (*P*) (Fig. 11).

Two furnaces (*F₁* and *F₂*) heat to 600°C. tubes that contain magnesium shavings.   This removes nitrogen and the last traces of oxygen; the last traces of water are removed by two additional traps (*P₁* and *P₂*) held at -196°C.

The reactor *Q* is a silica tube (diam., 3 cm.; length, 1 m.) sealed by epoxy resin at *a* and *b* to two Pyrex-glass caps.   The condenser *C* (outside diam., 8 cm.; inside diam., 6 cm.; height, 60 cm.) is connected at the bottom to a U-tube capillary that permits the recycling of the BBr_3.   The tubing, without ground-glass joints and stopcocks between the evaporator *V* and the condenser *C*, is heated to 80°C. by an electrical resistance heater.   The hydrogen bromide formed is removed by passing the exit gases from the P_2O_5 tower through several bubblers filled with an aqueous solution of sodium hydroxide. The hydrogen gas should be led into the base of an operating Meker burner or otherwise destroyed. ■ *Warning*. *Hydrogen gas mixed with air is an explosion hazard.*

Procedure

Before the reaction is carried out, all traces of air and water vapor are removed by prolonged pumping at R_1 and R_3 (15 hours, 5×10^{-3} torr) while the furnaces F, F_1, and F_2 are heated to 900, 600, and 600°C., respectively.

An ampule of BBr_3 (430 g.) is opened and connected rapidly to the apparatus at d by a short length of tubing. The BBr_3 is then distilled under reduced pressure (20 mm. Hg) into the evaporator V which is maintained at -40°C. The apparatus is then sealed at d. A slow stream of hydrogen (■ *Caution. Adequate provision to burn or remove exhausting hydrogen must be made.*) is introduced at R_2 and passed through the entire apparatus for 15 hours (P, P_1, P_2 at -196°C.; F_1, F_2 at 600°C.; V at -40°C.; F and C at room temperature).

The deposition of boron is then started by elevating the temperature of the reactor Q to 700°C. and of the evaporator V to 40°C. ($p_{BBr_3} = 130$ mm. Hg). To trap and recycle BBr_3, the condenser C is cooled to -40°C. The flow rate of hydrogen is adjusted to 200 ml./minute. The boron deposits in the form of dendritic granules. The furnace F is displaced periodically from right to left so that the boron already deposited acts as a nucleus for the new deposit. If the boron remains in the hot zone for more than 15 hours at 700°C., the product will be contaminated with small quantities of β-rhombohedral boron.[6]

When all the boron tribromide has been consumed, C and V are brought back to room temperature, and the total deposit of boron is subjected to prolonged degassing by pumping through R_3 (700°C., 5×10^{-3} torr). After the reactor is cooled, the apparatus is slowly filled with dry argon and opened at XX'

and YY'. The boron is removed from the reactor (in a dry argon atmosphere), sifted by a screen (100 mesh) under argon, and sealed in tubes.

The apparatus described permits the preparation of 15 g. of boron per run (duration 90 hours; yield on basis of BBr_3 is 80%).

Properties

The boron thus obtained is a very finely divided powder with red-brown color. The analysis of boron is better than 99%. Except for small quantities of Br_2, no impurities with atomic number greater than that of sodium were detected by use of x-ray fluorescence analysis. The x-ray pattern and the density ($d_{obs.} = 2.44$) permit the identification of the α-rhombohedral variety ($a = 5.057 \pm 0.003$ A., $\alpha = 58.06 \pm 0.05°$; $Z = 12$, space group $R\bar{3}m$; $d_{th} = 2.46^2$).

The α-rhombohedral variety is stable up to 1200°C. Above 1200°C., it transforms to the β variety. The transformation is irreversible.[9] The large chemical reactivity of α-boron, due to its crystalline structure and its state of division, makes this variety of boron an ideal material for the synthesis of borides.

References

1. L. V. McCarty, J. S. Kasper, F. H. Horn, B. F. Decker, and A. E. Newkirk, *J. Am. Chem. Soc.*, **80**, 2592 (1958).
2. B. F. Decker and J. S. Kasper, *Acta Cryst.*, **12**, 503 (1959).
3. L. V. McCarty and D. R. Carpenter, *J. Electrochem. Soc.*, **107**, 38 (1960).
4. Von E. Amberger and W. Dietze, *Z. Anorg. Allgem. Chem.*, **332**, 131 (1964).
5. P. Hagenmuller and R. Naslain, *Rev. Hautes Tempér. Réfract.*, **2**, 225 (1965).
6. R. Naslain, Thèse n° 188 Bordeaux (1967), Centre de Doc. CNRS n° A.O. 1332 (1967).
7. E. Gamble, *Inorganic Syntheses*, **3**, 27 (1950).
8. G. Urry, Boron Halides, in "The Chemistry of Boron and Its Compounds," E. L. Muetterties (ed.), John Wiley & Sons, Inc., New York, 1967.
9. R. Naslain, J. Etourneau, and P. Hagenmuller, "Proceedings of the Third International Symposium on Boron," Warsaw, Poland, 1968.

27. SODIUM TUNGSTEN BRONZES

$$Na_2WO_4 + WO_3 \rightarrow 2Na_xWO_3 + O_2\uparrow$$

Submitted by C. T. HAUCK,* A. WOLD,* and E. BANKS†
Checked by R. SEIVER‡ and H. A. EICK‡

The sodium tungsten bronzes were first reported by F. Wohler in 1824.[1] The synthesis of these compounds has been reported by many workers, including Brown and Banks.[2] The various syntheses include reduction of a sodium tungstate melt by hydrogen,[1] as well as by tin, zinc, iron, and phosphorus.[3,4] Bronzes can also be formed by reduction of a melt of sodium tungstate and tungsten(VI) oxide with tungsten metal,[5] and by the electrolytic reduction of tungstate melts.[6] The bronzes, Na_xWO_3, have metallic properties and form solid solutions of varying sodium content. The products range from a blue tetragonal bronze, in which $x = 0.3$ to 0.4, to a yellow bronze of cubic structure, for which x can be as high as 0.9. In the method discussed in this synthesis, the bronzes are formed at elevated temperatures (780–800°C.) by cathodic reduction of a sodium tungstate–tungsten(VI) oxide melt.§

Procedure

Preparation of Reactants. About 300 g.‖ of Na_2WO_4 is placed in a ceramic crucible and fused at 800°C. overnight. The dehydrated tungstate is then ground to a fine powder with a

---

* Brown University, Providence, R.I. 02912.

† Polytechnic Institute of Brooklyn, N.Y. 11201.

‡ Department of Chemistry, Michigan State University, East Lansing, Mich. 48823.

§ Simpler electrolytic procedures will result in the formation of bronzes, but the crystals will not be as large, homogeneous, or well formed.

‖ This weight is sufficient for several preparations.

porcelain mortar and pestle and stored in a desiccator. Reagent-grade tungsten(VI) oxide was used in these preparations. The components were intimately ground together in the following proportions:

Na_2WO_4 (g.)	WO_3 (g.)	Mole ratio ($Na_2WO_4:WO_3$)
83.06	65.56	10:10
117.14	76.84	12:10
153.57	46.43	26:10

Preparation of the Electrolytic Cell. The cell (Fig. 12) consists of a 50-ml. alumina crucible fitted with a platinum cap. The anode* is a rectangular platinum strip $\frac{1}{2} \times 2\frac{3}{4} \times 0.01$ in. suspended in the middle of the melt, approximately 1 in. from the bottom of the crucible. The cathode is a 1-in.-diam. platinum disk, 0.01 in. thick, placed on the bottom of the cell.

* The Na_2WO_4-WO_3 melt is corrosive, and both electrodes as well as the crucible are subject to considerable attack.

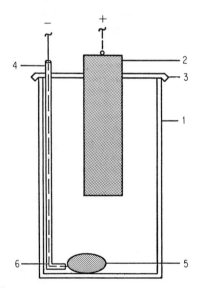

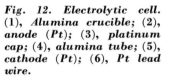

Fig. 12. Electrolytic cell. (1), Alumina crucible; (2), anode (Pt); (3), platinum cap; (4), alumina tube; (5), cathode (Pt); (6), Pt lead wire.

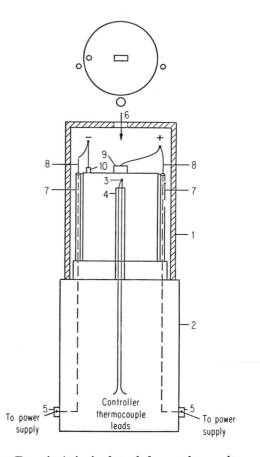

Fig. 13. Electrolysis apparatus. (1), Inconel heat shield; (2), ceramic base; (3), monitor thermocouple; (4), alumina tube (two-hole); (5), terminal posts; (6), insert for calibration thermocouple; (7), alumina tubes; (8), Pt leads; (9), anode (Pt); (10), alumina tube for cathode lead wire.

Its platinum lead (0.016-in. Pt wire) is isolated from the melt by means of an alumina tube placed snugly against the cell wall. It is essential that the cell be packed very tightly to the top with the finely ground mixture of Na_2WO_4-WO_3. Both platinum electrodes protrude a few inches out of the cell.

The electrolytic cell is placed on a stand (Fig. 13) which is made of Alsimag 222 ceramic.* The anode and cathode are connected to platinum wires that run down the inside of the stand and are attached to terminal posts near the base of the stand.

* American Lava Corp., 219 Kruesi Building, Chattanooga, Tenn. 37405.

Electrolytic Procedure. The cell is placed inside a vertical-column, Hevi-Duty, multiple furnace* and is surrounded by an Inconel heat shield (Fig. 13). The temperature-control thermocouple (Pt vs. 90% Pt − 10% Rh) is located between the crucible and the heat shield and is protected by an outer ceramic tube. A calibration run is initially made without a charge in the crucible. A second thermocouple is inserted through the heat shield and centered on the cover of the electrolytic cell. This thermocouple is read directly by means of a potentiometer; the temperature of this thermocouple corresponds closely to that of the melt. The temperature controller is adjusted until the second thermocouple indicates the desired temperature for the electrolysis.†

An electrolysis can be carried out by removing the second thermocouple, filling the crucible with a charge, and connecting the leads from the power supply to the cell. An Inconel plug is used to close the thermocouple hole in the heat shield. The d.-c. constant-current power supply is turned on to warm up without any current fed to the cell. The temperature is then raised to the operating temperature of 780°C. at a rate of 200°C. per hour. When the temperature has equilibrated, the voltage is applied, and the current is adjusted to 45 ma. The electrolysis is carried out for 5 days.

On completion of the run, the furnace is shut down, but the current from the power supply to the cell is left on. The resistance of the cell increases when the melt begins to solidify. This is indicated by a fluctuation of the current, and the power is then turned off. The reaction mixture is allowed to cool to 100°C. in the furnace, and the cell is disconnected. The crucible walls above the melt are broken off with a hammer, and the product is extracted with boiling water. This process is

---

* Type MK-2012, Fisher Scientific Co., 711 Forbes Ave., Pittsburgh, Pa. 15219.

† This calibration must be done since a thermocouple cannot be placed directly into the melt, and the heat shield prevents the control thermocouple from giving an accurate temperature reading.

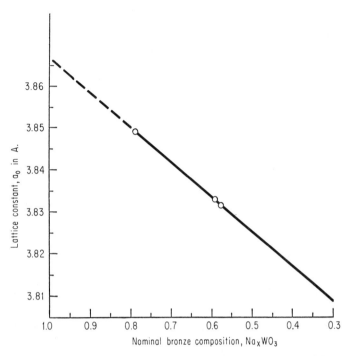

Fig. 14. Lattice constant vs. nominal bronze composition.

repeated until the flux has dissolved; the clean crystals are recovered and allowed to dry in air.

Preparation for X-ray Analysis. Lattice constants are calculated from patterns obtained on powder samples with a Norelco diffractometer using monochromatic radiation (AMR-202 Focusing Monochromator) from a high-intensity copper source. The crystals are powdered with a diamond* mortar and pestle, and the powder passed through a 74-μ sieve. Accurate lattice constants are calculated from the x-ray data.

Properties

The following bronzes·were obtained:

---

* Plattner's diamond mortar and pestle of hard tool steel.

Na_xWO_3, x	Lattice constant, a_0 (A.)	Mole ratio of reactants, $Na_2WO_4:WO_3$	Color
0.58	3.832 ± 0.005	10:10	Magenta
0.59	3.833 ± 0.005	12:10	Red
0.79	3.849 ± 0.005	26:10	Yellow

The values for x are obtained from a lattice constant vs. nominal composition curve (Fig. 14) reported by Brown and Banks.[2] The linear portion of the curve can be represented by:

$$a_0(\text{A.}) = 0.0819x + 3.7846$$

References

1. F. Wohler, *Ann. Chim. Phys.*, [2], **29**, 43 (1823).
2. B. W. Brown and E. Banks, *J. Am. Chem. Soc.*, **76**, 963 (1954).
3. H. Wright, *Liebigs Ann. Chem.*, **79**, 221 (1851).
4. J. Philip, *Berlk.*, **15**, 499 (1882); *Jahrb. Ber. Forte-Chem.*, **28**, 444 (1882).
5. M. E. Straumanis, *J. Am. Chem. Soc.*, **71**, 679 (1949).
6. L. D. Ellerbeck, H. R. Shanks, P. H. Sidles, and G. C. Danielson, *J. Chem. Phys.*, **35**(1), 298 (1961).

28. GROUP IV SULFIDES

$$M + 2S \rightarrow MS_2$$
$$MS_2 + 2I_2 \xrightarrow{900°C} MI_4 + 2S$$
$$MI_4 + 2S \xrightarrow{800°C} MS_2 + 2I_2$$

Submitted by LAWRENCE E. CONROY*
Checked by R. J. BOUCHARD†

Titanium(IV) sulfide, zirconium(IV) sulfide, hafnium(IV) sulfide, and tin(IV) sulfide constitute a group of isostructural compounds that can be prepared by similar methods. A

* Department of Chemistry, University of Minnesota, Minneapolis, Minn. 55455.
† E. I. du Pont de Nemours & Company, Wilmington, Del. 19898.

procedure for preparing polycrystalline titanium(IV) sulfide has appeared in this series.[1]   A review of methods, covering the literature to 1956, is contained in that article.   Titanium(IV) sulfide can also be prepared directly from the elements[2] and by using the chemical transport technique[3] described below. Zirconium(IV) sulfide can be prepared from reaction of the elements[4] at 1400°C.; by heating zirconium(IV) chloride, hydrogen, and sulfur vapor;[5] by heating Zr_3S_5 in hydrogen sulfide gas[6] at 900–1300°C.; and by treating zirconium(IV) chloride with hydrogen sulfide[7] at 500°C.   A procedure utilizing reaction of the elements and chemical transport[3] is described here.   Hafnium(IV) sulfide can be prepared from reaction of the elements alone[2] or with chemical transport.[3]

Tin(IV) sulfide can be prepared by hydrogen sulfide precipitation of Sn(IV) from solution, to produce a microcrystalline material that is contaminated with oxide.   *Mosaic gold* is a crystalline form of tin(IV) sulfide prepared by high-temperature sublimation procedures.   Mosaic gold is the reported product of heating mixtures of (1) tin and sulfur;[8] (2) tin, sulfur, and ammonium chloride;[8,9] (3) tin, sulfur, mercury, and ammonium chloride;[9] (4) tin(II) oxide, sulfur, and ammonium chloride;[9] (5) tin(II) chloride and sulfur;[9] (6) tin(II) sulfide, tin(II) chloride, and sulfur.[9]

Many of the transition-metal chalcogenides can be prepared as large crystals of high purity by means of high-temperature chemical transport reactions.[10]   The syntheses of TiS_2, ZrS_2, HfS_2, and SnS_2 that are described below adapt the procedure of Greenaway and Nitsche.[3]   The polycrystalline disulfide is synthesized from the elements and transported through a temperature gradient in the presence of iodine vapor.   Both the synthesis and transport reaction are carried out in the same evacuated ampule.   Upon heating, the polycrystalline disulfide and iodine equilibrate with the volatile metal tetraiodide and sulfur.   The equilibria favor the disulfide and iodine more in the cooler zone than in the hotter zone.   The disulfide may thus be transported from a hotter to a cooler zone under conditions that yield very pure, and often rather large, crystals. Because the iodine is regenerated in the cooler zone, the ratio of iodine to disulfide that is required is quite small.

A. TITANIUM(IV) SULFIDE

$$Ti + 2S \rightarrow TiS_2$$

$$TiS_2 + 2I_2 \underset{800°C.}{\overset{900°C.}{\rightleftarrows}} TiI_4 + 2S$$

Procedure

The synthesis and transport reactions are carried out in silica or Vycor ampules (Fig. 15).   Convenient dimensions are 1.5–2.5 cm. i.d. and 10–15 cm. long.   Titanium sponge* (0.96 g., 0.020 mole) and resublimed sulfur (1.34 g., slight excess over 0.04 mole) are placed in the tube along with a weighed quantity of the iodine transporting agent corresponding to 5 mg./ml. of the tube volume.   Two relatively simple procedures for introducing iodine into the ampule are available.   (1) The reactants are added to the transport tube through a long-stemmed

* The sponge form of titanium, zirconium, or hafnium is preferable in these syntheses because the tendency for explosive reaction is much less than with the powdered metals.   More massive forms, such as wire or shot, react too slowly and less completely.   Suitable sources are City Chemical Corp., 132 W. 22d St., New York, N.Y. 10011, and Electronic Space Products, Inc., 854 S. Robertson Blvd., Los Angeles, Calif. 90035.

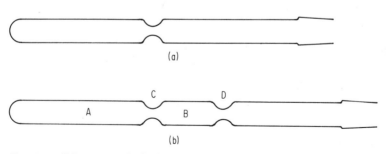

Fig. 15.   *Silica ampule for transport reactions.*

funnel, and the tube is attached to a vacuum line and evacuated at 10^{-5} torr. A tube containing the calculated weight of purified iodine is attached to the vacuum line and evacuated. The iodine is then sublimed over into the transport tube which is then sealed off.* (2) A transport tube of type *b* (Fig. 15) is prepared. The required weight of iodine (for the volume of zone *A*) is loaded into a *soft-glass* ampule (5 mm. o.d.) that is evacuated and sealed off at a length of approximately 3 cm. The reactants are loaded into zone *A* of the silica tube, and the iodine ampule is inserted into zone *B*. The silica tube is then evacuated, outgassed, and sealed off at *D*. The iodine ampule is opened by clamping the silica tube in a horizontal position and heating zone *B* at one point with a fine oxygen-gas flame until the iodine ampule melts at the hot spot and liberates the iodine. The iodine is then sublimed over into zone *A*, and the silica ampule is sealed off at *C*. Other more elaborate procedures for adding the transport agent have been described.[12,13]

The transport ampule is placed in a horizontal two-zone furnace in which different temperatures may be maintained at either end of the ampule. Such a furnace may be constructed by bolting together two laboratory tube furnaces end to end or by winding a center-tapped heating element on a refractory tube-furnace core. Commercial two-zone furnaces are available. For short ampules the normal center-to-end gradient of an ordinary tube furnace may be used. Because the reaction of titanium, zirconium, or hafnium with sulfur is highly exothermic, slow heating of the reactants is necessary to avoid explosion. After the ampule is inserted in the cold furnace (■ *Caution. Do not insert in a hot furnace!*), the furnace is *slowly* heated to transport temperatures, 900°C. at the reactant end and 800°C. at the growth end, over a period of 5 hours.†

---

* The checkers pointed out that commercially available titanium, zirconium, and hafnium (even the ultrapure variety) are usually contaminated with oxides, in which case the high-vacuum procedure suggested here is not justified.

† The checkers suggested an alternative heating procedure to avoid explosions. The reactant end of the ampule is inserted into the cold furnace, leaving the empty

(*footnote continues on page 162*)

Crystals are visible in the growth zone within 6 hours after the reactant zone reaches 900°C. The 900 → 800°C. gradient is maintained for at least 70 hours or until no further transport is evident. The ampule is allowed to cool and is cracked open. (■ *Caution.* *Wrap the tube in several layers of cloth before attempting to open.*) The crystals of TiS_2 are washed successively with CCl_4 and CS_2 to remove any surface iodine and sulfur.

If large crystals are desired, some modifications will promote the growth of a smaller number of large crystals. An ampule of at least 2.5 cm. i.d. is desirable. Before the reaction zone is heated above 800°C., the growth zone is heated to 950°C. for 3 hours to minimize seed sites. The growth zone is then allowed to cool to 800°C., the reaction zone is heated to 900°C., and transport is allowed to proceed at 900 → 800°C. for 70 to 80 hours. The yield of transported crystals depends upon tube dimensions, purity of reactants, and duration of the transport process.

B. ZIRCONIUM(IV) SULFIDE

$$Zr + 2S \rightarrow ZrS_2$$
$$ZrS_2 + 2I_2 \underset{800°C.}{\overset{900°C.}{\rightleftharpoons}} ZrI_4 + 2S$$

Procedure

Zirconium sponge (1.82 g., 0.020 mole) and resublimed sulfur (1.34 g., a slight excess over 0.040 mole) are charged into the transport tube. The procedures for adding iodine (5 mg./ml. of ampule volume) and heating that are described in the TiS_2

_____

growth end projecting. (■ *Caution.* *Provide a protective shield.*) The furnace is slowly heated to transport temperatures over a period of at least an hour, and the ampule is then slowly moved, over a period of 4 to 5 hours, into the heated zones. During this procedure most of the sulfur will sublime into the cool projecting portion of the ampule and will be removed from the reaction zone. As its vapor pressure increases with the temperature, reaction with the metal can proceed gradually.

synthesis (Sec. A above) are followed. Transport is carried out at 900 → 800°C. for at least 70 hours. The yield of transported crystals depends upon tube dimensions, purity of reactants, and duration of the transport process.

C. HAFNIUM(IV) SULFIDE

If hafnium sponge (3.57 g., 0.020 mole) and resublimed sulfur (1.34 g., a slight excess over 0.040 mole) are used in the above procedure in place of zirconium sponge and sulfur, HfS_2 is obtained.

D. TIN(IV) SULFIDE

$$Sn + 2S \rightarrow SnS_2$$
$$SnS_2 + 2I_2 \underset{600°C.}{\overset{700°C.}{\rightleftharpoons}} SnI_4 + 2S$$

Procedure

Tin metal (2.37 g., 0.020 mole) and resublimed sulfur (1.34 g., a slight excess over 0.040 mole) are charged into the transport tube. The procedures for adding iodine (5 mg./ml. of ampule volume) and heating that are described in the TiS_2 synthesis (Sec. A above) are followed. Because of the low melting point of tin, this reaction proceeds more smoothly at lower temperature with less danger of explosions than in the case of the transition metals. Transport is carried out at 700 → 600°C. for 25 to 30 hours. An alternative procedure is to prepare polycrystalline SnS_2 by the usual solution precipitation procedures. This material is also available commercially. Precipitated SnS_2 contains a considerable fraction of SnO_2, but it may be purified by the same transport procedure described above. Precipitated SnS_2 should be dried at 350°C. for at least a day before sealing in a transport ampule. Crystalline SnS_2 (mosaic gold) is usually prepared from tin amalgam and contains appreciable mercury contamination. This reagent is to be avoided in high-purity preparations. The yield depends upon tube dimensions, purity of reactants, and duration of the transport process.

Properties

The group IV disulfides form platelike crystals that are stable in air* and are insoluble in water, dilute acids and bases, and most organic solvents. Titanium(IV) sulfide is gold-colored, zirconium(IV) sulfide is brown-violet, hafnium(IV) sulfide is violet, and tin(IV) sulfide is golden yellow. The transition-metal compounds all exhibit a high metallic luster even in very thin sections. The tin(IV) sulfide appears metallic in thick section, but thin crystals are a transparent yellow color. All four compounds crystallize in the hexagonal cadmium iodide structure (C6 in the Strukturbericht classification) in which the sulfur atoms are arranged in a hexagonal close-packed array and the metal atoms reside in octahedral holes between alternate sulfur layers. The lattice parameters are: TiS_2[3], a = 3.405 A., c = 5.687 A.; ZrS_2,[11] a = 3.661 A., c = 5.825 A.; HfS_2,[11] a = 3.625 A., c = 5.846 A.; SnS_2,[3] a = 3.639 A., c = 5.884 A. All four compounds decompose or sublime, without melting, at temperatures above 800°C. They are all n-type semiconductors* when prepared by the known procedures, indicating a slight excess of metal atoms in the lattice[11] (of the order of 0.1% excess in TiS_2, 0.01% in ZrS_2, but less than 10^{-6}% in HfS_2 or SnS_2 when prepared by the procedure given here).

References

1. R. C. Hall and J. P. Mickel, *Inorganic Syntheses*, **5**, 82 (1957).
2. F. K. McTaggart and A. D. Wadsley, *Australian J. Chem.*, **11**, 445 (1958).
3. D. L. Greenaway and R. Nitsche, *J. Phys. Chem. Solids*, **26**, 1445 (1965).
4. R. Vogel and A. Hartung, *Arch. Eisenhuettenw.*, **15**, 413 (1941–1942).
5. A. E. van Arkel, *Physica*, **4**, 286 (1924).
6. M. Picon, *Compt. Rend.*, **196**, 2003 (1933).

* *Notes added in proof:* Checker reports TiS_2 and ZrS_2 changed color on standing in closed vials at 25°C. Authors report that TiS_2 is a metallic conductor down to 1°K.

7. E. F. Strotzer, W. Biltz, and K. Meisel, *Z. Anorg. Allgem. Chem.*, **242**, 249 (1939).
8. P. Woulfe, *Phil. Trans.*, **61**, 114 (1771).
9. J. W. Mellor, "A Comprehensive Treatise on Inorganic and Theoretical Chemistry," Vol. 7, p. 469, Longmans, Green & Co., Ltd., London, 1927, contains an extensive list of early references.
10. H. Schafer, "Chemical Transport Reactions," Academic Press, Inc., New York, 1961.
11. L. E. Conroy and K. C. Park, *Inorg. Chem.*, **7**, 459 (1968).
12. A. G. Karipides and A. V. Cafiero, *Inorganic Syntheses*, **11**, 5 (1968).
13. R. Kershaw, M. Vlasse, and A. Wold, *Inorg. Chem.*, **6**, 1599 (1967).

II. PREPARATION OF SOME METAL HALIDES— ANHYDROUS MOLYBDENUM HALIDES AND OXIDE HALIDES—A SUMMARY

Submitted by MELVIN L. LARSON*

A. BINARY MOLYBDENUM HALIDES

Molybdenum forms a group of fluorides of general formula MoF_n, where n takes all integral values from 6 to 3. With the less electronegative chlorine, the highest well-characterized compound is $MoCl_5$, although a recent report[1] suggests the existence of thermally unstable $MoCl_6$. Other well-established chlorides are $MoCl_4$, $MoCl_3$, and $MoCl_2$. Bromides of formula $MoBr_4$, $MoBr_3$, and $MoBr_2$ are known to exist, whereas with iodine only two lower-valent iodides, MoI_3 and MoI_2, are satisfactorily characterized.

In the case of fluorine and chlorine, molybdenum metal powder reacts with the elemental halogen to give the higher halides in accordance with the general equation

$$Mo + \frac{n}{2} X_2 \rightarrow MoX_n$$

where $n = 6$ for F and $n = 5$ for Cl. Because of the relative thermal instability of higher bromides and iodides, however, this direct reaction is efficient only for the synthesis of $MoBr_3$

* Research Laboratory, Climax Molybdenum Company, Ann Arbor, Mich. 48106.

and MoI_3.* Although some $MoBr_4$ can form at 400°C. with bromine pressure,[3] the compound decomposes thermally at this temperature in the absence of excess bromine.[4]

The lower molybdenum halides are prepared by the reduction or thermal decomposition of the appropriate higher halide. A variety of reducing agents (examples: H_2, $SnCl_2$, PF_3) have proved useful.

This review will cite only the established and more specific syntheses without attempting to include those of lesser efficiency.

1. Molybdenum Fluorides

a. MoF_6—The direct reaction of molybdenum with fluorine initiated around 100°C. in a copper tube[5-7] is specific for this highest fluoride. The method is relatively inconvenient in most laboratories because of experimental difficulties and hazards associated with the handling of elemental fluorine. A more recent procedure,[8] which avoids elemental F_2, involves the reaction between MoO_3 and SF_4 in a bomb at 350°C. Sulfur tetrafluoride, although expensive, is commercially available in cylinders.

b. MoF_5—Molybdenum(VI) fluoride can be reduced to MoF_5 by PF_3,[6] $Mo(CO)_6$,[9] or molybdenum powder.[9] The preferred reducing agent appears to be phosphorus(III) fluoride at room temperature because of the ease of removal of both the volatile by-product PF_5 (b.p. −85°C.) and the excess PF_3 (b.p. −102°C.).

c. MoF_4—Several unsatisfactory methods have been reported. Molybdenum(IV) fluoride is reported as one of several products of the low-temperature (−75°C.) reaction of $Mo(CO)_6$ with elemental fluorine. From this reaction mixture MoF_5 is vacuum-distilled, leaving a solid residue of composition approximating MoF_4.[10] Another method involves disproportionation of MoF_5 above 150°C. to MoF_6 and a solid phase believed to be

---

* The early literature contains an unverified and questionable report of MoI_4.[2]

MoF_4.[9] The fluorination of MoS_2 with SF_4 gave MoF_4 which was isolated as the $MoF_4 \cdot 2(CH_3)_2SO$ complex.[8]

d. MoF_3—The lowest reported molybdenum fluoride, MoF_3, was obtained by the reduction of MoF_5 with elemental molybdenum. It was found that the color of the product varied with the experimental conditions.[11] Because of varying compositions, the properties of the compound are not well defined.

2. Molybdenum Chlorides

Specific procedures for the synthesis of $MoCl_4$, $MoCl_3$, and $MoCl_2$ are described in the sections that follow. Other methods of synthesis are also summarized in these sections. Procedures for the synthesis of $MoCl_5$ have been given in *Inorganic Syntheses* **7**, 167 (1963) and **9**, 135 (1967). The industrial method is the direct chlorination of metal and is recommended in some laboratories.[12, 13] A convenient laboratory technique is the liquid-phase reaction system that uses refluxing thionyl chloride and molybdenum(VI) oxide.[14]

3. Molybdenum Bromides

A direct method of synthesis of molybdenum(IV) bromide is the reaction between $MoBr_3$ and warm (60°C.) liquid bromine.[4] If restricted amounts of bromine vapor and molybdenum metal powder are used ($T = 600°C.$), $MoBr_3$ is the product.[4] At still higher temperatures (650–700°C.) $MoBr_2$ is reported to form along with $MoBr_3$.[15] Described in a following section is the furnace reaction of $MoCl_2$ with excess LiBr which allows the formation of $MoBr_2$. (Both $MoCl_2$ and $MoBr_2$ are cluster compounds. See synthesis 29.)

4. Molybdenum Iodides

The reaction of molybdenum metal powder with an excess of iodine in a sealed tube at 300°C. gives MoI_3.[16] An earlier preparation which gives a less pure product is the reaction of

$Mo(CO)_6$ with iodine in a sealed tube at 105°C.[17]   The thermal dissociation of MoI_3 at 100°C. in a vacuum is reported to give MoI_2.[18]   The sealed-tube furnace reaction of iodine with molybdenum metal powder at 300–400°C.[19] as well as the thermally induced reaction of $MoCl_2$ with fused lithium iodide[19] are mentioned.

B. MOLYBDENUM OXIDE HALIDES

The known oxide halides of molybdenum are listed below:

TABLE I

	Mo(VI)		Mo(V)		Mo(IV)
Halogen	MoO_2X_2	$MoOX_4$	MoO_2X	$MoOX_3$	$MoOX_2$
Fluorine	MoO_2F_2	$MoOF_4$			
Chlorine	MoO_2Cl_2	$MoOCl_4$	MoO_2Cl	$MoOCl_3$	$MoOCl_2$
Bromine	MoO_2Br_2	. . . .	. . . .	$MoOBr_3$	

Presumably because of its lower electronegativity, bromine can stabilize Mo(V) as $MoOBr_3$ but not Mo(VI) as $MoOBr_4$. There are no reported molybdenum oxide iodides and no lower-valent [molybdenum(V) or molybdenum(IV)] oxide fluorides. The combination of oxygen plus chlorine can stabilize Mo(VI) in both MoO_2Cl_2 and $MoOCl_4$, but $MoCl_6$ is apparently unstable at room temperature.

1. Molybdenum(VI) Oxide Halides

a. MoO_2X_2—A suggested synthesis of MoO_2F_2 involves the substitution reaction of MoO_2Cl_2 with hydrogen fluoride.[20] The simplest and best method for the synthesis of MoO_2Cl_2 is the chlorination of molybdenum dioxide with dry chlorine.[21] One synthesis of MoO_2Br_2 involves passing a mixture of oxygen and bromine diluted with nitrogen over heated (300°C.) molybdenum metal.[13]

b. MoOX$_4$—An older synthesis of MoOF$_4$ involves the substitution reaction of MoOCl$_4$ with anhydrous hydrogen fluoride.[22] A more recent procedure involves the direct fluorination of molybdenum metal powder in the presence of oxygen.[23] The preferred method of synthesis of MoOCl$_4$ is the reaction of dry oxygen with molten molybdenum(V) chloride.[24]

$$MoCl_5 + \tfrac{1}{2}O_2 \rightarrow MoOCl_4 + \tfrac{1}{2}Cl_2$$

2. Molybdenum(V) Oxide Halides

a. MoO$_2$X—A concentrated solution of MoCl$_5$ in dry ethanol is evaporated and the residue heated to approximately 150°C. under vacuum to form MoO$_2$Cl.[25]

b. MoOX$_3$—The most specific and efficient synthesis of MoOCl$_3$ is the reduction of MoOCl$_4$ with refluxing chlorobenzene.[26] For MoOBr$_3$ the parent compound MoO$_2$Br$_2$ is allowed to react with PBr$_5$ in refluxing carbon tetrachloride solution.[13]

3. Molybdenum(IV) Oxide Halides

a. MoOX$_2$—The recently reported MoOCl$_2$ has been prepared by various chemical transport reactions.[27] (See synthesis 28.) For example, it was obtained by the sealed-tube-furnace gradient (400–300°C.) reaction of molybdenum(VI) oxide with molybdenum(III) chloride in a transporting medium of Cl$_2$, the essential equation being:

$$MoO_3 + 2MoCl_3 \rightarrow 3MoOCl_2$$

Another route involved the following equation:

$$5MoO_3 + 6MoCl_5 + 4Mo \rightarrow 15MoOCl_2$$

References

1. M. Mercer, *Chem. Commun.*, 119 (1967).
2. M. Guichard, *Ann. Chem. Phys.*, **23**(7), 569 (1901).

3. W. Klemm and H. Steinberg, *Z. Anorg. Allgem. Chem.*, **227**, 193 (1936).
4. P. J. H. Carnell, R. E. McCarley, and R. D. Hogue, *Inorganic Syntheses*, **10**, 49 (1967).
5. T. A. O'Donnell, *J. Chem. Soc.*, 4681 (1956).
6. T. A. O'Donnell and D. F. Stewart, *J. Inorg. Nucl. Chem.*, **24**, 309 (1962).
7. T. A. O'Donnell and D. F. Stewart, *Inorg. Chem.*, **5**, 1434 (1966).
8. A. L. Oppegard, W. C. Smith, E. L. Muetterties, and V. A. Engelhardt, *J. Am. Chem. Soc.*, **82**, 3835 (1960).
9. A. J. Edwards, R. D. Peacock, and R. W. H. Small, *J. Chem. Soc.*, 4486 (1962).
10. R. D. Peacock, *Proc. Chem. Soc.*, 59 (1957).
11. D. E. LaValle, R. M. Steele, M. K. Wilkinson, and H. L. Yokel, Jr., *J. Am. Chem. Soc.*, **82**, 2433 (1960).
12. G. Brauer, "Handbook of Preparative Inorganic Chemistry," Vol. 2, p. 1405, Academic Press, Inc., New York, 1965.
13. R. Colton and J. B. Tomkins, *Australian J. Chem.*, **18**, 447 (1965).
14. H. J. Siefert and H. P. Quak, *Angew. Chem.*, **73**, 621 (1961).
15. C. Durand, R. Schaal, and P. Souchay, *Compt. Rend.*, **248**, 979 (1959).
16. J. Lewis, D. J. Machlin, R. S. Nyholm, P. Pauling, and P. W. Smith, *Chem. Ind. (London)*, 259 (1960).
17. C. Djordjevic, R. S. Nyholm, C. S. Pande, and M. H. B. Stiddard, *J. Chem. Soc.*, **A**, 16 (1966).
18. V. F. Klanberg and H. W. Kohlschütter, *Z. Naturforsch.*, **15b**, 616 (1960).
19. J. C. Sheldon, *J. Chem. Soc.*, 410 (1962).
20. O. Ruff and F. Eisner, *Ber.*, **40**, 2926 (1907).
21. R. L. Graham and L. G. Hepler, *J. Phys. Chem.*, **63**, 723 (1959).
22. O. Ruff, F. Eisner, and W. Heller, *Z. Anorg. Allgem. Chem.*, **52**, 256 (1907).
23. G. H. Cady and G. B. Hargreaves, *J. Chem. Soc.*, 1568 (1961).
24. A. K. Mallock, *Inorganic Syntheses*, **10**, 54 (1967).
25. R. Colton and I. B. Tomkins, *Australian J. Chem.*, **21**, 1975 (1968).
26. M. L. Larson and F. W. Moore, *Inorg. Chem.*, **5**, 801 (1966).
27. H. Schäfer and J. Tillock, *J. Less-Common Metals*, **6**, 152 (1964).

29. MOLYBDENUM(II) HALIDES

Submitted by PIERO NANNELLI* and B. P. BLOCK*
Checked by D. A. EDWARDS† and A. K. MALLOCK‡

Molybdenum(II) halides, which can be formulated $(Mo_6X_8)X_4$ (X=Cl, Br, or I), are typical metal-cluster compounds.   The

* Technological Center, Pennwalt Corporation, King of Prussia, Pa. 19406.
† School of Chemistry, Bath University of Technology, Bath, Somerset, England.
‡ Research Laboratory, Climax Molybdenum Company, Ann Arbor, Mich. 48106.

six molybdenum atoms are arranged in an octahedron with eight halogens forming bridges between them. In order to satisfy the coordination requirements of the cluster, the remaining four halogens function as ligands, bridging between clusters.[1]

Blomstrand[2] prepared molybdenum(II) chloride either by first reducing molybdenum(V) chloride to molybdenum(III) chloride with hydrogen and then heating the trichloride in a stream of dry carbon dioxide or by passing chlorine, diluted with a large quantity of carbon dioxide, over gently heated molybdenum in the absence of oxygen. Lindner and co-workers[3] prepared it by passing phosgene vapor over molybdenum powder heated between 600 and 620°C. Helriegel[4] converted molybdenum(V) or molybdenum(III) chloride to the dichloride by fusion with molybdenum powder in an inert atmosphere. Senderoff and Brenner[5] made molybdenum(III) chloride from molybdenum powder and excess molybdenum(V) chloride in a sealed tube at 350°C., removed the excess molybdenum(V) chloride, and effected disproportionation to molybdenum(II) chloride at 650°C. More recently Sheldon[6] prepared molybdenum(II) chloride by first heating molybdenum powder in chlorine to form molybdenum(V) chloride. The molybdenum(V) chloride was then reduced to molybdenum(III) chloride by excess molybdenum at red heat, and the latter was disproportionated to molybdenum(II) chloride and volatile molybdenum(IV) chloride upon heating in a stream of nitrogen. Cotton and Curtis[7] improved the yield by including additional molybdenum in the reaction tube. The molybdenum(II) chloride prepared by these methods is usually amorphous; however, Schäfer and co-workers[1] prepared crystalline samples by careful disproportionation of molybdenum(III) chloride. Our method of preparation involves heating a mixture of molybdenum(V) chloride and excess molybdenum powder in a stream of nitrogen in such a way that the yield of pure product is near theoretical. The molybdenum(II) chloride is purified by conversion to

$(H_3O)_2(Mo_6Cl_8)Cl_6 \cdot 6H_2O$ which is then heated under a vacuum. The entire operation requires no more than 2 days.

Molybdenum(II) bromide was prepared first by Blomstrand[8] by passing bromine vapor over heated molybdenum. Lindner et al.[9] then improved the method by using bromine vapor diluted with nitrogen. More recently Sheldon[10] converted molybdenum(II) chloride to the bromide by fusion with lithium bromide. Like the chloride, molybdenum(II) bromide is usually an amorphous powder; however, crystalline samples have been prepared by disproportionation of molybdenum(III) bromide under vacuum at 600°C.[1] Our method is substantially that of Sheldon and consists of heating an intimate mixture of molybdenum(II) chloride and a large excess of lithium bromide under vacuum. The crude product that results is dissolved in dilute sodium hydroxide and precipitated in pure form with concentrated hydrobromic acid.

A. MOLYBDENUM(II) CHLORIDE

$$3MoCl_5 + 2Mo \rightarrow 5MoCl_3$$
$$12MoCl_3 \rightarrow (Mo_6Cl_8)Cl_4 + 6MoCl_4$$
$$3MoCl_4 + 3Mo \rightarrow (Mo_6Cl_8)Cl_4$$

Procedure

The reaction tube (see Fig. 16) is a Vycor tube about 125 cm. long and 2.5 cm. in diameter. Both ends of the tube are fitted with male ground-glass joints for connection to stopcocks. The tube is dried thoroughly by heating while a slow stream of nitrogen is passed through it.

An intimate mixture of 50 g. of molybdenum(V) chloride with 150 g. of molybdenum powder (mesh size: 100)* prepared under anhydrous conditions is distributed uniformly along the length of the tube. Two glass-wool plugs (*A* and *B* in the figure) are placed about 15 cm. from the joints at the ends of the tube to

---

* Available from Climax Molybdenum Company, 1270 Avenue of the Americas, New York, N.Y. 10020.

contain the reaction mixture. Since molybdenum(V) chloride is very sensitive to moisture and oxygen, care should be taken not to expose it to atmospheric moisture either during mixing or loading. A dry-box or dry-bag of sufficient size to hold the tube should be used.

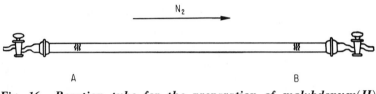

Fig. 16. Reaction tube for the preparation of molybdenum(II) chloride.

Heating is then started at the left end (A) of the tube with a tube furnace about 33 cm. long. A slow stream (about two bubbles per second) of dry, oxygen-free nitrogen is passed through the reaction tube in the direction indicated in Fig. 16. This section of the tube is kept at red heat (600–650°C.)* until the reaction mixture is yellow-brown with black spots due to excess molybdenum. During this operation a sizable amount of a very dark mass (mixture of higher molybdenum chlorides) collects in the colder part of the tube next to the right end of the furnace. At the beginning of this operation a temporary increase in the rate of flow of the nitrogen may be necessary in order to keep these volatile chlorides from also collecting to the left of the furnace. The furnace is then moved slowly toward the right end of the tube in order to convert the rest of the reaction mixture to yellow-brown. The dark mass of volatile chlorides will also automatically move toward the right end. When the front line of this mass has reached plug B, heating is discontinued. It takes about half an hour to reach this point. The tube is allowed to cool a little, both stopcocks are momentarily closed, and the direction of flow of the stream of nitrogen through the reaction mass is reversed (from B toward A).

* Disproportionation of $MoCl_2$ occurs above 700°C.

Heating is then resumed, starting from the extreme right (B) and moving toward the left (A). This second heating cycle is carried out in a manner analogous to the first except that the nitrogen, furnace, and volatile chlorides are now all moving in the opposite direction. When the front line of the volatile chlorides has reached the plug at the extreme left, heating is again discontinued, again the nitrogen stream is reversed, and another heating cycle from A toward B is started.

After five cycles the quantity of volatile chlorides is negligible, and in the fifth cycle the furnace can be brought up to the end of the tube. The reaction mass is allowed to cool to room temperature in the stream of nitrogen. Very volatile brownish-white cyrstals (probably MoO_2Cl_2) due to oxygen-containing impurities in the reagents may have collected at both ends of the tube. They are removed with a spatula, together with any unreacted higher molybdenum chlorides. The remaining content of the tube, a yellow mass, is transferred into a 1-l. beaker and extracted with 600, 200, and 100-ml. portions of hot 25% hydrochloric acid. The resulting solutions are filtered hot through a sintered-glass filter funnel of medium porosity and then cooled with ice. All the unreacted molybdenum powder is transferred into the filter, washed well with water, and then with alcohol. After it has been dried at 110°C., it can be reused.

The hydrochloric acid solutions, on cooling, deposit beautiful yellow crystals of $(H_3O)_2(Mo_6Cl_8)Cl_6\cdot6H_2O$ which are collected on fritted glass, washed with 30 ml. of cold 25% hydrochloric acid, and then transferred into a 250-ml. flask. The flask containing the product is evacuated (about 0.01 mm.) continuously and heated by means of an oil bath whose temperature is raised slowly (over a 2-hour period) to 200°C. A pair of traps cooled with liquid nitrogen or Dry Ice–acetone should be inserted in the vacuum line. When the oil-bath temperature reaches 100°C., the compound begins to lose both water and hydrogen chloride and is transformed into amorphous molyb-

denum(II) chloride. Especially at the beginning of the decomposition, when copious quantities of volatile products are being evolved, it is advisable not to move the flask because movement may cause some of the solids to be carried into the vacuum line.*
Heating at 200°C. under vacuum is continued to constant weight.

The yield ranges from 65 to 70 g.† [85–91%, based on molybdenum(V) chloride for the over-all reaction $2MoCl_5 + 3Mo \rightarrow 5MoCl_2$]. *Anal.* Calcd. for $(Mo_6Cl_8)Cl_4$: Cl, 42.5; Mo, 57.5.
Found: Cl, 42.4; Mo, 57.7. By checker Edwards: Cl, 42.3; Mo, 57.2. By checker Mallock: Cl, 42.00; Mo, 59.22.

Properties

Molybdenum(II) chloride is an infusible, amorphous hygroscopic powder which is yellow-brown when hot but deep yellow at room temperature. It is soluble in ethanol and ether but insoluble in water, in which it is slowly hydrolyzed. Although it is unaffected by dilute acids, it is soluble in dilute strong bases and yields solutions from which crystalline $(Mo_6Cl_8)(OH)_4 \cdot nH_2O$ can be obtained.[6] It is completely decomposed by concentrated strong bases with evolution of hydrogen and formation of higher molybdenum hydroxides. Treatment with concentrated hydrobromic or hydriodic acid gives crystals of $(H_3O)_2(Mo_6Cl_8)Br_6 \cdot 6H_2O$ or $(H_3O)_2(Mo_6Cl_8)I_6 \cdot 6H_2O$, respectively, which decompose, when heated under vacuum, to $(Mo_6Cl_8)Br_4$ or $(Mo_6Cl_8)I_4$.[6] Fusion with excess lithium bromide or iodide yields molybdenum(II) bromide or iodide.[10] Molybdenum(II) chloride is diamagnetic.

* The checker placed a glass-wool plug in the vacuum line above the 250-ml. flask to prevent solid from being carried into the vacuum line, should the flask be moved.

† Mallock reported the yield of $MoCl_2$ is dependent upon exposure time at elevated temperature. Agitation of reactants and/or a greater number of sublimations over molybdenum powder increases the yield.

B. MOLYBDENUM(II) BROMIDE

$$(Mo_6Cl_8)Cl_4 + 12LiBr \rightarrow (Mo_6Br_8)Br_4 + 12LiCl$$

Procedure

An intimate mixture of 30 g. of molybdenum(II) chloride and 120 g. of LiBr·nH$_2$O* is placed in a Vycor tube about 35 cm. long and 5 cm. in diameter, one end of which is closed and the other fitted with a ground-glass joint to permit connection to a vacuum system protected by a trap cooled with liquid nitrogen. After a small glass-wool plug is placed in the upper part of the reaction vessel, the tube is put in a pot furnace and continuously evacuated (about 0.01 mm.). The furnace temperature is then raised to the 650–700°C. range over a period of $1\frac{1}{2}$ hours. The temperature is maintained in that range for a half hour, or until the reaction mixture is a very dark brown, almost black, mass much reduced in volume. The tube is then allowed to cool to room temperature under vacuum. The reaction product is stirred well with about 300 ml. of water, and the resulting suspension is filtered through a fritted glass of medium porosity. The brown product is washed with water on the filter until the washing liquor is colorless.

To a suspension of this product in 200 ml. of water, magnetically stirred and heated at 65°C., a 10% sodium hydroxide solution is added dropwise. The compound gradually dissolves in the alkaline solution, but it is very important to add the sodium hydroxide solution slowly, especially toward the end of the operation when the product is almost all dissolved. The pH of the suspension should never go above 10. At higher pH the compound may readily decompose to a very dark, soft precipitate of higher molybdenum hydroxides with evolution of hydrogen. From 60 to 70 ml. of sodium hydroxide solution is usually needed to dissolve the compound completely. The

---

* N. F. Grade. Available from Fisher Scientific Co., Chemical Manufacturing Division, Fair Lawn, N.J.

warm solution is then quickly filtered directly into 200 ml. of 48% hydrobromic acid through a fritted glass of coarse porosity covered by a layer of diatomaceous earth (Hyflo Super-Cel, Johns-Manville Corp., New York) about 0.5 cm. thick after some of this filter aid is suspended in the solution. The resulting brown precipitate is filtered onto a medium-fritted-glass filter and washed well with 24% hydrobromic acid. The product is then dried under vacuum (about 0.01 mm.) at 200°C. to constant weight. There should be two traps cooled with liquid nitrogen in the vacuum line.

The yield ranges from 35 to 40 g. (76–87%). *Anal.* Calcd. for $(Mo_6Br_8)Br_4$: Br, 62.5; Mo, 37.5. Found: Br, 62.3; Mo, 37.7. By checker Edwards: Br, 61.9; Mo, 37.6. By checker Mallock: Br, 62.38; Mo. 38.84.

Properties

Molybdenum(II) bromide is an infusible, amorphous, hygroscopic, yellow-brown powder. It is insoluble in water and unaffected even by strong acids, but it is soluble in boiling dimethyl sulfoxide or hot concentrated sulfuric acid. It dissolves in warm dilute strong bases to give solutions from which $(Mo_6Br_8)(OH)_4 \cdot nH_2O$ can be obtained.[10] Addition of concentrated hydrochloric or hydriodic acid to an alkaline solution of molybdenum(II) bromide gives $(Mo_6Br_8)Cl_4$ or $(Mo_6Br_8)I_4$, respectively.[10] Concentrated strong bases decompose the compound completely. Molybdenum(II) bromide is diamagnetic. The structure of the crystalline form is isomorphous with that of the crystalline chloride (reference 1, p. 310).

References

1. H. Schäfer, H. G. v. Schnering, J. Tillack, F. Kuhnen, H. Wöhrle, and H. Baumann, *Z. Anorg. Allgem. Chem.*, **353**, 281 (1967).
2. C. W. Blomstrand, *J. Prakt. Chem.*, **77**, 96 (1859).

3. K. Lindner, E. H. Haller, and H. Helwig, *Z. Anorg. Allgem. Chem.*, **130**, 209 (1923).
4. W. Helriegel, German patent 703,895 (1941).
5. S. Senderoff and A. Brenner, *J. Electrochem. Soc.*, **101**, 28 (1954).
6. J. C. Sheldon, *J. Chem. Soc.*, 1007 (1960).
7. F. A. Cotton and N. F. Curtis, *Inorg. Chem.*, **4**, 241 (1965).
8. C. W. Blomstrand, *J. Prakt. Chem.*, **77**, 89 (1859).
9. K. Lindner and H. Helwig, *Z. Anorg. Allgem. Chem.*, **142**, 181 (1925).
10. J. C. Sheldon, *J. Chem. Soc.*, 410 (1962).

30. MOLYBDENUM(III) CHLORIDE

$$MoCl_5 + SnCl_2 \rightarrow MoCl_3 + SnCl_4$$

Submitted by ALAN K. MALLOCK*
Checked by D. A. EDWARDS†

Molybdenum(III) chloride has been prepared in limited quantities by the reduction of molybdenum(V) chloride with hydrogen or molybdenum metal powder.[1,2] The procedures give low yields and require several steps. The new procedure described here for the preparation of molybdenum(III) chloride gives a high yield and involves reduction of pure molybdenum(V) chloride, using anhydrous tin(II) chloride as a reducing agent.

Procedure

Because of the high reactivity of both the materials and the resulting product, the entire preparation must be conducted under a moisture-free and oxygen-free atmosphere of prepurified nitrogen. The conventional glass apparatus and the necessary working tools are degassed in an oven at 150°C. for 48 hours. The reaction chamber consists of one 200-ml., round-bottomed, single-necked flask with a 24/40 ground-glass joint equipped with a Teflon-coated magnetic impeller, Friedrichs condenser,

* Research Laboratory, Climax Molybdenum Company, Ann Arbor, Mich. 48106.
† School of Chemistry, Bath University of Technology, Bath, Somerset, England.

and glass fittings to facilitate a continuous flow of prepurified nitrogen under slight pressure. A dry trap and bubbler are connected at the exhaust end of the condenser to prevent suck-back into the reaction vessel and also to allow a flow of nitrogen through the system.

The method of Seifert and Quak[3] was used quite successfully to prepare a moisture-free and oxygen-free molybdenum(V) chloride. (See also synthesis 32.) *Anal.:* Mo, 36.92; Cl, 63.39; m.p. 202°C., yield 65%. The elemental ratio of molybdenum to chlorine, 4.66, is somewhat low; however, ratios of 4:60 to 4:80 are typical for molybdenum(V) chloride unless it is stored under some chlorine pressure. The Seifert and Quak preparation can be scaled-up twenty times without difficulty. Tin(II) chloride was dehydrated by heating[4] it to redness in a stream of hydrogen chloride. *Anal.:* Sn, 62.46; Cl, 36.96. Commercial sources of the $MoCl_5$* and $SnCl_2$† are available as noted below.

Excess molybdenum(V) chloride (13.66 g., 0.049 mole) and anhydrous tin(II) chloride (9.03 g., 0.047 mole) are ground to a fine powder with a mortar and pestle and then placed in the reaction vessel with the Teflon impeller.

The apparatus is then completely assembled, using a Teflon sleeve between the condenser and reaction-vessel neck. Upon removal from the dry-box, the loaded apparatus is immersed in a silicone-oil bath at room temperature and secured with a sturdy clamp. The nitrogen line is connected to the apparatus, and nitrogen is introduced at a minimal flow rate. The silicone-oil bath is heated by a magnetic-stirrer hot plate. The contents of the reaction vessel are then agitated at a high rate, and the temperature of the oil bath is slowly elevated to 290–300°C.

A very small amount of tin(IV) chloride begins to reflux between 85 and 100°C. Reflux should continue for 30 minutes after a temperature of 290°C. is reached. Then an oil-bath temperature of 290–300°C. should be maintained without water

* Climax Molybdenum Company, 1270 Avenue of the Americas, New York, N.Y. 10020.

† Stannochlor, M and T Chemicals, Rahway, N.J.

circulation in the condenser to allow tin(IV) chloride to distill into the dry trap.   Upon removal of the bulk of tin(IV) chloride, the nitrogen flow is increased, and the Teflon sleeve is quickly removed; the male joint of the condenser is coated with a high-temperature silicone grease and reinserted.   The trap containing the tin(IV) chloride and the bubbler are removed from the apparatus, and the flow of nitrogen is quickly replaced by a vacuum system containing one cold-finger trap that collects the small amount of residual tin(IV) chloride while the excess molybdenum(V) chloride is sublimed into the condenser. Evacuation of the vessel at 290–300°C. is continued until no additional volatile material is collected.   The apparatus is removed from the oil bath while under vacuum and allowed to cool to room temperature before being placed in the dry-box and before samples are taken for infrared studies and elemental analysis.   The bulk of the product is stored in flame-sealed evacuated ampules.

Molybdenum(III) chloride can be prepared in considerably larger quantities by simply modifying the size of the equipment. *Anal.:* Calcd. for $MoCl_3$: Mo, 47.42; Cl, 52.58.   Found: Mo, 46.56; Cl, 52.51.   Checker's analysis: Mo, 47.00; Cl, 52.60.

Infrared Data.   The presence of moisture and/or oxygen bonds is easily detected by infrared analysis, and the spectra serve partly to indicate the type of contamination that might be present in anhydrous or oxygen-free compounds.   Tin(II) chloride was studied as a hexachlorobutadiene mull, and molybdenum(V) chloride and molybdenum(III) chloride were studied as Nujol mulls.   All starting materials and the final product were free of moisture, OH^- stretching, and molybdenum-to-oxygen bonding, as indicated by the absence of absorptions at 3500, 1800, and 900–1020 cm^{-1}, respectively.

Properties

Molybdenum(III) chloride, $MoCl_3$, has a molecular weight of 202.32.   The dark red compound is stable when protected from the air.   It is gradually hydrolyzed as well as oxidized by moist air.   The trichloride is insoluble in water, dilute hydrochloric acid, acetone, carbon tetrachloride, and benzene.   Minimal solubility occurs in ether, isopropyl alcohol, and pyridine, but the compound is completely dissolved by dilute nitric acid, probably as some more highly oxidized species.   DTA studies show that molybdenum(III) chloride disproportionates to $MoCl_2$ and $MoCl_4$ ($MoCl_3 + MoCl_5$ upon cooling) at temperatures above 410°C.[6] α-$MoCl_3$ and β-$MoCl_3$ structures (both monoclinic) have been established.[7]

References

1. D. E. Couch and A. Brenner, *J. Res. Natl. Bur. Std., A,* **63,** 185 (September–October, 1959).
2. "Gmelins Handbuch der anorganischen Chemie," System-Nummer 67, Achte Auflage, p. 158, Verlag Chemie GmbH, Weinheim, Germany, 1935.
3. H. J. Seifert and H. P. Quak, *Angew. Chem.,* **73,** 621 (1961).
4. H. Remy, "Treatise on Inorganic Chemistry," Vol. 1, p. 535, American Elsevier Publishing Company, Inc., New York, 1956.
5. N. A. Lange, "Handbook of Chemistry," 9th ed., p. 276, McGraw-Hill Book Company, New York, 1956.
6. A. K. Mallock, Climax Molybdenum Company, unpublished results.
7. H. Schäfer, H. G. Schnering, J. Tillack, F. Kuhnen, H. Wohrle, and H. Baumann, *Z. Anorg. Allgem. Chem.,* **353,** 281 (1967).

31. MOLYBDENUM(IV) CHLORIDE

$$2MoCl_5 + C_2Cl_4 \rightarrow 2MoCl_4 + C_2Cl_6$$

Submitted by E. L. McCANN III* and T. M. BROWN*
Checked by ALAN K. MALLOCK†

The preparation of pure molybdenum(IV) chloride is complicated by its extreme ease of disproportionation and oxidative instability.   Formerly, the preparation of the tetrachloride was

* Arizona State University, Tempe, Ariz. 85281.
† Research Laboratory, Climax Molybdenum Company, Ann Arbor, Mich. 48106.

accomplished by hydrogen reduction of molybdenum(V) chloride[1,2] or thermal disproportionation of the trichloride.[3] The reactions between molybdenum(III) chloride and molybdenum(V) chloride,[4] molybdenum(IV) oxide and carbon tetrachloride[5,6] or hexachloro-1,3-butadiene,[7] or molybdenum(V) chloride and benzene[8] also yield molybdenum(IV) chloride.

The following procedure[9,10] takes advantage of the ability of tetrachloroethylene to remove chlorine from solutions of molybdenum(V) chloride and in this manner bring about a convenient photochemical synthesis of very pure molybdenum(IV) chloride.

Procedure

All operations must be performed in an inert atmosphere or *in vacuo* to prevent contamination by oxychlorides. The apparatus used for the preparation of molybdenum(IV) chloride consists of an all-Pyrex-glass vessel which can be attached to a standard vacuum line for easy manipulation of the reactants and products. This is shown in Fig. 17.

In a dry-box, 5 g. of freshly sublimed and powdered molybdenum(V) chloride is placed in chamber *A* of the vessel through sidearm *B*. The vessel is evacuated on the vacuum line, and the sidearm is sealed off with a torch at point *C*. This is necessary since tetrachloroethylene will attack stopcock grease. Thirty milliliters of purified[11] and dried tetrachloroethylene is vacuum-distilled onto the molybdenum(V) chloride by cooling bulb *A* with liquid nitrogen or Dry Ice and acetone. The evacuated apparatus is detached from the vacuum line and warmed to 25°C.; then chamber *A* is immersed in a clear oil bath at 130°C. The reactants are stirred for 48 hours, with illumination from a 100-watt light bulb placed approximately 1 foot from the reaction vessel. ■ *Caution. The apparatus should be placed behind a safety shield.*

During the reaction the solution changes from deep red to

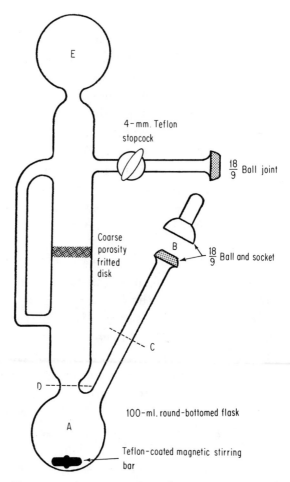

Fig. 17. Apparatus for the preparation of molybdenum(IV) chloride.

essentially colorless; a fine powder appears upon completion of the reaction.   The products are cooled to ambient temperature, and the solution is filtered by inverting the vessel.   A continuous extraction of the impurities from the insoluble molybdenum(IV) chloride with the unreacted tetrachloroethylene is effected by heating chamber *E*.*   This results in the solvent being driven

* The checker recommended stirring the liquid in *E* during extraction to minimize bumping.

into chamber *A* where it condenses and subsequently passes through the frit.   The extraction is continued until the effluent solution is colorless.   In this manner any excess molybdenum(V) chloride and hexachloroethane are removed from the desired product.   The volatile organic materials are removed by vacuum distillation and the molybdenum(IV) chloride dried under vacuum for several hours.

In the dry-box, the stopcock is opened and bulb *A* is removed from the vessel by scoring and breaking the neck at *D*.   The product is removed from the glass frit and stored in an air-tight container.   The yield is approximately 100% based on the molybdenum(V) chloride starting material.   *Anal.* Calcd. for $MoCl_4$: Mo, 40.35; Cl, 59.65.   Found: Mo, 40.34; Cl, 59.86.   Checker reported: Mo, 40.40; Cl, 58.28.   The amount of hexachloroethane formed can be determined by gas chromatographic analysis of the organic distillate.

Infrared spectral analysis[9] is employed as a check for other impurities which might be present in the molybdenum(IV) chloride.   Molybdenum oxychlorides can be easily identified by their characteristic absorptions in the 900–1000-cm.$^{-1}$ region, which is typical of Mo—O groups.   In addition, an examination of the far infrared spectrum should show only the characteristic bands of molybdenum(IV) chloride at 404, 350, and 268 cm.$^{-1}$.   These can be distinguished from those of other molybdenum halide species.   A test for the presence of molybdenum(III) chloride, which often occurs as a side product in other preparations of molybdenum(IV) chloride, is treatment with 6 *N* hydrochloric acid.   Complete solubility indicates the absence of this chloride.

Properties

The chemical and physical properties of molybdenum(IV) chloride are well established.[9, 10, 12, 13]   It is a black, nonvolatile compound which readily disproportionates to molybdenum(III) chloride and molybdenum(V) chloride at moderate temperatures. At less than 150°C. *in vacuo* (10^{-5} torr), molybdenum-(IV) chloride undergoes thermal decomposition to molybdenum-(III) chloride and chlorine. It is insoluble in nonpolar solvents and reacts slowly with most polar solvents.

The form of molybdenum(IV) chloride obtained by the above method has been labeled the low-temperature α-isomer because of its structural similarity with the other tetrachlorides of tungsten, niobium, and tantalum and to distinguish it from the higher-temperature β-isomer prepared from molybdenum(III) chloride and molybdenum(V) chloride. The β-isomer may be obtained by heating the low-temperature form at 250°C. in a sealed tube with molybdenum(V) chloride for 24 hours. The excess molybdenum(V) chloride is sublimed away at 125°C. Because of its structural difference, the β-isomer possesses a considerably higher paramagnetism and thermal stability than the α-isomer.

Discussion

With relatively little attention, high yields of very pure molybdenum(IV) chloride can be obtained. Although in well-lighted areas, the use of a 100-watt light may not be necessary for the preparation of molybdenum(IV) chloride, its presence ensures completion of the reaction within the allotted period. The organic product, hexachloroethane, is quite volatile so that contamination with carbonaceous materials is avoided. Because of the completeness of the reaction, normally no molybdenum(V) chloride will be found in the product. This method also avoids further reduction to molybdenum(III) chloride, a contaminant so often found in the other preparations of molybdenum(IV) chloride.

Extension of Method—Tungsten(IV) Chloride and Niobium(IV) Chloride

Application of this method to other transition-metal systems has been examined, with only minor procedural changes.

The synthesis of tungsten(V) chloride from tungsten(VI) chloride has been accomplished in a 24-hour reaction at 100°C. when illuminated by a 100-watt light bulb. To purify the tungsten(V) chloride, the dry product is transferred to a tube which is then sealed and placed part way in a tube furnace so that a 180/25°C. gradient is maintained. The volatile tungsten(V) chloride, which does not contain any unreacted tungsten(VI) chloride, is sublimed to the cooler zone, leaving behind a small amount of nonvolatile, black tungsten(IV) chloride powder. Yields greater than 90% based on the weight of tungsten(VI) chloride used are obtained, depending on the amount of tungsten(IV) chloride formed.

The method has also been examined for the production of other nonvolatile compounds. In these cases, the purification process involves subliming excess starting materials away from the product. For example, greater than 95% yields of tungsten(IV) chloride from tungsten(VI) chloride and 85% yields of niobium-(IV) chloride from niobium(V) chloride can be obtained by using a 500-watt light bulb at 150°C. and a reaction period of 3 days.

References

1. W. Klemm and H. Steinberg, *Z. Anorg. Allgem. Chem.*, **227**, 193 (1936).
2. C. W. Blomstrand, *J. Prakt. Chem.*, **71**, 453 (1857); **77**, 97 (1873).
3. L. P. Liechti and B. Kempe, *Liebigs Ann. Chem.*, **169**, 350 (1873).
4. H. Schäfer, H. G. v. Schnering, J. Tillack, F. Kuhnen, H. Wöhrle, and H. Baumann, *Z. Anorg. Allgem. Chem.*, **353**, 281 (1967).
5. E. R. Epperson and H. Frye, *Inorg. Nucl. Chem. Letters*, **2**, 223 (1966).
6. S. A. Schukarev, G. I. Novikova, I. V. Vasilkova, A. V. Surorov, A. K. Baev, N. V. Andreeva, and B. N. Shapurin, *Russ. J. Inorg. Chem. (English Transl.)*, **5**, 802 (1960).
7. T. E. Austin and S. Y. Tyree, *J. Inorg. Nucl. Chem.*, **14**, 141 (1960).
8. M. L. Larson and F. W. Moore, *Inorg. Chem.*, **3**, 285 (1964).
9. T. M. Brown and E. L. McCann III, *ibid.*, **7**, 1227 (1968).
10. D. L. Kepert and R. Mandyczewsky, *ibid.*, **7**, 2091 (1968).
11. D. D. Perrin, W. L. F. Armarego, and D. R. Perrin, "Purification of Laboratory Chemicals," p. 260, Pergamon Press, New York, 1966.
12. J. E. Fergusson, "Halogen Chemistry," V. Gutman (ed.), Vol. 3, pp. 259–262, Academic Press, Inc., New York, 1967.
13. "Nouveau Traité de Chimie Minérale," P. Pascal (ed.), Vol. 14, Masson et Cie, Paris, 1958.

32. ANHYDROUS METAL HALIDES, NbBr₅, TaBr₅, WCl₆, MoCl₅

$$M + nX_2 = MX_{2n}$$

Submitted by AMOS J. LEFFLER* and ROBERT PENQUE*
Checked by G. W. A. FOWLES†

Anhydrous metal halides have been synthesized by a number of different reactions,[1-4] but the direct reaction of the metal and halogen is most convenient for larger amounts of product. The method described here involves a modification of earlier equipment;[2] the modified apparatus is simpler to use and is designed for easy removal of the product from the apparatus. If the reactants are pure and if moisture and oxygen are rigorously excluded, pure products will be obtained. With the equipment described here, it is readily possible to make of the order of 500 g. of product during an 8-hour period.

Procedure

The apparatus is shown in Fig. 18. The high-temperature portion is constructed of Vycor‡ or quartz with graded seals to borosilicate glass. Since it is necessary to maintain the outlet of the reactor above the boiling point of the product in order to prevent condensation in the outlet of the reactor, the cyclone must be directly sealed to the reactor. The cyclone is constructed from a 1-l. Erlenmeyer flask with the inlet sealed tangentially to the bottom of the flask. The gas exit consists of 15-mm. tubing that extends to a distance of 10 cm. into the flask. It is convenient to wrap the exit of the reactor with Nichrome wire, using asbestos insulation, to serve as a heating

---

* Villanova University, Villanova, Pa. 19085.
† The University, Whiteknights, Reading, England.
‡ Trademark of Corning Glass Works.

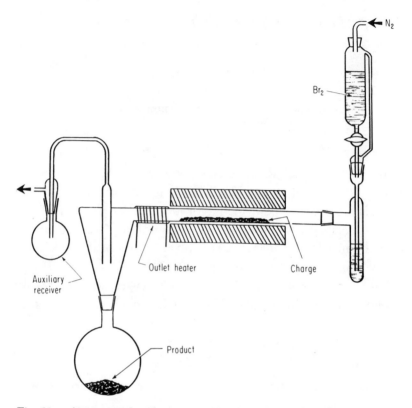

Fig. 18.   Apparatus for the preparation of anhydrous metal halides.

source, although it is possible to flame this portion of the reactor frequently. The former procedure is more desirable since the graded seal should be heated slowly and evenly. In order to avoid greasing the receiver flask joint, Teflon sleeves* are convenient.

The reactor as shown is set up for bromination reactions. The desired amount of metal powder is placed in the center of the reactor, and the system is purged with nitrogen or argon. Experimental conditions including any pretreatment of the metal can be obtained from earlier work.[2] After any pretreat-

* Available from Arthur F. Smith Co., Inc., 201 S.W. 12th Ave., Pompano Beach, Fla.

ment, the addition funnel is removed, bromine is added in a hood, the outlet of the reactor is heated above the boiling point of the product, and the funnel is set back in place.   During the removal of the addition funnel, the inert-gas flush can be continued through the trap.   Enough bromine is added to half-fill the trap, and the purge gas bubbling through the liquid will carry vapor into the reactor.   The flow of nitrogen or argon is held at 500–600 ml./min. as determined by a flowmeter in the inlet line (not shown).   Since the bromine evaporation cools the liquid, it is useful to place a beaker of warm water around the trap.   As the bromine is depleted, more can be added periodically from the addition funnel.   At the completion of the reaction, the excess bromine is flushed from the system, and the product receiver is quickly removed and stoppered.

If it is desired to carry out a chlorination, the apparatus is modified by replacing the dropping funnel and trap with a simple Y inlet tube attached directly to the reactor.   One side of the Y is attached to the purge gas and the other to the chlorine source.   If there is any question about the purity of the chlorine, it can be passed through a sulfuric acid wash.   The chlorine flow rate during the reaction is about 300 ml./min. with no purge-gas flow.   At the end of the reaction, the chlorine is purged from the system, and the product is removed as before.

This apparatus has been successfully used in the preparation of niobium(V) bromide, tantalum(V) bromide, tungsten(VI) chloride, and molybdenum(V) chloride.*

References

1. M. Chaigneau, *Compt. Rend.*, **248**, 3173 (1959); **243**, 957 (1956).
2. K. M. Alexander and F. M. Fairbrother, *J. Chem. Soc.*, S223 (1949); D. H. Nowicky and I. E. Campbell, *Inorganic Syntheses*, **4**, 130 (1953).

* Niobium and tantalum powders can be obtained from the Rare Metals Division of the Ciba Company, Fair Lawn, N.J., or from the Kawecki Chemical Co., Boyertown, Pa.   Molybdenum and tantalum used in this study were supplied by Alfa Inorganics, Inc., P.O. Box 159, Beverly, Mass. 01915.   Metals are also available from Climax Molybdenum Company, P.O. Box 1568, Ann Arbor, Mich. 48106.

3. Teclu, *Ann.*, **187**, 255 (1877); J. A. Ketelaar and van Oosterhut, *Rec. Trav. Chim.*, **62**, 197, 597 (1943); E. R. Epperson, S. M. Horner, K. Knox, and S. Y. Tyree, Jr., *Inorganic Syntheses*, **7**, 163 (1963); W. W. Porterfield and S. Y. Tyree, Jr., *ibid.*, **9**, 133 (1967); M. H. Lietzke and M. L. Holt, *ibid.*, **3**, 183 (1950).
4. E. R. Epperson, S. M. Horner, K. Knox, and S. Y. Tyree, Jr., *Inorganic Syntheses*, **7**, 163 (1963); W. W. Porterfield and S. Y. Tyree, Jr., *ibid.*, **9**, 133 (1967).

33. MOLYBDENUM OXIDE TRICHLORIDE

$$2MoOCl_4 + C_6H_5Cl \rightarrow 2MoOCl_3 + C_6H_4Cl_2 + HCl$$

Submitted by M. L. LARSON* and F. W. MOORE*
Checked by D. A. EDWARDS†

Molybdenum oxide trichloride has been prepared by heating molybdenum(VI) oxide or molybdenum dioxide dichloride with molybdenum(V) chloride. Molybdenum(VI) oxide tetrachloride is a by-product of the latter reaction.[1] It was also prepared by the sealed-tube reaction of liquid sulfur dioxide with molybdenum(V) chloride[2] and by the thermal decomposition of molybdenum oxide tetrachloride in a stream of nitrogen.[3] In the following procedure, it is prepared by reducing molybdenum oxide tetrachloride with refluxing chlorobenzene.[4]

Procedure

In all the following operations, an atmosphere of prepurified nitrogen must be maintained throughout the reaction, and all glassware must be flame-dried or oven-dried and purged with nitrogen to remove traces of moisture. Transferals of molyb-

* Research Laboratory, Climax Molybdenum Company, Ann Arbor, Mich. 48106.
† School of Chemistry, Bath University of Technology, Claverton Down, Bath, England.

denum oxide chlorides were conducted in a glove box having a measured dew point of −55 to −60°C.

A 500-ml., three-necked, round-bottomed flask with ground-glass joints is equipped with two stopcock valves and with a mechanically operated stirring assembly consisting of a precision-ground glass stirrer bearing and a stirrer shaft with an attached Teflon blade. All the ground-glass connections are lubricated with silicone high-vacuum grease.*

In a glove box, 40.0 g. (0.158 mole) of pure, freshly ground molybdenum oxide tetrachloride,[5] m.p. 100–101°C., is transferred to the reaction flask. Chlorobenzene that has been stored over Drierite is filtered and distilled into an addition funnel. This funnel is attached to the reaction flask, and 160 ml. (177.6 g., 1.58 moles) of the chlorobenzene is added. The flask is equipped with a reflux condenser, and the reaction mixture is stirred vigorously in an oil bath at 138–143°C. for $3\frac{1}{2}$ hours with refluxing. During the end of this reflux period, the rate of HCl evolution becomes quite slow. The cooled reaction mixture contains fine, brown powder and is filtered through a medium-fritted filter tube.† Filtration is accomplished by using ground-glass joints to connect a short length of Tygon tubing from the reaction flask to the filter tube. The filter tube is attached to a 500-ml., two-necked, round-bottomed flask equipped with a valve connected to a vacuum pump. The shaken reaction flask is lifted to transfer the product to the filter tube, and the valve of the filter flask is opened to allow vacuum filtration. Dry chlorobenzene is added to the reaction flask as a wash liquid to effect complete transfer of the solid product to the filter tube.‡ The closed filter tube is then connected to a dry flask equipped with a valve for oil-pump evacuation to remove excess chlorobenzene. Residual chlorobenzene

---

* Product of Dow Corning Corporation, Midland, Mich.

† Corning Glass Works, Corning, N.Y., Catalog No. 99220.

‡ Laboratory air and moisture are excluded from flask and filter system during this operation by keeping valves from the reactor flask to the outside closed.

is removed by transferring the product to a beaker inside a vacuum desiccator placed in the glove box. The oil pump is used to evacuate the desiccator until no further condensate is collected in a cold trap (Dry Ice–trichloroethylene slush).

The yield of brown powder is 32.32 g. or 93.7% of theoretical. *Anal.* Calcd. for $MoOCl_3$: Cl, 48.72; Mo, 43.45. Found: Cl, 48.76; Mo, 43.16. Checker reported: Cl, 48.6; Mo, 43.9. Of this product 0.069% is insoluble in 1:1 aqueous hydrochloric acid; m.p. (sealed capillary) 301.5–303.5°C. (literature,[1] m.p. 296 ±1°C.).

Properties

Molybdenum oxide trichloride gives a black melt that freezes to black needles. The magnetic moment is 1.62 B.M. at 22°C.[2] A Nujol-mull infrared absorption spectrum shows only one peak in the sodium chloride region at 1007(s) cm.$^{-1}$.[4] $MoOCl_3$ is rapidly hydrolyzed by atmospheric moisture and is soluble in water to give a brown solution. Methanol and acetone are reactive solvents.[2] Molybdenum oxide trichloride is insoluble in nonhydroxylic organic solvents, such as chloroform, carbon tetrachloride, 1,2-dichloroethane, diethyl ether, and benzene. Reaction of a benzene suspension of $MoOCl_3$ with 2,4-pentanedione (acetylacetone) yields the benzene-soluble acetylacetonate, $MoOCl(C_5H_7O_2)_2$.[4] Dimethyl sulfoxide and pyridine react with $MoOCl_3$ to form $MoOCl_3 \cdot 2(CH_3)_2SO$ and $MoOCl_3 \cdot 2C_5H_5N$,[4] respectively.

References

1. I. A. Glukhov and S. S. Eliseev, *Russ. J. Inorg. Chem. (English Transl.)*, **7**, 40 (1962).
2. D. A. Edwards, *J. Inorg. Nucl. Chem.*, **25**, 1198 (1963).
3. R. Colton and I. B. Tomkins, *Australian J. Chem.*, **18**, 447 (1965).
4. M. L. Larson and F. W. Moore, *Inorg. Chem.*, **5**, 801 (1966).
5. A. K. Mallock, *Inorganic Syntheses*, **10**, 54 (1967).

34. NONACHLOROTRIRHENIUM (III)

$$3ReCl_5 \rightarrow Re_3Cl_9 + 3Cl_2$$

Submitted by H. GEHRKE, Jr.,* and D. BUE*
Checked by B. M. FOXMAN† and F. A. COTTON‡

With the current interest in the trinuclear species nonachloro-trirhenium(III) and coordination compounds derived from it,[1] a good method for the preparation of nonachlorotrirhenium(III) is necessary. The only practical method is the thermal decomposition of either rhenium(V) chloride[2] or silver hexachloro-rhenate(IV).[3]   The *Inorganic Syntheses* procedure of Hurd and Brimm employing rhenium(V) chloride is the most convenient,[2] but the design of the apparatus allows considerable amounts of easily sublimed rhenium(V) chloride to be carried by the nitrogen stream through the decomposition train without undergoing decomposition. Therefore, the yield is substantially reduced.

The method to be described is a significant modification of the synthesis outlined by Hurd and Brimm[2] and is derived from the apparatus of Geilmann and Wrigge.[5] This procedure allows for the decomposition of large quantities of rhenium(V) chloride in a short period (approximately an hour for a 14-g. sample), in high yield (of the order of 90%) and by an inexperienced person.

Procedure

The apparatus illustrated in Fig. 19 is utilized for the synthesis of nonachlorotrirhenium(III). The bulbs are constructed from

* Chemistry Department, South Dakota State University, Brookings, S.D. 57006.
† Australian National University, Canberra, Australia.
‡ Department of Chemistry, Massachusetts Institute of Technology, Cambridge, Mass. 02139.

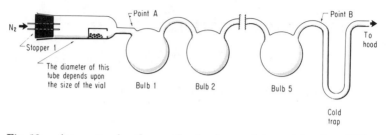

Fig. 19.   *Apparatus for the synthesis of nonachlorotrirhenium(III).*

250-ml. distilling flasks connected by 10-mm. tubing.* A reaction train of this size has been found to be satisfactory for the preparation of 10-g. quantities. For larger amounts, the size of the flasks should be increased. The use of smaller flasks results in loss of rhenium(V) chloride into the cold trap and the necessity of isolating small quantities of nonachlorotrirhenium(III) from the latter bulbs.

The rhenium(V) chloride can be prepared by standard methods[6,7] or purchased commercially† and then stored in a screw-capped vial until needed. The movement of the rhenium(V) chloride into bulb 1 and subsequent decomposition are accomplished by the heat of a medium Bunsen flame. The temperature is not critical, but a very hot flame should be avoided, as some of the unreacted rhenium(V) chloride will be lost and some decomposition of the nonachlorotrirhenium(III) will occur. The nitrogen flow rate during the decomposition should be two to three bubbles per second. The flow rate can be doubled during the movement of the sample into bulb 1 to aid in the transfer process.

In a typical run, the decomposition apparatus is flushed with dry nitrogen, and then 14.0 g. of rhenium(V) chloride is transferred to the apparatus by removal of stopper 1 and quick insertion of the opened vial.‡ Upon heating the tube directly

* The use of 10-mm. tubing alleviates obstruction of the connecting tubes. Elevation of the connecting tube in the fashion illustrated in Fig. 19 allows condensation of the rhenium(V) chloride vapor in the first bulb and facilitates the decomposition.

† $ReCl_5$ can be purchased from Shattuck Chemical Co., Denver, Colo.

‡ The very brief exposure of the rhenium(V) chloride to the atmosphere does not result in a measurable amount of reaction with air or water.

under the vial to move the sample into bulb 1, a mixture of liquid and gaseous rhenium(V) chloride is formed. The vapor condenses in the cooler parts of the tube, and the liquid decomposes to nonachlorotrirhenium(III). Close proximity of the vial to bulb 1 and gentle heating will facilitate movement of most of the solid into this bulb with a minimum of decomposition.* Since decomposition occurs in the liquid state, the solid in bulb 1 is melted, and heating of the mixture of nonachlorotrirhenium(III) and rhenium(V) chloride is continued until the mass no longer evolves the red-brown vapor.† At this point, the solid which has condensed in the connecting tube between bulb 1 and bulb 2 is moved into bulb 2 by heating, and the decomposition process is repeated. When 250-ml. bulbs are used, 90–95% of the rhenium(V) chloride is decomposed in the vial and the first bulb. After the decomposition is complete, 10.8 g. (95.6%) of crude nonachlorotrirhenium(III) is isolated from the vial and bulbs by blowing holes in the sides of the bulbs and removing the product.‡ The crude nonachlorotrirhenium(III) is purified by vacuum sublimation at 450°C. and 1-mm. pressure to yield 9.75 g. (86.3%) of pure nonachlorotrirhenium(III). *Anal.* Calcd. for Re_3Cl_9: Re, 63.6; Cl, 36.4. Found: Re, 63.5; Cl, 36.5. Checkers reported 90.3% yield of crude product.

If the procedure and apparatus of Hurd and Brimm[6] are used for the production of the rhenium(V) chloride,§ the decomposition train, from point A to point B, is attached to the combustion tube in place of tubes D to I. The bottom of the combustion tube and the connecting tube of bulb 1

* Decomposition in the tube makes the isolation of the product more difficult, and the yield will be decreased slightly.

† Heating should be discontinued occasionally in order to see how the reaction is progressing and to move all the condensed rhenium(V) chloride back to the bottom of the bulb.

‡ The holes can be sealed and the decomposition train used again.

§ As most laboratories use commercially obtained rhenium metal rather than a material prepared *in situ*, the following modifications are used. The rhenium metal is heated in a stream of hydrogen at 1000°C. for an hour to remove surface oxides, and the reaction of rhenium metal and chlorine is carried out at 650°C. Because of these temperatures, a Vycor combustion tube is necessary.

(point *A*) should be level at their junction to allow the easy movement of the rhenium(V) chloride into the first bulb. After the rhenium(V) chloride has been moved into bulb 1, the decomposition process is carried out in the manner described previously. A yield of 70–80% based on the weight of rhenium metal is obtained, but it is dependent upon the extent of formation of rhenium oxide chloride impurities.

Properties

Nonachlorotrirhenium(III) is a dark red crystalline solid. When heated strongly it gives a green vapor. It is reduced by hydrogen to the metal at 250–300°C. and reacts with oxygen to yield perrhenyl chloride and rhenium oxide tetrachloride.[4]

It is soluble with decomposition in water, and forms stable, deep-red solutions in acetone, ethanol, and concentrated hydrochloric acid. Early molecular-weight determinations in glacial acetic acid indicated a dimeric structure,[8] but recent molecular-weight studies in Sulfolane have definitely established the existence of the trinuclear structure. X-ray crystallography also shows a trimer.[9] Re_3Cl_9 reacts with a number of ligands to form complexes of the type $L_3Re_3Cl_9$, where L = triphenylphosphine, pyridine, etc.[10]

References

1. F. A. Cotton, N. F. Curtis, and W. R. Robinson, *Inorg. Chem.*, **4**, 1696 (1965), and references cited therein.
2. L. C. Hurd and E. Brimm, *Inorganic Syntheses*, **1**, 182 (1939).
3. R. B. Bevan, Jr., R. A. Gilbert, and R. H. Busey, *J. Phys. Chem.*, **70**, 147 (1966).
4. R. Colton, "The Chemistry of Rhenium and Technetium," Interscience Publishers, a division of John Wiley & Sons, Inc., New York, 1965.
5. W. Geilmann and F. W. Wrigge, *Z. Anorg. Allgem. Chem.*, **214**, 248 (1933).
6. L. C. Hurd and E. Brimm, *Inorganic Syntheses*, **1**, 180 (1939).
7. E. R. Epperson, S. M. Horner, K. Knox, and S. Y. Tyree, *ibid.*, **7**, 163 (1963).
8. F. W. Wrigge and W. Biltz, *Z. Anorg. Allgem. Chem.*, **228**, 372 (1936).
9. F. A. Cotton and J. T. Mague, *Inorg. Chem.*, **3**, 1402 (1964).
10. F. A. Cotton, S. J. Lippard, and J. T. Mague, *ibid.*, **4**, 508 (1965).

Chapter Five

METAL COORDINATION COMPOUNDS

35. BINUCLEAR COMPLEXES OF COBALT(III)

Submitted by R. DAVIES,* MASAYASU MORI,† A. G. SYKES,*
and J. A. WEIL†
Checked by L. CENTOFANTI‡ and by H. OGINO§ and J. C. BAILAR, JR.§

The first dicobalt(III) complex was reported by Fremy in 1852. Later, in 1910, Werner described the preparation of a wide range of ammine and ethylenediamine complexes with as many as three bridging groups.[1]   A review summarizing the literature dealing with binuclear complexes of cobalt(III) has been written.[2]   Ligands which are known to bridge two cobalt(III) atoms are NH_2^-, OH^-, O_2^-, O_2^{2-}, O_2H^-, SO_4^{2-}, SeO_4^{2-}, NO_2^-, $CH_3CO_2^-$, and $C_2O_4^{2-}$, and more recently Cl^- and possibly Br^-.   With the exception of the hydroxo-bridged species, complexes having more than one bridging group of the same kind are rare.   Details of the preparation of $[(NH_3)_5Co(O_2)$-

* Department of Inorganic and Structural Chemistry, The University, Leeds, LS2 9JT, England.   Inquiries should be addressed to either A. G. Sykes or J. A. Weil.

† Argonne National Laboratory, Argonne, Ill. Present address of M. Mori is Osaka City University, Osaka, Japan.

‡ Department of Chemistry, University of Utah, Salt Lake City, Utah 84112; current address: Emory University, Atlanta, Ga. 30322.

§ Department of Chemistry, University of Illinois, Urbana, Ill. 61801.

$Co(NH_3)_5]^{4+}$, $[(NH_3)_5Co(O_2)Co(NH_3)_5]^{5+}$, $[(CN)_5Co(O_2)Co-(CN)_5]^{5-}$, $[(NH_3)_4Co(NH_2,O_2)Co(NH_3)_4]^{3+}$, $[(NH_3)_4Co(NH_2,O_2)Co(NH_3)_4]^{4+}$, $(en)_2Co(NH_2,O_2)Co(en)_2]^{4+}$, $[(NH_3)_4Co(NH_2,Cl)Co(NH_3)_4]^{4+}$, $[(NH_3)_4Co(NH_2,OH)Co(NH_3)_4]^{4+}$, and $[(NH_3)_5Co(NH_2)Co(NH_3)_5]^{5+}$ are described here.

Two oxidation states of dicobalt complexes in which there is an O_2 bridging ligand are known. Examples of these are $[(NH_3)_5Co(O_2)Co(NH_3)_5]^{4+}$ and $[(NH_3)_5Co(O_2)Co(NH_3)_5]^{5+}$ and, with a second bridging ligand, $[(NH_3)_4Co\cdot(NH_2,O_2)Co(NH_3)_4]^{3+}$ and $[(NH_3)_4Co(NH_2,O_2)Co(NH_3)_4]^{4+}$, respectively. Recent work including x-ray,[3] e.p.r.,[4] and infrared[5] studies has indicated that the two types are best referred to as peroxo and superoxo complexes, respectively. Alternative names which have been used are diamagnetic peroxo and paramagnetic peroxo complexes, respectively.[6] The peroxocobalt(III,IV) terminology for the superoxo complexes (implying nonequivalent cobalt atoms) is no longer satisfactory.[7] The peroxo complexes are generally red-brown in color, and the superoxo complexes generally, but not always, are green in color.

A. μ-PEROXO-BIS[PENTAAMMINECOBALT(III)] TETRANITRATE

$$2Co^{2+} + 10NH_3 + O_2 \rightarrow [(NH_3)_5Co(O_2)Co(NH_3)_5]^{4+}$$

Procedure

A 50-g. quantity of cobalt(II) nitrate hexahydrate is dissolved in 100 ml. of water and filtered. Aqueous ammonia (15 M, 250 ml.) is added, and after the mixture has been cooled to 5–15°C., a current of oxygen (1000 ml./minute, i.e., a brisk stream) is passed through the cooled solution for an hour, the mixture being stirred with a magnetic stirrer. Oxygen may be replaced by air, in which case longer bubbling (2–3 hours) at 0–5°C. is recommended. A stream of air can be drawn through the solution contained in a Büchner flask, using the suction provided by a water tap. The suction is applied to the sidearm of the flask, the air entering by a piece of glass tubing (held in place by a rubber stopper) extending into the solution. Sodium nitrate (20 g.) dissolved in 50 ml. of water is then added; oxygen (or air) is passed through the solution for another hour, with the mixture cooled in ice toward the end of this period. The dark brown crystals are gathered on a glass filter and washed with a small quantity of 15 M aqueous ammonia and then with ethanol. The dihydrate thus prepared can be kept in an evacuated desiccator overnight in order to obtain the anhydrous nitrate. The yield is about 44 g. of $[(NH_3)_5Co(O_2)Co(NH_3)_5]$-$(NO_3)_4$.[5] *Anal.* Calcd.: Co, 20.7; N(ammoniacal), 24.7. Found: Co, 20.8; N(ammoniacal), 24.4.

Properties

The peroxo complex is brown in color and is diamagnetic. It decomposes slowly in the solid state and is stable in solution only in *ca.* 7 M ammonia. In acidic solutions it decomposes with the evolution of oxygen:

$$[(NH_3)_5Co(O_2)Co(NH_3)_5]^{4+} + 10H^+ \rightarrow 2Co^{2+} + 10NH_4^+ + O_2$$

Impurities which are sometimes present in such samples can be detected most easily by adding acid and identifying the products.[6] The O—O bond distance has been shown to be 1.47 A., and the Co—O—O angle is 113°.[3d] A red hydroperoxo complex $[(NH_3)_5Co(O_2H)Co(NH_3)_5]H_3(SO_4)_4$ has been prepared by treating the sulfate salt of the peroxo complex with ice-cold 3 M sulfuric acid.[6]

B. μ-SUPEROXO-BIS[PENTAAMMINECOBALT(III)] PENTACHLORIDE MONOHYDRATE

Procedure

The peroxo complex is first prepared as in Sec. A, but the product is not isolated from the final solution. The peroxo complex is then oxidized with (in this instance) peroxodisulfate, acid being added to neutralize the excess ammonia:[9]

$$2[(NH_3)_5Co(O_2)Co(NH_3)_5]^{4+} + S_2O_8^{2-} \rightarrow$$
$$2[(NH_3)_5Co(O_2)Co(NH_3)_5]^{5+} + 2SO_4^{2-}$$

The chloride salt is prepared here, since it is more soluble than the nitrate or sulfate salts.   In the presence of chloride a lower temperature (0°C.) is desirable for the first stage, so that side reactions in which chloropentaamminecobalt(III) is produced are minimized.

Ten grams of cobalt(II) chloride hexahydrate is dissolved in 100 ml. of water at 0°C., and 50 ml. of 15 M (0.880 sp. gr.) ammonia at 0°C. is added.   A brisk stream of oxygen (or air) is sucked through the resultant cooled solution as described in Sec. A, in this case for 45 minutes (or 2 hours if air is used). The solution is kept at 0°C. while ammonium peroxodisulfate (6 g.) dissolved in a minimum amount of ice-cold water is added, and the solution is allowed to stand for up to 10 minutes. Ice-cold concentrated hydrochloric acid is slowly added with stirring (the temperature being kept very close to 0°C.) until the solution is acidic, as indicated by pH papers (30–40 ml. of acid is required).   The solution is kept ice-cold until precipitation of the product is complete.   The blue-green solid is filtered off and washed with ethanol and then with diethyl ether.

To remove the sulfate introduced during the peroxodisulfate oxidation, the crude product is dissolved in a minimum amount of 1 M hydrochloric acid ($\sim$160 ml.)* at 40°C., and barium chloride (3 g.) dissolved in a minimum amount of hot water is added.   The solution is allowed to stand for 20–30 minutes, rewarmed to 50°C., and the barium sulfate filtered off on a medium-porosity sintered-glass filter.   Concentrated hydrochloric acid (20 ml.) is added to the filtrate, which is cooled until precipitation is complete.   The solid is recrystallized by dissolving it in a minimum amount of 0.1 M hydrochloric acid

---

* Checkers found a 400-ml. quantity was required.

at 40°C. and allowing the solution to stand. Solutions of the complex are sensitive to light. Light green or dark green crystals are obtained, depending on the speed of recrystallization. The crystals are filtered off and air-dried, since washing with ethanol and ether appears to speed up decomposition. The solid should be stored in a lightproof desiccator; it loses water to yield a monohydrate form of the pentachloride salt.[9] Yields of $[(NH_3)_5Co(O_2)Co(NH_3)_5]Cl_5 \cdot H_2O$ are variable but are generally *ca.* 5 g. (50%). *Anal.* Calcd.: N, 27.2; Co, 22.9; Cl, 34.4; H, 6.3. Found: N, 26.7; Co, 22.7; Cl, 34.5; H, 6.2. Checkers Ogino and Bailar reported: N, 27.2; H, 6.25.

Properties

The complex has a green color with a characteristic absorption peak at 670 nm. ($\epsilon \cong 890$ M^{-1} cm.$^{-1}$). It is paramagnetic (one unpaired electron), and e.p.r. studies indicate the equivalence of the cobalt atoms and delocalization of the odd electron.[7] Recent x-ray crystallographic studies[3b] have shown that the bridging oxygen atoms are σ-bonded (Co—O—O bond angle 118°) and that the O—O bond distance is 1.31 A. This is shorter than that normally found in peroxides (1.48 A.) and is close to that found for the superoxide ion in KO_2. Whereas in the peroxo complex there is a torsion angle of 146° about the O—O bond,[3a] the Co—O—O—Co atoms are coplanar in the superoxo complex.[3b]

The complex is reasonably stable in acidic solution, decomposition (i.e., aquation):

$$[(NH_3)_5Co(O_2)Co(NH_3)_5]^{5+} + H_2O \rightarrow$$
$$[Co(NH_3)_5(H_2O)]^{3+} + O_2 + Co^{2+} + 5NH_3$$

being about 10^2 times slower than that of chloropentaamminecobalt(III) ion.[10] At pH > 3–4 there is a much more rapid decomposition of the complex. With one-electron donors (for example, Fe^{2+}, I^-), there is a reduction[11] to the peroxo complex, which decomposes rapidly in acidic solution with the evolution of oxygen:

$$[(NH_3)_5Co(O_2)Co(NH_3)_5]^{5+} + e^- \rightarrow [(NH_3)_5Co(O_2)Co(NH_3)_5]^{4+}$$

$$[(NH_3)_5Co(O_2)Co(NH_3)_5]^{4+} \xrightarrow{\text{H+}} 2Co^{2+} + 10NH_4^+ + O_2$$

With sulfite a different stoichiometry is observed:

$$[(NH_3)_5Co(O_2)Co(NH_3)_5]^{5+} + 2SO_3^{2-} \rightarrow$$
$$[Co(NH_3)_5(SO_4)]^+ + Co^{2+} + 5NH_3 + SO_4^{2-}$$

and cobalt(III) and cobalt(II) ions are produced in equal amounts. The mechanism in this case involves atom transfer.[12]

C. POTASSIUM μ-SUPEROXO-BIS[PENTACYANOCOBALT(III)] MONOHYDRATE

$$[(NH_3)_5Co(O_2)Co(NH_3)_5]^{5+} + 10CN^- \rightarrow$$
$$[(CN)_5Co(O_2)Co(CN)_5]^{5-} + 10NH_3$$

Procedure

This entire process must be carried out in a good hood with proper gas disposal. It is essential that the airflow be sufficient to carry away all the hydrogen cyanide.

Three grams of $[(NH_3)_5Co(O_2)Co(NH_3)_5]Cl_5 \cdot H_2O$ is added, a little at a time, to a stirred solution of 5 g. of KCN in 50 ml. of water.[13] A brown precipitate is formed. The mixture is filtered and the precipitate discarded. The reddish-brown solution is acidified with 2 *M* nitric acid using pH papers; ■ *HCN is given off on addition of the acid.* A solution of zinc sulfate is added until most of the red-orange complex precipitates. This is allowed to stand 5 minutes and is then filtered. The zinc salt is washed with water and dissolved in a 10% solution of KCN, whereupon five times the volume of methanol and then an equal* volume of ethanol are added. The product is precipitated once more by dissolving in a minimum quantity of

---

* Equal to methanol added.

water and again adding methanol and ethanol in equal amounts. The precipitate is washed with ethanol and finally with ether. The yield is about 50%. *Anal.* Calcd. for $K_5[(CN)_5Co(O_2)$-$Co(CN)_5]\cdot H_2O$: C, 19.2; Co, 18.9; H, 0.32; N, 22.5. Found: C, 19.1; Co, 19.1; H, 0.90; N, 22.4.

Properties

The superoxo complex is red (and not green) with an intense charge transfer band at 310 nm. ($\epsilon = 1.15 \times 10^4$ M^{-1} cm.$^{-1}$).[14] The e.p.r. characteristics are very similar to those of the $[(NH_3)_5Co(O_2)Co(NH_3)_5]^{5+}$ ion.[13]   The solid should be stored in a desiccator.   It is soluble in water (>1 g./ml. at 25°C.) and quite stable in both acidic and basic solutions.   The complex is reduced by iodide, the reaction being essentially 3:1 in iodide to complex.[15]   The corresponding μ-peroxo-bis[penta-cyanocobalt(III)](6−) ion is obtained by the action of oxygen on $[Co(CN)_5]^{3-}$.[15]

D. μ-AMIDO-μ-PEROXO-BIS[TETRAAMMINECOBALT(III)] TRINITRATE MONOHYDRATE

$$[(NH_3)_5Co(O_2)Co(NH_3)_5]^{4+} \rightarrow$$
$$[(NH_3)_4Co(NH_2,O_2)Co(NH_3)_4]^{3+} + NH_4^+$$

Procedure

Until recently the starting material for all amido-bridged complexes of cobalt was Vortmann's sulfate.[1]   In the much improved procedure described here, the monobridged μ-peroxo complex is first prepared and then converted into the μ-peroxo-μ-amido complex.[5]

This μ-peroxo-μ-amido complex can also be prepared by the reduction of the μ-superoxo-μ-amido compound (Sec. E) in aqueous ammonia.   The latter method is useful in the preparation of the corresponding ethylenediamine complex.[6] A 10-g. quantity of the product from the procedure of Sec. A is added, a little at a time, to 100 ml. of 15 M aqueous ammonia while the latter is constantly stirred. Potassium hydroxide (1 g.) dissolved in a little water is added, and the mixture is slowly warmed up to and kept at 32–35°C. for 90 minutes. It is then cooled to below 10°C., and ammonium nitrate (70 g.) is added, followed by 200 ml. of methanol which is slowly added to the stirred mixture. If oily material separates out, the mixture is stirred until it becomes crystalline. The precipitate is washed with methanol and redissolved in 100 ml. of 6 M aqueous ammonia; the filtered solution is stirred with 100 g. of sodium perchlorate monohydrate. The precipitate of crude double perchlorate containing sodium perchlorate is filtered and washed with ethanol and ethyl ether.

A 5- to 6-g. sample of the product is dissolved in 100 ml. of 6 M ammonia at 20°C. and, after filtering, 40 g. of ammonium nitrate and 200 ml. of methanol are added to the stirred filtrate, and the mixture is cooled with ice. The precipitate is filtered, washed with methanol, and redissolved in 100 ml. of 6 M aqueous ammonia; after filtration the complex salt is crystallized by addition of 40 g. of ammonium nitrate and 100 ml. of methanol and by cooling in ice. For every gram of the product a 23-ml. aliquot of 15 M aqueous ammonia is added and, after filtering, a 23-ml. quantity of methanol is slowly added. The mixture is cooled in ice, and the precipitate is filtered and washed with methanol. The product is $[(NH_3)_4Co(NH_2,O_2)Co(NH_3)_4]$-$(NO_3)_3 \cdot H_2O$. *Anal.* Calcd.: Co, 23.3; N(ammoniacal), 24.9; N (in NO_3), 8.30; H, 5.58. Found: Co, 23.5; N(ammoniacal), 24.5; N (in NO_3), 8.45; H, 5.75.

Properties

In solution, the μ-peroxo-μ-amido-bis[tetraamminecobalt(III)] complex is not as stable as the μ-peroxo-μ-amido-bis[bis(ethyl-enediamine)cobalt(III)] complex. Both solids have a brown color. In acidic solutions, protonation occurs and there is isomerization to the corresponding red hydroperoxo complex.[6] Details of the equilibria involved have been studied in the case of the ethylenediamine complex:

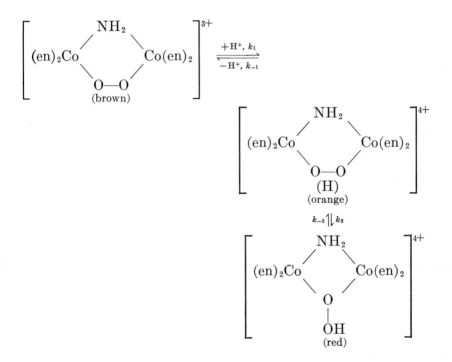

At 25°C. and ionic strength equal to 0.25 M, K_1 (defined k_{-1}/k_1) is 0.15 mole/l., and K_2 (defined k_2/k_{-2}) is 1.1. Whereas both the protonation reaction and its reverse are rapid, the isomerization reactions are much slower, $k_2 = 0.060$ sec.$^{-1}$ and $k_{-2} = 0.053$ sec^{-1} at 25°C.[6] The five-atom ring in the brown μ-amido-μ-peroxo ethylenediamine complex is not planar (compare the μ-amido-μ-superoxo complex in Sec. E); the distance O—O is 1.48 A., and the angle Co—O—O is *ca.* 110°.[3e] The structure of the red hydroperoxo complex has been determined.[3a] The O—O bond here has a length of 1.41 A. and lies at an angle of *ca.* 41° out of the plane of the CoNCoO ring.

E. μ-AMIDO-μ-SUPEROXO-BIS[TETRAAMMINECOBALT(III)] TETRANITRATE

$$[(NH_3)_5Co(O_2)Co(NH_3)_5]^{4+} \rightarrow$$
$$[(NH_3)_4Co(NH_2,O_2)Co(NH_3)_4]^{3+} + NH_4^+$$
$$[(NH_3)_4Co(NH_2,O_2)Co(NH_3)_4]^{3+} + Ce^{4+} \rightarrow$$
$$[(NH_3)_4Co(NH_2,O_2)Co(NH_3)_4]^{4+} + Ce^{3+}$$

Procedure

If the monobridged μ-peroxo-bis[pentaamminecobalt(III)] nitrate is freshly obtained (Sec. A), the brownish-black precipitate of the dihydrate just separated from the mother liquor may be used without washing. A 45-g. quantity is added in several portions to 500 ml. of 15 M aqueous ammonia with stirring. Potassium hydroxide, (4.2 g.) dissolved in a minimum quantity of water, is added, and the mixture is heated to and then kept at 32–35°C. for 90 minutes. After cooling to below 5°C., the product is poured, a little at a time and with vigorous stirring, into a mixture of 700 ml. of concentrated nitric acid (14–15 M, which has been cooled to 5°C.), 100 g. of diammonium hexanitratocerate(IV),* and 1200 g. of ice. The temperature of the mixture must be kept below 10°C., by means of an ice-salt bath. The mixture is stirred for 60 minutes in an ice bath to assist crystallization. It is convenient at this stage to leave the mixture in a refrigerator overnight.

The gray-green precipitate is filtered on a glass filter and washed twice with a little 2 M nitric acid. The precipitate is then stirred mechanically at 20°C. with 2000 ml. of water containing 5 ml. of 15 M nitric acid until all the crystal blocks are dispersed, and the suspension is filtered. To the filtrate

---

* The oxidation to the superoxo complex can also be effected with 10 g. of potassium permanganate, completely dissolved in 700 ml. of nitric acid together with 1200 g. of ice. The yield in this case is reduced by *ca.* 20% as compared with that obtained with the Ce(IV) salt. Care must be taken to avoid excess permanganate, since the solid superoxo dicobalt permanganates are explosive. Chlorine water is also a possible oxidizing agent.

a 200-ml. quantity of 3 M sulfuric acid is added and the mixture stirred for 30 minutes. The gray crystalline precipitate of the crude μ-amido-μ-superoxo-bis[tetraamminecobalt(III)] disulfate dihydrate is filtered on a glass filter and washed with a little water and ethanol. The yield of crude sulfate is 16–17 g.

A 10-g. mass of the crude sulfate is stirred with 500 ml. of water, and a 100-ml. quantity of concentrated nitric acid is added to this suspension little by little under constant agitation. The mixture is cooled with running water, or in ice, and filtered. The crude nitrate is then dissolved in a minimum quantity (a little more than 1 l.) of 0.5% nitric acid and the solution filtered. The pure nitrate can be precipitated by adding concentrated nitric acid of about one-fifth the volume of the solution and cooling to 5°C. A light-colored, fluffy precipitate may form in the first instance but will change to dark green needles on standing. The crystals are filtered and washed with dilute nitric acid and ethanol. The yield is *ca.* 9 g. from 10 g. of crude sulfate or 14–15 g. (28–30%) from 50 g. of cobalt(II) nitrate hexahydrate. Solutions of the complex are sensitive to light. The solid should be stored in a lightproof container. *Anal.* Calcd. for $[(NH_3)_4Co(NH_2,O_2)Co(NH_3)_4]$- $(NO_3)_4$: Co, 21.4; N(ammoniacal), 22.9. Found: Co, 21.5; N(ammoniacal), 22.8.

Properties

The complex is green ($\epsilon = 306$ M^{-1} cm.$^{-1}$ at 700 nm.) and paramagnetic.[4,7] From an x-ray crystallographic study,[3c] the O_2 bridge is known to be σ-bonded to the cobalt atoms (Co—O—O angles 120°), and the O—O distance is 1.32 A. The five atoms in the ring are very nearly coplanar. The complex is stable in acidic solutions, but in neutral and alkaline solutions it is reduced to the corresponding peroxo complex, with subsequent decomposition of the latter. Ferrous ion also reduces the complex, in the first place to the peroxo complex.

With iodide the reaction proceeds to the μ-amido-μ-hydroxo-bis[tetraaminecobalt(III)] complex:[11a]

$$3I^- + [(NH_3)_4Co(NH_2,O_2)Co(NH_3)_4]^{4+} + 2H_2O \rightarrow$$
$$\tfrac{3}{2}I_2 + [(NH_3)_4Co(NH_2,OH)Co(NH_3)_4]^{4+} + 3OH^-$$

F. μ-AMIDO-μ-SUPEROXO-BIS[DI(ETHYLENEDIAMINE)-COBALT(III)] TETRANITRATE

$$[(NH_3)_4Co(NH_2,O_2)Co(NH_3)_4]^{4+} + 4 \text{ en} \rightarrow$$
$$[(en)_2Co(NH_2,O_2)Co(en)_2]^{4+} + 8NH_3$$

Procedure

Four grams of the nitrate salt of the amine complex is treated with 25 g. of 10% ethylenediamine solution at 60°C. until the smell of ammonia is weak (1–2 hours are sufficient). After it has cooled to room temperature, concentrated nitric acid is slowly added to the brown solution while the temperature is maintained at *ca.* 25°C. until the solution turns olive-green. The solution is ice-cooled, and the green crystals are filtered off; ethanol may be used to aid crystallization. The compound is recrystallized from hot ($\sim$70°C.) 5% nitric acid. Yield is $\sim$2 g. of $[(en)_2Co(NH_2,O_2)Co(en)_2](NO_3)_4$. *Anal.* Calcd.: H, 5.2; C, 14.7. Found: H, 5.4; C, 14.7.

The complex may contain some (10–20%) $[(en)_2Co(NH_2, NO_2)Co(en)_2](NO_3)_4$ as an impurity unless contact with nitrate ions is avoided.[17] To avoid nitrate contamination, chlorine gas can be used to oxidize the peroxo complex.[17] The complex is difficult to purify by recrystallization. The μ-amido-μ-nitrito complex is extremely unreactive.

Properties

General properties and reactions of the cation are similar to those of [μ-amido-μ-superoxo-bis{tetraamminecobalt(III)}]$^{4+}$.

There is a characteristic absorption peak at 687 nm. ($\epsilon = 485$ M^{-1} cm.$^{-1}$). The structure of the complex has been determined.[3a]

G. μ-AMIDO-μ-CHLORO-BIS[TETRAAMMINECOBALT(III)] TETRACHLORIDE TETRAHYDRATE

a. $2[(NH_3)_4Co(NH_2,O_2)Co(NH_3)_4]^{4+} + 3SO_3^{2-} + 2H^+ \rightarrow$
   $2[(NH_3)_4Co(NH_2,SO_4)Co(NH_3)_4]^{3+} + SO_4^{2-} + H_2O$

b. $[(NH_3)_4Co(NH_2,SO_4)Co(NH_3)_4]^{3+} + HCl + H_2O \rightarrow$
   $[(NH_3)_4(H_2O)Co(NH_2)CoCl(NH_3)_4]^{4+} + HSO_4^{-}$

c. $[(NH_3)_4(H_2O)Co(NH_2)CoCl(NH_3)_4]^{4+} \rightarrow$
   $[(NH_3)_4Co(NH_2,Cl)Co(NH_3)_4]^{4+} + H_2O$

Procedure

The starting material is the nitrate salt of the μ-amido-μ-superoxo-bis[tetraamminecobalt(III)] complex as prepared in Sec. E. The crude salt obtained following treatment with diammonium hexanitratocerate(IV) may be used. The gray-green precipitate is filtered on a glass filter and washed with a little 2 M nitric acid. For every 0.8 g. of the complex, a 50-ml. quantity of 0.1 M nitric acid is added, and the complex is dissolved by heating to 50–60°C. A vigorous stream of sulfur dioxide is passed through the solution for about 3 minutes or for such time as is required for the green color to disappear.[17] On cooling to 0°C., a red precipitate of the μ-amido-μ-sulfato-bis[tetraamminecobalt(III)] complex is obtained. This is filtered, washed with ethanol and ether, and allowed to dry; yield is 90%.

For every gram of the product, 15 ml. of concentrated hydrochloric acid is added, and the mixture is shaken or stirred for 24 hours. A red precipitate is obtained; this is filtered off by suction and washed carefully with ethanol until acid-free. For recrystallization, the precipitate is dissolved in a minimum amount of ice-cold water and quickly filtered into an equal volume of ice-cold concentrated hydrochloric acid. Dark red crystals are obtained when the solution is allowed to stand for several hours at 0°C. These are filtered, washed with ethanol, and dried by suction. *Anal.* Calcd. for [(NH₃)₄Co(NH₂, Cl)Co(NH₃)₄]Cl₄·4H₂O: H, 6.6; Cl, 34.1; N, 24.2. Found: H, 6.3; Cl, 34.0; N, 24.0.

Properties

The chloride salt of the solid complex has recently been shown to have a chloride bridge.[18] In solution, the μ-amido-μ-chloro and μ-amido-aquo-chloro complexes appear to equilibrate rapidly (see Eq. *c*, p. 209). It is possible that the μ-amido-μ-chloro complex is never present in solution in large amounts but that it is much the least soluble form. The aquo-chloro complex also equilibrates with the μ-amido-μ-hydroxo complex in aqueous solution:

$$[(NH_3)_4(H_2O)Co(NH_2)Co(Cl)(NH_3)_4]^{4+} \rightleftharpoons$$
$$[(NH_3)_4Co(NH_2,OH)Co(NH_3)_4]^{4+} + H^+ + Cl^-$$

When the μ-amido-μ-chloro complex is dissolved in 0.1 M perchlorate solution, pH measurements (extrapolated back to the time of dissolution) indicate the initial presence of the aquo-chloro-μ-amido complex.[19]

H. μ-AMIDO-μ-HYDROXO-BIS[TETRAAMMINECOBALT(III)] TETRACHLORIDE TETRAHYDRATE

$$[(NH_3)_4Co(NH_2,Cl)Co(NH_3)_4]^{4+} + H_2O \rightarrow$$
$$[(NH_3)_4(H_2O)Co(NH_2)CoCl(NH_3)_4]^{4+}$$
$$[(NH_3)_4(H_2O)Co(NH_2)CoCl(NH_3)_4]^{4+} \rightarrow$$
$$[(NH_3)_4Co(NH_2,OH)Co(NH_3)_4]^{4+} + H^+ + Cl^-$$

Procedure

The procedure for converting the μ-amido-μ-chloro (or aquo-chloro) complex to the μ-amido-μ-hydroxo complex is: For every gram of the μ-amido-aquo-chloro complex, a 7-ml. quantity of distilled water is added, and the solution is left to stand for 6–8 hours. The more insoluble chloride salt of the hydroxo complex $[(NH_3)_4Co(NH_2,OH)Co(NH_3)_4]^{4+}$ crystallizes out as a dark red solid. This is filtered off, washed with alcohol, and recrystallized from hot (~70°C.) dilute acetic acid. The yield is 0.6 g. of $[(NH_3)_4Co(NH_2,OH)Co(NH_3)_4]Cl_4·4H_2O$ per gram of starting material. *Anal.* Calcd.: H, 6.99; Cl, 28.35. Found: H, 7.00; Cl, 28.2.

Properties

The complex absorbs at 520 nm. ($\epsilon = 149$ M^{-1} cm.$^{-1}$). In aqueous solutions there is no acid dissociation of the NH$_2^-$ or OH$^-$ bridging ligands (pH measurements). The complex can be used as starting material for the preparation of a wide range of other dicobalt complexes. With strong acids there is cleavage of the hydroxo bridge, for example:

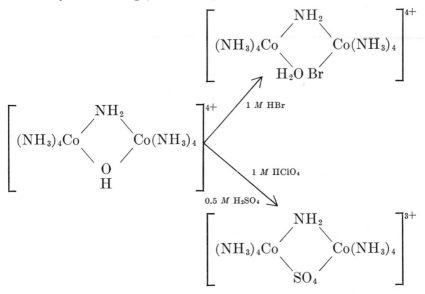

The reactions with HCl, HBr, H$_2$SO$_4$, HNO$_3$, and HClO$_4$ can be followed spectrophotometrically.

I. μ-AMIDO-BIS[PENTAAMMINECOBALT(III)] PENTAPERCHLORATE MONOHYDRATE

$$[(NH_3)_4(H_2O)Co(NH_2)CoCl(NH_3)_4]^{4+} + 2NH_3 \rightarrow$$
$$[(NH_3)_5Co(NH_2)Co(NH_3)_5]^{5+}$$

Procedure

Three grams of the μ-amido-μ-chloro complex

$$[(NH_3)_4Co(NH_2,Cl)Co(NH_3)_4]Cl_4 \cdot 5H_2O$$

as prepared in Sec. G is dissolved in a minimum amount (about 50 ml.) of ice-cold water, and immediately one-third the volume of concentrated nitric acid (also at 0°C.) is added. The magenta-colored crystals obtained are filtered off, washed with ethanol and ether, and allowed to dry. About a 2.5-g. quantity of the nitrate salt of the aquo-chloro complex is obtained. This is added a little at a time to 500–600 ml. of liquid ammonia in a 1-l. beaker. The beaker is immersed in an ice-salt mixture, and the ammonia is allowed to evaporate overnight under a fume hood. The red solid which is obtained is placed on a sintered-glass filter funnel, just covered with water, mixed well, and then sucked dry. This process is repeated until the washings change from brown to red. The yield is around 70%.

Two grams of the product is dissolved in a minimum amount of water at room temperature, and a slight excess (~1.75 g.) of ammonium bromide is added. After it has cooled in ice, the bromide salt of the complex is filtered off and recrystallized. To convert to the perchlorate salt, a slight excess (by weight) of silver perchlorate is added to a saturated solution of the bromide salt in *ca.* 0.3 *M* perchloric acid at 40°C. (small quantities of perchloric acid only are required). The silver bromide is filtered off, and a third of the volume of ice-cold 72% perchloric acid is added.* The sample obtained is recrystallized twice. *Anal.* Calcd. for $[(NH_3)_5Co(NH_2)Co(NH_3)_5](ClO_4)_5 \cdot H_2O$: Cl, 21.7; H, 4.15; N, 18.8. Found: Cl, 21.2; H, 4.3; N, 18.6.

* ■ *Caution. Ammine perchlorates are explosive. Do not increase scale.*

Properties

The complex has a red color with characteristic absorption peaks at 360 nm. ($\epsilon = 708$ M^{-1} cm.$^{-1}$) and 505 nm. ($\epsilon = 428$ M^{-1} cm.$^{-1}$). Decomposition of the complex in the solid phase and in solution occurs at the rate of *ca.* 2% per day at 0–20°C. In perchloric acid solutions, equivalent amounts of hexaamminecobalt(III) and aquopentaamminecobalt(III) are produced in 2–3 hours at about 85°C.:[10]

$$[(NH_3)_5Co(NH_2))Co(NH_3)_5]^{5+} + H_2O + H^+ \rightarrow$$
$$[Co(NH_3)_6]^{3+} + [Co(NH_3)_5(H_2O)]^{3+}$$

From an x-ray crystallographic study, the Co—N—Co bond angle is known to be 153°, and the distance Co—N is 2.06 A.[20] These values show that the bridge of the cation in the solid is highly strained.

References

1. A. Werner, *Ann.*, **375**, 1 (1910).
2. A. G. Sykes and J. A. Weil, *Progr. Inorg. Chem.*, to be published.
3a. U. Thewalt and R. E. Marsh, *J. Am. Chem. Soc.*, **89**, 6364 (1967).
3b. R. E. Marsh and W. P. Schaefer, *Acta Cryst.*, **B24**, 246 (1968).
3c. G. G. Christoph, R. E. Marsh, and W. P. Schaefer, *Inorg. Chem.*, **8**, 291 (1969).
3d. W. P. Schaefer, *ibid.*, **7**, 725 (1968).
3e. U. Thewalt, unpublished work, 1969.
4. J. A. Weil and J. K. Kinnaird, *J. Phys. Chem.*, **71**, 3341 (1967).
5. M. Mori, J. A. Weil, and M. Ishiguro, *J. Am. Chem. Soc.*, **90**, 615 (1968).
6. M. Mori and J. A. Weil, *ibid.*, **89**, 3732 (1967).
7. E. A. V. Ebsworth and J. A. Weil, *J. Phys. Chem.*, **63**, 1890 (1959).
8a. F. H. Rohm and C. J. Nyman, *J. Inorg. Nucl. Chem.*, **32**, 165 (1970).
8b. R. G. Charles and S. Barnartt, *J. Inorg. Nucl. Chem.*, **22**, 69 (1961).
9. M. Linhard and M. Weigel, *Z. Anorg. Allgem. Chem.*, **308**, 254 (1961).
10. R. D. Mast and A. G. Sykes, *J. Chem. Soc.*, **A**, 1031 (1968).
11a. R. Davies and A. G. Sykes, *ibid.*, **A**, 2237 (1968).
11b. R. Davies and A. G. Sykes, *ibid.*, **A**, 2831 (1968).
12. R. Davies, A. K. E. Hagopian, and A. G. Sykes, *ibid.*, **A**, 623 (1969).
13. M. Mori, J. A. Weil, and J. K. Kinnaird, *J. Phys. Chem.*, **71**, 103 (1967).
14. J. Barrett, *Chem. Commun.*, 874 (1968).
15. A. Haim and W. K. Wilmarth, *J. Am. Chem. Soc.*, **83**, 509 (1961).
16. L. R. Thompson and W. K. Wilmarth, *J. Phys. Chem.*, **56**, 5 (1952).

17. M. B. Stevenson and A. G. Sykes, *J. Chem. Soc.*, A, 2293, 2979 (1970).
18. R. Barro, R. E. Marsh, and W. P. Schaefer, *Inorg. Chem.*, to be published.
19. M. B. Stevenson, R. D. Mast, and A. G. Sykes, *J. Chem. Soc.*, A, 937 (1969).
20. W. P. Schaefer, A. W. Cordes, and R. E. Marsh, *Acta Cryst.*, B24, 283 (1968).

36. PENTAAMMINEPERRHENATOCOBALT(III) SALTS

Submitted by ELIZABETH LENZ* and R. KENT MURMANN*
Checked by MARY F. SWINIARSKI,† MANFRED FILD,† and RONALD D. ARCHER†

Complexes of cobalt(III) containing coordinated perrhenate ion have not previously been reported.‡   They are of special interest because of their close relationship to the unstable perchlorate complexes and are of use in studies on the mechanism of hydrolysis.   Since the equilibrium constant of formation is small and $[Co(NH_3)_5(H_2O)](ReO_4)_3 \cdot 2H_2O$ is relatively insoluble, the present procedure is similar to that used previously[2] for more common complex ions.

A. $[Co(NH_3)_5(H_2O)](ReO_4)_3 \cdot 2H_2O$

$$[Co(NH_3)_5(H_2O)](ClO_4)_3 + 3ReO_4^- + 2H_2O \rightarrow$$
$$[Co(NH_3)_5(H_2O)](ReO_4)_3 \cdot 2H_2O + 3ClO_4^-$$

Procedure

In a 100-ml. beaker is placed 4.60 g. (0.010 mole) of $[Co(NH_3)_5(H_2O)](ClO_4)_3$§ and 60 ml. of pure water.   To this solution 5.76 ml. of a $HReO_4$ solution (0.301 mole) containing

---

* Chemistry Department, University of Missouri, Columbia, Mo. 65201.

† University of Massachusetts, Amherst, Mass. 01002.

‡ Since submission of this synthesis, the complexes have been reported.   See reference 1.

§ Prepared as described in reference 3, and recrystallized from dilute $HClO_4$ three times as discussed in reference 1.   An equivalent amount of $[Co(NH_3)_5Cl]Cl_2$ which has been allowed to aquate for 4 days at room temperature can be used, but lower yields are obtained.   The purity of the initial pentaammine complex is important.   Hexaammine impurities are carried through the reaction sequence.

1 g. of Re per milliliter* is quickly added with stirring. An immediate precipitation of $[Co(NH_3)_5(H_2O)](ReO_4)_3 \cdot 2H_2O$ occurs which is nearly quantitative. After 10 minutes at 0°C. the solid is collected on a sintered-glass filter, washed with two 50-ml. portions of methanol and five 20-ml. portions of acetone, and dried in air. The yield of $[Co(NH_3)_5(H_2O)](ReO_4)_3 \cdot 2H_2O$ is 9.17 g. {97% of theory based on $[Co(NH_3)_5(H_2O)](ClO_4)_3$}. *Anal.* Calcd. for $CoRe_3N_5H_{21}O_{15}$: Co, 6.21; Re, 58.88; N, 7.38; lattice water, 3.8; coordinated water, 1.9; formula weight, 948.7. Found: Co, 6.18; Re, 59.00; N, 7.29; lattice water, 3.8; coordinated water, 1.9; formula weight, 948.1.†

B. $[Co(NH_3)_5(OReO_3)](ReO_4)_2$

$$[Co(NH_3)_5(H_2O)](ReO_4)_3 \cdot 2H_2O \xrightarrow[50°C.]{vac.} [Co(NH_3)_5(H_2O)](ReO_4)_3 + 2H_2O\uparrow$$

$$[Co(NH_3)_5(H_2O)](ReO_4)_3 \xrightarrow[115°C.]{vac.} [Co(NH_3)_5(OReO_3)](ReO_4)_2 + H_2O\uparrow$$

Procedure

The $[Co(NH_3)_5(H_2O)](ReO_4)_3 \cdot 2H_2O$ is placed in a 3-cm.-diam. tube sealed at one end and having a ground-glass joint at the other. The tube is slowly evacuated on a vacuum system to a pressure of 1 μ. A Dry-Ice trap protects the pump. A silicone-oil bath is placed around the tube so that the oil is 1 cm. above the complex. At 50°C. the lattice water is removed at a reasonable rate. After 2 hours at 50°C the bath temperature is raised to 115–120°C. over a period of an hour. At this temperature the coordinated water is removed, and the solid slowly becomes purple. The reaction is complete in 4–5 hours as

* This solution may be of any concentration above 0.5 M, provided the volume of it is changed accordingly. $HReO_4$ solution (5.38 M) is available from S. W. Shattuck Corp., Denver, Colo. Alternatively the checkers found $NaReO_4$ (Alfa Inorganics, Inc., Beverly, Mass.) to be satisfactory for precipitation of the perrhenate.

† All formula weights were determined as described in the analysis subsection, assuming three equivalents of H^+ liberated from the resin per formula weight.

shown by the vacuum approaching the value obtained in the absence of a sample. After cooling to room temperature, the product $[Co(NH_3)_5(OReO_3)](ReO_4)_2$ is removed and maintained in a dry atmosphere. Yield is 8.63 g., quantitative except for transfer losses. *Anal.* Calcd. for $CoRe_3N_5H_{15}O_{12}$: Co, 6.59; Re, 62.5; N, 7.83; formula weight, 894.1. Found: Co, 6.62; Re, 62.3; N, 7.79; formula weight, 893.4.

C. $[Co(NH_3)_5(OReO_3)]X_2$

(Z = resin cation; X = ClO_4^-, NO_3^-, Cl^-)

$$[Co(NH_3)_5(OReO_3)](ReO_4)_2 + 2ZX \rightarrow$$
$$[Co(NH_3)_5(OReO_3)]X_2 + 2ZOReO_3$$

Procedure

Conversion of $[Co(NH_3)_5(OReO_3)](ReO_4)_2$ to $[Co(NH_3)_5(OReO_3)]X_2$, where X^- is ClO_4^-, NO_3^-, or Cl_2^- is accomplished by using an excess of Dowex 1-X4 (50–100 mesh) in the Cl^- form. One hundred milliliters of Dowex 1-X4 in the Cl^- form is placed in a 600-ml. sintered-glass filter and washed with pure water (about 1.5 l.) until the washings are neutral and free of Cl^-. (It has been found helpful to wash the resin with a 0.002 M 2,6-lutidine solution to remove traces of acid.*) The resin is placed in a large mortar with 100 ml. of water and the assembly cooled to 10°C. Then 4.00 g. of $[Co(NH_3)_5(OReO_3)](ReO_4)_2$ is added with vigorous stirring. (At this point it is necessary to carry out the operations quickly and to keep the solutions at 10–15°C. in order to prevent undue hydrolysis.) The solids are ground vigorously with a pestle for 5 minutes, and the mixture is added to a filter apparatus which drains into a 1-l. flask containing 150 ml. of a filtered, saturated, NaClO₄, LiNO₃, or LiCl solution at 0°C. The resin is again treated with a 50-ml. portion of cooled water; after 5 minutes of grinding, the liquid is decanted into the filter, and the filtrate is allowed to flow into the salt solution.

---

* 2,6-Lutidine does not increase the rate of hydrolysis as do most other bases. This step is not necessary but increases the yield.

The product precipitates shortly after contact with the salt solution, and crystallization is essentially complete in 5 minutes at 0°C.* The violet-red solid is collected on a filter and washed with three 10-ml. portions of methanol and four 10-ml. portions of acetone. It is then dried quickly under vacuum.

The yield of $[Co(NH_3)_5(OReO_3)](ClO_4)_2$ is 78%, that of $[Co(NH_3)_5(OReO_3)](NO_3)_2 \cdot H_2O$ is 69%, and that of $[Co(NH_3)_5-OReO_3]Cl_2$ is 64%. These appear to be optimum yields since time delays and slight temperature increases allow appreciable hydrolysis, which lowers the yield. *Anal.* $[Co(NH_3)_5(OReO_3)]-(ClO_4)_2$. Calcd.: Co, 9.94; Re, 31.4; N, 11.84; H, 2.55; Cl, 11.95; formula weight, 593.2. Found: Co, 9.87; Re, 31.6; N, 11.92; H, 2.60; Cl, 12.01; formula weight, 591.4. $[Co(NH_3)_5-(OReO_3)](NO_3)_2 \cdot H_2O$. Calcd.: Re, 34.72; N, 18.29; formula weight, 536.3. Found: Re, 34.51; N, 18.32; formula weight, 535.1. $[Co(NH_3)_5(OReO_3)]Cl_2$. Calcd.: Re, 40.03; N, 14.95; Cl, 15.24%; formula weight, 465.2. Found: Re, 39.94; N, 14.97; Cl, 15.31%; formula weight, 468.7.

Analysis

Since $[Co(NH_3)_5(OReO_3)]^{2+}$ aquates rapidly in neutral media to $[Co(NH_3)_5(H_2O)]^{3+}$ and ReO_4^-, the formula weight may be determined by measuring the amount of aquo complex formed. This is conveniently done by warming a solution of $[Co(NH_3)_5-(OReO_3)](ClO_4)_2$ until aquation is complete (10 minutes, 50°C.), absorbing it on a neutral resin (Dowex 50W-X2) in the H^+ form, eluting the H^+ quantitatively, and titrating it with 0.100 N NaOH, using phenolphthalein indicator. The calculated formula weight is based on $3H^+$/complex ion.

Co, C, H, N, and Cl are determined by conventional methods. If Cl^- is determined gravimetrically, care must be taken to remove, by careful washing with warm water, the slightly soluble $AgReO_4$ formed.

* The checkers had to add a few drops of methanol to initiate precipitation of the chloride in a reasonable time period.

Properties

Pentaammineperrhenatocobalt(III) perchlorate, nitrate, and chloride are red-violet solids slightly soluble in water and essentially insoluble in methanol or acetone. In water the solubilities increase in the following order: $ClO_4^- < NO_3^- \ll Cl^-$. With about 50% loss, the ClO_4^- and NO_3^- salts may be recrystallized from cold water by the addition of concentrated solutions of $NaClO_4$ or $LiNO_3$. With the Cl^- salt, however, losses are much greater and the purity is not increased. In pure water at 25°C. they aquate following first-order kinetics to the aquopentaamminecobalt(III) ion and ReO_4^- with a half-time of about 30 minutes. In 10^{-2} M OH^- or H^+ the aquation rate is almost instantaneous. The visible spectrum resembles that of $[Co(NH_3)_5OH]^{2+}$ and has an absorption maximum in the visible region at 525 nm. ($\epsilon = 63$ M^{-1} cm.$^{-1}$). The infrared spectra show a splitting of the ReO_4^- band at 920 cm.$^{-1}$ due to the loss of symmetry of ReO_4^- upon coordination.

References

1. E. Lenz and R. K. Murmann, *Inorg. Chem.*, **7**, 1880 (1968).
2. Fred Basolo and R. Kent Murmann, *Inorganic Syntheses*, **4**, 171–176 (1953).
3. A. C. Rutenberg and H. Taube, *J. Chem. Phys.*, **20**, 823 (1952).
4. G. G. Schlessinger, *Inorganic Syntheses*, **9**, 160–163 (1967).

37. LINKAGE ISOMERS OF METAL COMPLEXES

Submitted by JOHN L. BURMEISTER* and FRED BASOLO†
Checked by S. E. JACOBSON‡ and A. WOJCICKI‡

Linkage isomerism in metal complexes was discovered in 1894 by Jørgensen.[1] He prepared and characterized the nitrito

* University of Delaware, Newark, Del. 19711.
† Northwestern University, Evanston, Ill. 60201.
‡ Department of Chemistry, The Ohio State University, Columbus, Ohio 43210.

(M—ONO) and nitro (M—NO$_2$) isomers of some acidoammine-cobalt(III) complexes. Many years later Adell[2] made extensive studies of the kinetics of isomerization in these systems. However, it was not until 1962 that linkage isomers of other metal complexes of this type were reported.[3]

It is noteworthy that the successful synthesis of nitrito-nitro linkage isomers for other metal ammines was designed only after the mechanism of formation of $[Co(NH_3)_5ONO]^{2+}$ and its rearrangement had been elucidated.[4] It was found that the nitrito complex is readily formed without the cleavage of the Co—O bond:

$$[(NH_3)_5Co—OH]^{2+} \xrightarrow{\text{NOX}} \begin{bmatrix} (NH_3)_5Co—O \cdots H \\ \vdots \\ \vdots \\ O—N \cdots X \end{bmatrix}^{2+}$$

$$\downarrow$$

$$[(NH_3)_5Co—ONO]^{2+} + HX$$

The kinetic product then slowly rearranges by an intramolecular process to the stable isomer $[(NH_3)_5Co—NO_2]^{2+}$. On this basis it was reasoned that it should be possible to obtain other nitrito complexes by operating at an appropriate pH and under mild conditions.

By taking advantage of the observation that the other ligands in a metal complex may alter the type of metal-atom bonding to an ambidentate ligand,[5] it was possible to prepare thiocyanato (M—SCN) and isothiocyanato (M—NCS) linkage isomers.[6] The syntheses of two such palladium(II) complexes are described below.

For a period of about 68 years after the original work of Jørgensen, no new types of linkage isomers of metal complexes were prepared. However, in recent years many such compounds have been obtained or their presence has been detected.[7] The known examples are given in Table I. An examination of the

TABLE I   Known Linkage Isomers of Metal Complexes

M—X linkage	Complex	Reference
M—ONO and M—NO$_2$	$[Co(NH_3)_5X]^{2+}$	1[a]
	$[Co(NH_3)_2(py)_2X_2]^+$	8[a]
	$[Co(en)_2X_2]^+$	4,8[a]
	$[M(NH_3)_5X]^{m+}$	3[a, b]
	$[Co(CN)_5X]^{3-}$	9[c]
	$[Ni(Me_2en)_2X_2$	10[c]
	$[Ni(EtenEt)_2X_2]$	10[c]
M—SCN and M—NCS	$[Pd(As(C_6H_5)_3)_2X_2]$	6[a]
	$[Pd(bipy)X_2]$	6[a]
	$[CdX_4]^{2-}$	11[c]
	$[Mn(CO)_5X]$	12[c]
	$[Pd(As(n\text{-}C_4H_9)_3)_2X_2]$	13[d]
	$[M(NH_3)_5X]^{2+}$	14[a, e]
	$[Cr(H_2O)_5X]^{2+}$	15[c]
	$[Pd(Et_4dien)X]^+$	16[a]
	$[Pd(4,7\text{-}diphenylphen)X_2]$	17[a]
	$[Cu(tripyam)X_2]$	18[a]
	$[Cu(dppa)X_2]$	18[a]
	$[(C_5H_5)Fe(CO)_2X]$	19[a]
	$[(C_5H_5)Mo(CO)_3X]$	19[a]
	$[Pd(P(OCH_3)_3)_2X_2]$	20[f]
	$[Pt_2(P(n\text{-}C_3H_7)_3)_2Cl_2X_2]$	21[a]
M—SSO$_3$ and M—OS$_2$O$_2$	$[Co(NH_3)_5X]^+$	22[f]
M—NC and M—CN	$[Co(CN)_5X]^{3-}$	9[c]
	$KFe[CrX_6]$	23[a]
	$[Cr(H_2O)_5X]^{2+}$	24[c]
	$cis\text{-}\alpha\text{-}[Co(trien)X_2]^+$	25[a]
M—SeCN and M—NCSe	$[Pd(Et_4dien)X]^+$	26[a]

[a] Both isomers isolated.
[b] M=Rh(III), Ir(III), Pt(IV).
[c] Unstable isomer detected in solution but not isolated.
[d] Partial isomerization to N-bonded isomer in molten state.
[e] M=Rh(III), Ir(III).
[f] Mixture of isomers initially isolated in solid state.

Abbreviations: py = pyridine; en = ethylenediamine; Me$_2$en = N,N-dimethyl-ethylenediamine; EtenEt = N,N'-diethylethylenediamine; bipy = 2,2'-bipyridine; Et$_4$dien = N,N,N',N'-tetraethyldiethylenetriamine; 4,7-diphenylphen = 4,7-diphenyl-1,10-phenanthroline; tripyam = tri(2-pyridyl)amine (bidentate); dppa = phenyldi(2-pyridyl)amine; trien = triethylenetetramine.

original references will show that the modern approaches to the syntheses of these isomers were based primarily on the current theories of bonding and/or mechanisms of reaction in these systems.

A. DITHIOCYANATO-S-BIS(TRIPHENYLARSINE)PALLADIUM(II)

$$K_2[Pd(SCN)_4] + 2As(C_6H_5)_3 \xrightarrow{0°C.} [Pd\{As(C_6H_5)_3\}_2(SCN)_2] + 2KSCN$$

Procedure

Solutions of 1.03 g. (0.0025 mole) of $K_2[Pd(SCN)_4]$,[27] dissolved in 25 ml. of absolute ethanol and 5 drops of water, and 1.52 g. (0.0050 mole) of triphenylarsine,* dissolved in 25 ml. of absolute ethanol and 5 drops of ethyl ether, are precooled to 0°C. and then mixed in a beaker fitted with a magnetic stirrer and surrounded with an ice bath.   The solution is stirred for 1 minute, whereupon the yellow-orange S-bonded isomer is precipitated by the addition of 50 ml. of ice water, isolated by filtration, washed with 50 ml. each of absolute ethanol and ethyl ether, both precooled to 0°C., and dried *in vacuo* over P_2O_5.   Yield is 1.9 g. (91%).   After isomerizing, the compound melts at 195°C. with decomposition.   *Anal.* Calcd. for $[Pd\{As(C_6H_5)_3\}_2(SCN)_2]$: C, 54.66; H, 3.62; N, 3.36.   Found: C, 54.50; H, 3.80; N, 3.27.

B. DITHIOCYANATO-N-BIS(TRIPHENYLARSINE)PALLADIUM(II)

$$K_2[Pd(SCN)_4] + 2As(C_6H_5)_3 \xrightarrow{25°C.} [Pd\{As(C_6H_5)_3\}_2(NCS)_2] + 2KSCN$$

Procedure

The S-bonded isomer slowly isomerizes to the bright yellow N-bonded isomer in the solid state at room temperature.   The

---

* The toxicity of arsenic should be recalled.

rate of isomerization may be accelerated by raising the temperature; however, this also has the effect of increasing the rate of decomposition of the sample.

A pure sample of the N-bonded isomer can be prepared by mixing 2.1 g. (0.005 mole) of $K_2[Pd(SCN)_4]$,[27] dissolved in 50 ml. of absolute ethanol and 10 ml. of distilled water, with 3.1 g. (0.010 mole) of triphenylarsine, dissolved in 50 ml. of boiling absolute ethanol. The solution is stirred for 6 hours without further heating and then poured into 200 ml. of ice water. The resulting solid is isolated by filtration and washed with 50 ml. each of distilled water, absolute ethanol, and ethyl ether. It is finally stirred in 50 ml. of ethyl ether for 30 minutes to remove the last traces of any unreacted triphenylarsine and/or S-bonded isomer, isolated by filtration, and dried *in vacuo*. Yield is 3.4 g. (74%); m.p. 195°C. (decomp.). *Anal.* Calcd. for $[Pd(As(C_6H_5)_3)_2(NCS)_2]$: C, 54.66; H, 3.62; N, 3.36. Found: C, 54.74; H, 3.63; N, 3.45.

C. DITHIOCYANATO-S-(2,2'-BIPYRIDINE)PALLADIUM(II)

$$K_2[Pd(SCN)_4] + C_{10}H_8N_2 \xrightarrow{-78°C.} [Pd(C_{10}H_8N_2)(SCN)_2] + 2KSCN$$

Procedure

Solutions of 0.42 g. (0.0010 mole) of $K_2[Pd(SCN)_4]$,[27] dissolved in 10 ml. of absolute ethanol and 5 drops of water, and 0.16 g. (0.0010 mole) of 2,2'-bipyridine, dissolved in 10 ml. of absolute ethanol and 5 drops of ethyl ether, are precooled to -78°C. in a Dry Ice–acetone bath and then mixed in a beaker fitted with a magnetic stirrer and surrounded by a Dry Ice–acetone bath. The solution is stirred for 1 minute, then removed from the bath, and allowed to stand for 20 minutes at room temperature. The unidentified orange solid which precipitates is removed by filtration, whereupon the desired orange-yellow S-bonded isomer immediately precipitates in the filtrate. This is isolated, washed with 50 ml. each of absolute ethanol and ethyl ether, both pre-cooled to $-78°C$., and dried *in vacuo* over magnesium perchlorate. Yield is 0.081 g. (21%). After isomerizing, the compound melts at 270°C. with decomposition. *Anal.* Calcd. for $[Pd(C_{10}H_8N_2)(SCN)_2]$: C, 38.04; H, 2.13; N, 14.79. Found: C, 38.17; H, 2.48; N, 14.95.

D. DITHIOCYANATO-*N*-(2,2′-BIPYRIDINE)PALLADIUM(II)

$$[Pd(C_{10}H_8N_2)(SCN)_2] \xrightarrow{156°C.} [Pd(C_{10}H_8N_2)(NCS)_2]$$

or

$$K_2[Pd(SCN)_4] + C_{10}H_8N_2 \xrightarrow{25°C.} [Pd(C_{10}H_8N_2)(NCS)_2] + 2KSCN$$

Procedure

The *S*-bonded isomer can be quantitatively converted to the light yellow *N*-bonded isomer by heating the former *in vacuo* at 156°C. for 30 minutes. *Anal.* Found: C, 38.56; H, 2.43; N, 14.81.

Alternatively, the pure *N*-bonded isomer may also be prepared by mixing, at 25°C., solutions of 0.42 g. (0.0010 mole) of $K_2[Pd(SCN)_4]$,[27] dissolved in 5 ml. of water, and 0.16 g. (0.0010 mole) of 2,2′-bipyridine, dissolved in 5 ml. of absolute ethanol. The resulting mixture is stirred for 10 minutes, after which the product is isolated by filtration, washed with 50 ml. each of water, absolute ethanol, and ethyl ether, and dried *in vacuo* over magnesium perchlorate. Yield is 0.36 g. (94%); m.p. 270°C. (decomp.). *Anal.* Found: C, 38.01; H, 2.34; N, 14.86.

Properties

The linkage isomers may be conveniently distinguished from each other by comparing their infrared spectra. The *N*-bonded isomers (Secs. B and D) exhibit strong, broad C—N stretching bands at, respectively, 2089 and 2100 cm.$^{-1}$ and C—S stretching bands of medium intensity at, respectively, 854 and 842 (849 sh) cm.$^{-1}$.   The *S*-bonded isomers (Secs. A and C) exhibit sharp C—N stretching bands which are shifted to higher frequencies [2119 and 2117, 2108 (doublet) cm.$^{-1}$, respectively].   Their C—S stretching bands are shifted to lower frequencies.   That of Sec. A is hidden by phenyl ring absorption, whereas that of Sec. C is found at 700 cm.$^{-1}$ and is of very low intensity.

References

1. S. M. Jørgensen, *Z. Anorg. Chem.*, **5**, 169 (1894).
2. B. Adell, *ibid.*, **252**, 272 (1944); **279**, 219 (1955).
3. F. Basolo and G. S. Hammaker, *Inorg. Chem.*, **1**, 1 (1962).
4. R. G. Pearson, P. M. Henry, J. G. Bergmann, and F. Basolo, *J. Am. Chem. Soc.*, **76**, 5920 (1954); R. K. Murmann and H. Taube, *ibid.*, **78**, 4886 (1956).
5. A. Turco and C. Pecile, *Nature*, **191**, 66 (1961).
6. J. L. Burmeister and F. Basolo, *Inorg. Chem.*, **3**, 1587 (1964).
7. J. L. Burmeister, *Coord. Chem. Rev.*, **3**, 225 (1968).
8. A. Werner, *Ber.*, **40**, 765 (1907).
9. J. Halpern and S. Nakamura, *J. Am. Chem. Soc.*, **87**, 3002 (1965).
10. D. M. L. Goodgame and M. A. Hitchman, *Inorg. Chem.*, **5**, 1303 (1966).
11. A. Tramer, in "Theory and Structure of Complex Compounds," B. Jezowska-Trzebiatowska (ed.), p. 225, Pergamon Press, New York, 1964; O. W. Howarth, R. E. Richards, and L. M. Venanzi, *J. Chem. Soc.*, 3335 (1964).
12. M. F. Farona and A. Wojcicki, *Inorg. Chem.*, **4**, 857 (1965).
13. A. Sabatini and I. Bertini, *ibid.*, **4**, 1665 (1965).
14. H. H. Schmidtke, *Z. Physik. Chem. (Frankfurt)*, **45**, 305 (1965); *J. Am. Chem. Soc.*, **87**, 2522 (1965); *Inorg. Chem.*, **5**, 1682 (1966).
15. A. Haim and N. Sutin, *J. Am. Chem. Soc.*, **87**, 4210 (1965); **88**, 434 (1966).
16. F. Basolo, W. H. Baddley, and J. L. Burmeister, *Inorg. Chem.*, **3**, 1202 (1964); F. Basolo, W. H. Baddley, and K. J. Weidenbaum, *J. Am. Chem. Soc.*, **88**, 1576 (1966).
17. I. Bertini and A. Sabatini, *Inorg. Chem.*, **5**, 1025 (1966).
18. G. C. Kulasingam and W. R. McWhinnie, *J. Chem. Soc.*, A, 254 (1968).
19. T. E. Sloan and A. Wojcicki, *Inorg. Chem.*, **7**, 1268 (1968).
20. J. L. Burmeister and H. J. Gysling, *Inorg. Chim. Acta*, **1**, 100 (1967).
21. J. Chatt and F. A. Hart, *J. Chem. Soc.*, 1416 (1961).
22. D. E. Peters and R. T. M. Fraser, *J. Am. Chem. Soc.*, **87**, 2758 (1965).
23. D. F. Shriver, S. A. Shriver, and S. E. Anderson, *Inorg. Chem.*, **4**, 725 (1965).
24. J. P. Birk and J. H. Espenson, *J. Am. Chem. Soc.*, **90**, 1153 (1968).
25. K. Kuroda and P. S. Gentile, *Inorg. Nucl. Chem. Letters*, **3**, 151 (1967).
26. J. L. Burmeister, H. J. Gysling, and J. C. Lim, *J. Am. Chem. Soc.*, **91**, 44 (1969).
27. "Gmelin's Handbuch der Anorganischen Chemie," Vol. 65, p. 319, 1942.

38. HEXAHALOGENO SALTS AND ALKYL NITRILE COMPLEXES OF TITANIUM(IV), ZIRCONIUM(IV), NIOBIUM(V), TANTALUM(V), PROTACTINIUM(IV) AND -(V), THORIUM(IV), AND URANIUM(IV)

Submitted by D. BROWN,* G. W. A. FOWLES,† and R. A. WALTON‡
Checked by LOUIS CENTOFANTI§

In a recent review[1] of the halide and oxyhalide complexes of elements of the titanium, vanadium, and chromium subgroups, the preparative methods discussed involved the use of protonic solvents such as aqueous and ethanolic hydrohalic acids and halogen-containing solvents such as thionyl chloride and iodine monochloride. An alternative approach makes use of the alkyl nitrile complexes of the metal halides as starting materials. Such complexes of several transition[2-6] and actinide[5,7-10] metal halides have been thoroughly characterized and are, in many instances, rather more stable toward hydrolysis than the simple halides themselves. Reaction of these complexes in either acetonitrile or chloroform solution with alkylammonium or tetraphenylarsonium halides to yield the appropriate hexahalogeno salt is particularly advantageous in those cases where the desired hexahalo complexes cannot be prepared from aqueous solution owing to their ready hydrolysis.

Since the alkyl nitrile (normally acetonitrile) is itself generally used as solvent, it is not always essential for the complexes to be isolated. Solutions formed by many halides in the alkyl nitrile can be used directly for the preparation of the halide complexes. It is often more convenient to store the complexes than the halides themselves, however, and in some instances [Th(IV)] it is advantageous if the complexes are isolated. Thus

* Chemistry Division, U.K.A.E.A., Harwell, England.
† Chemistry Department, University of Reading, Reading, England.
‡ Department of Chemistry, Purdue University, Lafayette, Ind. 47907.
§ Department of Chemistry, University of Utah, Salt Lake City, Utah; now at Emory University, Atlanta, Ga. 30322.

the preparation[8] of $\{(C_6H_5)_4As\}_2[ThI_6]$ is complicated if an excess of iodine is present because tetraphenylarsonium periodide then crystallizes from acetonitrile as large black plates; the use of $ThI_4 \cdot 4CH_3CN$ rather than ThI_4 ensures the absence of excess iodine.

Because the halides and the alkyl nitrile complexes are prone to hydrolysis, either a dry-box or a vacuum-line procedure must be used.

A. $ThX_4 \cdot 4CH_3CN$, $PaX_4 \cdot 4CH_3CN$, AND $UX_4 \cdot 4CH_3CN$

(X = Cl and Br for Pa and U; X = Cl, Br, and I for Th)

$$MX_4 + 4CH_3CN \rightarrow MX_4 \cdot 4CH_3CN$$

Procedure

1. $ThI_4 \cdot 4CH_3CN$

Anhydrous acetonitrile* (5 ml.) is added to thorium(IV) iodide† (2 g.) and the reaction mixture maintained at room temperature for 1–2 hours in the dry-box. The insoluble yellow complex, $ThI_4 \cdot 4CH_3CN$, is isolated by filtration, washed with acetonitrile (3 × 3 ml.), and vacuum-dried at room temperature. *Anal.* Calcd. for $ThI_4 \cdot 4CH_3CN$: Th, 25.8; I, 56.0. Found: Th, 25.7; I, 56.2.

2. Other Complexes

The following complexes can be prepared by the same procedure, although washing the insoluble complex with acetonitrile is unnecessary: $ThCl_4 \cdot 4CH_3CN$ (white), $ThBr_4 \cdot 4CH_3CN$ (white), $PaCl_4 \cdot 4CH_3CN$ (yellow-green), $PaBr_4 \cdot 4CH_3CN$

---

* Solvents are rigorously dried by repeated distillation *in vacuo* from phosphorus-(V) oxide. The alkyl nitriles are generally predried by refluxing them over calcium hydride for several days, then distilling several times from phosphorus(V) oxide, and finally from potassium carbonate; the liquid is then stored either over phosphorus(V) oxide on the vacuum line or in contact with molecular sieves in the dry-box.

† ThX₄ compounds are available from Alfa Inorganics, Inc., P.O. Box 159, Beverly, Mass. 01915. Also see reference 11.

(orange), $UCl_4 \cdot 4CH_3CN$ (gray-green), and $UBr_4 \cdot 4CH_3CN$ (blue-gray).

B. $MX_5 \cdot CH_3CN$

$$MX_5 + CH_3CN \rightarrow MX_5 \cdot CH_3CN$$

$NbBr_5 \cdot CH_3CN$ and Related Compounds

Niobium(V) bromide* (2 g.) is dissolved in anhydrous acetonitrile (10 ml.) and excess solvent removed *in vacuo* at room temperature to leave the dark-red solid complex (yield 100%).

Niobium and tantalum(V) chlorides and tantalum(V) bromide† yield the complexes $MX_5 \cdot CH_3CN$, and protactinium(V) bromide yields orange $PaBr_5 \cdot 3CH_3CN$ if the same procedure is used.

C. $ZrX_4 \cdot 2RCN$

$$ZrX_4 + 2RCN \rightarrow ZrX_4 \cdot 2RCN$$

A vacuum-line procedure is recommended for this preparation. The vacuum line can also be used for the preparations previously described in Secs. A and B.

Figure 20 shows the standard vacuum line used. The dry solvents are stored in traps A and B. The reaction of the metal halide with the appropriate alkyl nitrile is carried out in an ampule (C) which consists of a 200-ml. Pyrex vessel with a standard-taper joint ($B.19$). The ampule is sealed to the vacuum line and thoroughly "flamed" *in vacuo* to remove traces of moisture.

Zirconium chloride (2 g.), previously stored under nitrogen in a tube fitted with a standard-taper joint, is added through joint D (against a countercurrent of nitrogen) to the reaction

---

* NbX_5 compounds are available from Alfa Inorganics, Inc., P.O. Box 159, Beverly, Mass. 01915.

† For $TaBr_5$ see *Inorganic Syntheses,* **4,** 130 (1953) and synthesis 32 of this volume.

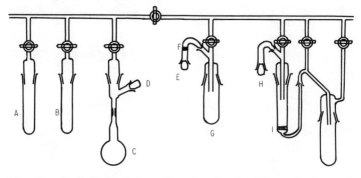

Fig. 20.   Standard vacuum line for the handling of air- and moisture-sensitive materials.

ampule.   The apparatus is then evacuated, and 10 ml. of anhydrous acetonitrile is distilled into ampule *C*, with liquid nitrogen or a Dry Ice–acetone slush bath used as coolant. The ampule is sealed at the constriction, above the extended joint, and removed from the line.   When the ampule and its contents are allowed to warm to room temperature and are shaken thoroughly, the zirconium(IV) chloride dissolves to give a colorless solution.   This solution is stored (almost indefinitely if necessary) until required.

The solid complex $ZrCl_4 \cdot 2CH_3CN$ is isolated as follows: The ampule is cooled to about 0°C., opened under nitrogen, and attached to the filtration section of the vacuum line at joint *E*. After the apparatus has been evacuated, the ampule and its contents are allowed to warm to room temperature, the colorless solution is filtered through the filter disk *F* into the trap *G*, and the excess acetonitrile is evaporated from the filtrate to leave the white crystalline complex, which is exposed to the vacuum pump for several hours.   (Yield 100%.)

With the procedure described above, the complexes $ZrCl_4 \cdot 2RCN$ (white), $ZrBr_4 \cdot 2RCN$ (off white), $NbCl_5 \cdot RCN$ (yellow), $NbBr_5 \cdot RCN$ (dark red), $TaCl_5 \cdot RCN$ (white), and $TaBr_5 \cdot RCN$ (yellow), where $R = CH_3$, C_2H_5, or C_3H_7, can be readily prepared.   If partial crystallization of the complex should occur prior to filtration, acetonitrile (or the appropriate nitrile) is distilled into the ampule after the solution has been filtered, and the crystalline complex is dissolved by washing while on the filter disk F.

The complexes $TiCl_4 \cdot 2RCN$ and $TiBr_4 \cdot 2RCN$, where R = CH_3, C_2H_5, or C_3H_7, require a slight modification of the above procedure. Titanium(IV) chloride (2 g.), being a liquid at room temperature, is distilled into the ampule C, with liquid nitrogen, used as coolant. Titanium(IV) bromide, which is stored in the usual type of tube fitted with a standard-taper joint, has to be melted (by gentle heating with a Bunsen flame) while in the storage tube which is attached to the vacuum line through joint D, and it is then distilled into C in the usual manner. The procedures are otherwise identical.

D. HEXAHALOGENO SALTS $[R_4N]_nMX_6$

(M = Nb, Ta, Pa, Ti, Zr, Th, and U; X = Cl and Br)

$$MX_4 \cdot nRCN + 2R_4NX \rightarrow [R_4N]_2MX_6 + nRCN$$
$$MX_5 \cdot RCN + R_4NX \rightarrow [R_4N]MX_6 + RCN$$

1. Tetraethylammonium Hexabromotantalate(V)

Acetonitrile (10 ml.) is added to $TaBr_5 \cdot CH_3CN$ (1.7 g.) and tetraethylammonium bromide (0.5 g.) in a dry-box. The stoppered reaction flask is removed from the dry-box and heated to dissolve the reactants. When the solution is cooled with ice, orange crystals of $[N(C_2H_5)_4][TaBr_6]$ deposit. The reaction flask is transferred to the dry-box, and the product is isolated by centrifugation and vacuum-dried at room temperature (yield 80%). *Anal.* Calcd. for $[(C_2H_5)_4N]TaBr_6$: Ta, 22.8; Br, 60.6%. Found: Ta, 22.7; Br, 60.3%.

2. Tetraphenylarsonium Hexaiodothorate(IV)

Procedures similar to those described in subsection 1 above are used. Starting materials are 1 g. of $ThI_4 \cdot 4CH_3CN$ and 1.15 g.

of $(C_6H_5)_4AsI$. Alternatively, the reaction may be carried out by using an excess of tetraphenylarsonium iodide; in this case the resulting pale yellow crystals are washed with ice-cold acetonitrile (3×5 ml.) prior to vacuum drying (yield 75%). *Anal.* Calcd. for $[(C_6H_5)_4As]_2ThI_6$: Th, 13.2; I, 43.3%. Found: Th, 13.35; I, 43.1%.

By an identical procedure, $[(C_2H_5)_4N]NbBr_6$ (dark red), $[(CH_3)_4N]_2ThCl_6$ (white), $[(C_2H_5)_4N]_2ThCl_6$ (white), $[(CH_3)_4N]_2$-$ThBr_6$ (white), $[(CH_3)_4N]PaCl_6$ (yellow), $[(C_2H_5)_4N]PaBr_6$ (orange), $[(CH_3)_4N]_2UCl_6$ (green), $[(C_2H_5)_4N]_2UCl_6$ (pale green), and $[(C_2H_5)_4N]_2UBr_6$ (pale blue-green) can be prepared.

A pure acetonitrile complex of uranium(IV) iodide cannot be isolated,[8] but the salt $[(C_6H_5)_4As]_2UI_6$ (red) can be readily prepared by the procedure described above in subsection 1, the pure halide being used instead of the acetonitrile complex. $[(C_2H_5)_4N]PaBr_6$ is extremely soluble in acetonitrile but it can be isolated in 100% yield by room-temperature evaporation of the solvent containing equimolar amounts of the component bromides.

3. Diethylammonium Hexachlorotitanate(IV), $[(C_2H_5)_2NH_2]_2TiCl_6$

The vacuum apparatus is convenient for this preparation. $TiCl_4 \cdot 2C_2H_5CN$ (3 g.) and diethylammonium chloride (2.2 g.) are added to an ampule through the sidearm D (Fig. 20) against a countercurrent of nitrogen in the usual way. A 30–40-ml. quantity of chloroform is distilled into the evacuated ampule which is then sealed at the constriction. The ampule and contents are shaken for several hours at room temperature to ensure complete precipitation of $[(C_2H_5)_2NH_2]_2TiCl_6$ which is insoluble in chloroform. The ampule is cooled, opened under nitrogen, and attached to the filtration section of the vacuum line at joint H. The precipitated salt is filtered off on the pad I, washed at least three times with dry chloroform to remove any unreacted materials, and finally exposed to the vacuum pumps for several hours to remove traces of solvent (yield 90%). *Anal.* Calcd. for $[(C_2H_5)_2NH_2]_2TiCl_6$: Ti, 11.7; Cl, 52.0. Found: Ti, 11.3; Cl, 51.2%. Other salts of the type $[(C_2H_5)_2NH_2]_2MX_6$ (M = Ti, Zr; X = Cl and Br) can be prepared from the propionitrile complexes by the same procedure.

Properties of Hexahalogeno Salts

Most of the salts are hydrolyzed in moist air but are stable indefinitely in a dry nitrogen atmosphere. A notable exception is tetraethylammonium hexabromotantalate(V), which is quite stable in the atmosphere and is only slowly hydrolyzed by concentrated aqueous ammonia. Complete hydrolysis is more readily achieved by the addition of acetone followed by aqueous ammonia, the mixture being gently warmed on a water bath. They are generally insoluble in nonpolar or slightly polar solvents but are fairly soluble in more polar solvents such as acetonitrile (methyl cyanide); in the latter solvent the solutions show the expected molar conductivities.

The salts of tetravalent titanium, zirconium, and thorium and of pentavalent niobium and tantalum are diamagnetic, as would be expected, and have colors ranging from white to orange, depending upon the halogen.

References

1. G. W. A. Fowles, in "Preparative Inorganic Reactions," W. L. Jolly (ed.) Vol. 1, p. 121, Interscience Publishers, a division of John Wiley & Sons, Inc. New York, 1964.
2. E. A. Allen, B. J. Brisdon, and G. W. A. Fowles, *J. Chem. Soc.*, 4531 (1964).
3. G. W. A. Fowles and R. A. Walton, *ibid.*, 2840 (1964).
4. K. Feenan and G. W. A. Fowles, *ibid.*, 2842 (1964).
5. D. Brown and P. J. Jones, *ibid.*, **A**, 247 (1967).
6. M. W. Duckworth, G. W. A. Fowles, and R. A. Hoodless, *ibid.*, 5665 (1963).
7. D. Brown, *ibid.*, **A**, 766 (1966).
8. K. W. Bagnall, D. Brown, P. J. Jones, and J. G. H. du Preez, *ibid.*, 350 (1965).
9. K. W. Bagnall, D. Brown, and P. J. Jones, *ibid.*, **A**, 1763 (1966).

10. D. Brown and P. J. Jones, *Chem. Commun.*, 280 (1966).
11. J. S. Anderson and R. W. M. D'Eye, *J. Chem. Soc.*, 5244 (1949).
12. A. Cowley, F. Fairbrother, and N. Scott, *ibid.*, 3133 (1958).

39. POTASSIUM TRICHLOROTRIFLUORO-PLATINATE(IV) AND POTASSIUM HEXAFLUOROPLATINATE(IV)

Submitted by K. R. DIXON,* D. W. A. SHARP,† and A. G. SHARPE‡
Checked by N. NGHI§ and N. BARTLETT§

Potassium hexafluoroplatinate(IV) was first prepared by the action of the complex fluoride $3KF \cdot HF \cdot PbF_4$ on platinum metal.[1] However, separation of the products of that interaction is not simple, and the salt is more conveniently obtained by the fluorination of potassium hexachloroplatinate(IV), with bromine trifluoride as both solvent and fluorinating agent.[2] This synthesis has been shown to involve the intermediate trichlorotrifluoroplatinate(IV) ion[3] as an isolable species, and the following procedure gives details of the isolation of both the intermediate and fully fluorinated products. Analogous procedures may be used for most preparations of simple and complex fluorides which involve bromine trifluoride as the fluorinating agent. Many examples of the importance and versatility of this reagent have been represented in reviews.[4] In most cases, preparations may be carried out in a manner similar to that described below for potassium hexafluoroplatinate(IV) except that dry-box techniques are often required in the handling of the products.

* Department of Chemistry, University of Western Ontario, London, Ontario, Canada; current address: University of Victoria, Victoria, British Columbia, Canada.
† Department of Pure and Applied Chemistry, University of Strathclyde, Glascow, C.I., Scotland.
‡ Chemical Laboratories, Lensfield Road, Cambridge, England.
§ Frick Chemical Laboratory, Princeton University, Princeton, N.J.; current address of N. Bartlett: Department of Chemistry, University of California, Berkeley, Calif. 94720.

The two syntheses described below each require 2–3 hours on the first day and about an hour on the second day, the vacuum pumping part of each being carried out overnight.

General Procedure

■ *Caution. Bromine trifluoride is an extremely corrosive liquid and ignites immediately and often explosively with most organic materials. The use of a fume hood for initial stages of the preparations is essential (i.e., up to transfer to the vacuum line), and if possible all operations should be carried out in a fume hood. The wearing of asbestos gloves, face mask, and polyvinyl chloride apron at all times is essential. Excess reagent should be destroyed (within a fume hood) by slow addition to a large volume of carbon tetrachloride, the ensuing reaction being vigorous but not dangerous. Very small quantities (e.g., residues on a glass rod used in pouring) may be destroyed by washing with carbon tetrachloride.*

Scale. ■ *Increasing the scale of these preparations is potentially hazardous because of the vigor of the reactions.* In the case of the trichlorotrifluoroplatinate(IV), increase of scale is not feasible since it becomes difficult to combine a short reaction time with effective temperature control. Larger quantities of the hexafluoroplatinate(IV) may be prepared provided the addition of bromine trifluoride is carried out very slowly.

Apparatus. The preparations may be carried out in quartz or Teflon vessels. A bottle of ~30-ml. capacity and internal diameter ~2.5 cm., provided with a standard-taper quartz joint, for attachment to a vacuum system is satisfactory. A small copper, nickel, Monel, or Kel-F funnel is needed to facilitate addition of liquid bromine trifluoride. Excess bromine trifluoride removed on the vacuum line must be collected in a quartz or Kel-F trap. ■ *Silicone or Kel-F vacuum grease must be employed in those parts of the vacuum system likely to come into contact with bromine trifluoride vapor.* Vacuum grease should be applied to the cone of the quartz bottle only *after* bromine trifluoride has been added. It is desirable that the vacuum path from reaction vessel to trap be free of stopcocks.

Reagents. Chloroplatinic acid may be purchased from Johnson Matthey and Mallory, Ltd. (Toronto or London), Freon 113 from E. I. du Pont de Nemours & Company, Wilmington, Del., and bromine trifluoride from the Matheson Co., East Rutherford, N.J., or may be prepared as previously described[5] and stored in stainless-steel or Monel-metal bottles. Sufficient potassium hexachloroplatinate(IV) for use in the following procedures is prepared by the dropwise addition of excess potassium chloride (0.8 g. in 5 ml. of water) to a solution of chloroplatinic acid, $H_2PtCl_6 \cdot 6H_2O$ (2 g. in 5 ml. of water). The yellow precipitate of K_2PtCl_6 is collected, washed with water and acetone, and dried under vacuum at 20°C. It is important that the potassium hexachloroplatinate(IV) be prepared in this way and not recrystallized, since the resulting small particle size is essential for smooth dissolution in bromine trifluoride.

A. POTASSIUM TRICHLOROTRIFLUOROPLATINATE(IV)

$$K_2PtCl_6 + BrF_3 \xrightarrow{Cl_2FCCClF_2} K_2PtCl_3F_3 + \text{other products}$$

Procedure

To 0.5 g. of potassium hexachloroplatinate(IV) in a quartz or Kel-F bottle, about 5 ml. of Freon 113 (CCl_2F—$CClF_2$) is added to act as moderator. Excess bromine trifluoride (1–2 ml.) is then added dropwise in the shortest possible time (a quartz or Kel-F rod may be used to assist pouring) until solution of the chloroplatinate is complete (bromine trifluoride and Freon 113 are immiscible). It is the aim of this procedure to combine minimum reaction time with minimum local heating of the mixture. It is usually necessary to shake the bottle (■ *care should be taken to avoid spills*) to aid the solution process. As soon as dissolution is complete, the mixture is rapidly frozen in liquid nitrogen and the bottle joined to a vacuum line. Freon and excess bromine trifluoride are removed by vacuum pumping at 20°C. for 18 hours. The resulting yellow powder still contains traces of bromine trifluoride (■ *caution*) which are removed by pouring the solid slowly into 20 ml. of acetone. (*Note:* addition of acetone to the solid usually results in local heating and charring.) The solid product is collected on a sintered-glass filter, washed with acetone, and dried under vacuum at 20°C. The yield is quantitative. *Anal.* Calcd. for $K_2PtCl_3F_3$: K, 17.9; Pt, 44.6; Cl, 24.4; F, 13.0. Found: K, 17.7; Pt, 44.8; Cl, 23.3; F, 12.3.

Properties

Potassium trichlorotrifluoroplatinate(IV) may be identified by its x-ray powder photograph and infrared* and ultraviolet spectra.[3] Possible impurities are potassium hexafluorosilicate† and hexafluoroplatinate(IV) which may be detected by their infrared spectra [see subsection on properties of potassium hexafluoroplatinate(IV) below]. Any unreacted potassium hexachloroplatinate(IV) may be detected by its very intense ($\epsilon = 24,500$) ultraviolet absorption at 38,200 cm.$^{-1}$ in aqueous solution.[3] The product cannot be purified by recrystallization from water, since it disproportionates in aqueous solution. A

---

* The x-ray powder photograph of $K_2PtCl_3F_3$ obtained by the checkers who used Kel-F vessels for their preparations showed no lines due to K_2PtF_6, K_2PtCl_6, or K_2SiF_6. The strong line pattern was indexable on a face-centered cubic cell $a = 9.23$ A. Since a number of weaker lines were not indexable on this or related cells, it seems probable that the solid is a mixture of two crystalline forms, the cubic phase being dominant. The infrared spectrum of $K_2PtCl_3F_3$ which showed traces of K_2PtF_6 to be present had bands at 570(m, broad), 517(vs), 378(s), and 260(w) cm.$^{-1}$, in fair agreement with the published spectrum.[3]

† The checkers did not detect hexafluorosilicate in either preparation and found infrared, visible, and ultraviolet spectra in close agreement with those reported for K_2PtF_6 and $K_2PtCl_3F_3$.

similar change of the solid probably occurs on prolonged storage. The salt is moderately soluble (0.1 M) in cold water, and the solution is stable for short periods.[3]

B. POTASSIUM HEXAFLUOROPLATINATE(IV)

$$K_2PtCl_6 + BrF_3 \xrightarrow{Br_2} K_2PtF_6 + \text{other products}$$

Procedure

To 1 g. of potassium hexachloroplatinate(IV) in the quartz or Kel-F bottle about 5 ml. of bromine is added to act as moderator. Excess bromine trifluoride (3–4 ml.) is then added dropwise, and the resulting red solution is boiled until the bromine is expelled (3–5 minutes), leaving a clear yellow solution. This is then frozen in liquid nitrogen and transferred to a vacuum line where excess bromine trifluoride is removed by pumping, initially at 20°C. and later for half an hour at 120°C. to remove the last traces of BrF_3. The product remains as a pale yellow powder (yield 0.8 g., 100%). It may be recrystallized from water, but it is best to extract with warm water, filter, and precipitate the salt from solution by addition of acetone. *Anal.* Calcd. for K_2PtF_6: K, 20.1; Pt, 50.4; F, 29.5. Found: K, 19.8; Pt, 50.0; F, 29.1.

Properties

Potassium hexafluoroplatinate(IV) may be identified by its x-ray powder diffraction pattern[6] and infrared spectrum[7] which in a Nujol mull has bands as follows: 583(vs), 282(s), 259(ms) cm.$^{-1}$. The crude product before recrystallization is sufficiently pure for most purposes but preparations carried out in quartz usually contain some potassium hexafluorosilicate, which may be detected by its very intense infrared absorptions at 741 and 483 cm.$^{-1}$.[8]

References

1. H. I. Schlesinger and M. W. Tapley, *J. Am. Chem. Soc.*, **46**, 276 (1924).
2. A. G. Sharpe, *J. Chem. Soc.*, 3444 (1950).
3. D. H. Brown, K. R. Dixon, and D. W. A. Sharp, *ibid.*, A, 1244 (1966).
4. R. D. Peacock, *Progr. Inorg. Chem.*, **2**, 193 (1960); A. G. Sharpe, *Advan. Fluorine Chem.*, **1**, 29 (1960); N. Bartlett in "Preparative Inorganic Reactions," W. L. Jolly (ed.), Vol. 2, p. 301, Interscience Publishers, a division of John Wiley & Sons, Inc., New York, 1965; A. G. Sharpe, in "Non-aqueous Solvent Systems," T. C. Waddington (ed.), Academic Press, Inc., New York, 1965.
5. J. H. Simons, *Inorganic Syntheses*, **3**, 184 (1950).
6. B. Cox and A. G. Sharpe, *J. Chem. Soc.*, 1783 (1953).
7. L. A. Woodward and M. J. Ware, *Spectrochim. Acta*, **19**, 775 (1963); R. D. Peacock and D. W. A. Sharp, *J. Chem. Soc.*, 2762 (1959); K. R. Dixon, Ph.D. thesis, Strathclyde University, 1966.
8. D. H. Brown, K. R. Dixon, C. M. Livingston, R. H. Nuttall, and D. W. A. Sharp, *J. Chem. Soc.*, A, 100 (1967).

40. TETRAKIS(TRIPHENYLPHOSPHINE)DICHLORO-RUTHENIUM(II) AND TRIS(TRIPHENYLPHOSPHINE)-DICHLORORUTHENIUM(II)

Submitted by P. S. HALLMAN,* T. A. STEPHENSON,* and G. WILKINSON*

The interaction of triphenylphosphine with methanolic solutions of commercial hydrated ruthenium trichloride† leads to mono-nuclear complexes $[RuCl_2\{P(C_6H_5)_3\}_4]$ and $[RuCl_2\{P(C_6H_5)_3\}_3]$, the product depending upon the reaction conditions.[1] However, similar reactions, involving mixed alkyl aryl tertiary phosphines, yield binuclear complexes of the type $[Ru_2Cl_3(PR_3)_6]Cl$ [R = $(C_2H_5)_2(C_6H_5)$, $(C_2H_5)(C_6H_5)_2$, etc.].[2]

* Imperial College of Science and Technology, London, S.W. 7, England.
† Johnson Matthey Ltd., London, supply "ruthenium trichloride trihydrate" with *ca.* 44.6% Ru; other suppliers' products are similar. These materials are mainly ruthenium(IV) complexes. On repeated evaporation almost to dryness with concentrated hydrochloric acid, a solution containing the ruthenium(III) complex ion $RuCl_6{}^{3-}$ is obtained. For the present purpose, the commercial material may be used; identical complexes are obtained by using the Ru(III) solutions, however.

A. TETRAKIS(TRIPHENYLPHOSPHINE)- DICHLORORUTHENIUM(II)

$$RuCl_3 \cdot 3H_2O + n(C_6H_5)_3P \xrightarrow[\substack{N_2 \text{ atm.} \\ 24 \text{ hours} \\ \text{in} \\ \text{methanol} \\ \text{solution}}]{25°C.} [RuCl_2\{P(C_6H_5)_3\}_4] + \text{other products}$$

Checked by RICHARD SINGLER* and ROBERT D. FELTHAM*

Procedure

Ruthenium trichloride trihydrate (0.6 g., 2.2 mmoles) is dissolved in methanol (150 ml.), filtered, and then refluxed for 5 minutes under nitrogen. The solution is allowed to cool in this inert atmosphere, and a sixfold excess (3.6 g., 13.7 mmoles) of triphenylphosphine is added. The solution, which becomes deep brown, is shaken at *ca.* 25°C. under nitrogen for 24 hours. The dark brown crystals which separate out are collected under nitrogen, washed with degassed methanol and diethyl ether, and dried under vacuum. The yield is approximately 1.85 g. (70% based on ruthenium); m.p. 130–132°C. *Anal.* Calcd. for $C_{72}H_{60}Cl_2P_4Ru$: C, 70.8%; H, 5.0%; Cl, 5.8%; P, 10.1%. Found: C, 70.0%; H, 5.2%; Cl, 6.2%; P, 10.5%.

B. TRIS(TRIPHENYLPHOSPHINE)DICHLORORUTHENIUM(II)

$$RuCl_3 \cdot 3H_2O + n(C_6H_5)_3P \xrightarrow[\substack{65°C., \ 3 \text{ hours } N_2 \text{ atm.}}]{\text{methanol soln.}}$$
$$[RuCl_2\{P(C_6H_5)_3\}_3] + \text{other products}$$

Checked by RICHARD HOLM†

Procedure

Ruthenium trichloride trihydrate (1.0 g., 3.8 mmoles) is dissolved in methanol (250 ·ml.) and the solution refluxed under

---

* Department of Chemistry, University of Arizona, Tucson, Ariz. 85721.

† Department of Chemistry, Massachusetts Institute of Technology, Cambridge, Mass. 02139.

nitrogen for 5 minutes. After cooling, triphenylphosphine (6.0 g., 22.9 mmoles) is added in the ratio of 6 moles of $(C_6H_5)_3P$ per mole of $RuCl_3 \cdot 3H_2O$, and the solution is again refluxed under nitrogen for 3 hours. The complex precipitates from the hot solution as shiny black crystals; on cooling, they are filtered under nitrogen, washed several times with degassed ether, and dried under vacuum. The yield is *ca.* 2.7 g. (74% based on ruthenium); m.p. 132–134°C. *Anal.* Calcd. for $C_{56}H_{45}Cl_2P_3Ru$: C, 67.6%; H, 4.7%; Cl, 7.4%; P, 10.3%; Ru, 10.6%. Found: C, 67.9%; H, 4.9%; Cl, 7.1%; P, 10.4%; Ru, 9.9%.

Properties

The tetrakis- and tris(triphenylphosphine)ruthenium(II) complexes are moderately soluble in warm chloroform, acetone, benzene, and ethyl acetate to give yellow-brown solutions. These solutions are air-sensitive, becoming green. Molecular-weight determinations[1] give low values indicating dissociation. The x-ray crystal structure of $[RuCl_2\{P(C_6H_5)_3\}_3]$[3] indicates a distorted octahedral structure with a vacant site which is occupied by an α-hydrogen atom of one of the phenyl rings of a phosphine ligand.

Solutions of the complexes behave similarly. Thus they react with carbon monoxide[1] at room temperature and pressure, and with norbornadiene[4] to give, respectively, *trans*-[RuCl$_2$-(CO)$_2\{P(C_6H_5)_3\}_2$] and [RuCl$_2$(C$_7$H$_8$)$\{P(C_6H_5)_3\}_2$]. The tris complex in ethanol–benzene has been shown[5] to be a homogeneous hydrogenation catalyst for reduction of 1-alkenes; however, the active catalytic species is chlorohydridotris(triphenylphosphine)ruthenium(III), formed by hydrogenolysis.[5, 6]

References

1. T. A. Stephenson and G. Wilkinson, *J. Inorg. Nucl. Chem.*, **28**, 945 (1966).
2. J. Chatt and R. G. Hayter, *J. Chem. Soc.*, 896 (1961).

3. S. J. LaPlaca and J. A. Ibers, *Inorg. Chem.*, **4**, 778 (1965).
4. S. D. Robinson and G. Wilkinson, *J. Chem. Soc.*, **A**, 300 (1966).
5. D. Evans, J. A. Osborn, F. H. Jardine, and G. Wilkinson, *Nature*, **208**, 1203 (1965).
6. P. S. Hallman, D. Evans, J. A. Osborn, and G. Wilkinson, *Chem. Commun.*, 305 (1967); P. S. Hallman, B. R. McGarvey, and G. Wilkinson, *J. Chem. Soc.*, **A**, 3143 (1968).

41. DI-μ-CHLORO-1,3-DICHLORO-2,4-BIS(CYCLO-HEXYLDIPHENYLPHOSPHINE)DIPLATINUM(II) AND RELATED COMPOUNDS

$$\textit{cis- and/or trans}[\{(C_6H_{11})(C_6H_5)_2P\}_2PtCl_2] + PtCl_2 \xrightarrow{C_2H_2Cl_4} [\{(C_6H_{11})(C_6H_5)_2P\}_2Pt_2Cl_4]$$

Submitted by A. C. SMITHIES,* P. SCHMIDT,* and MILTON ORCHIN*
Checked by LOUIS F. CENTOFANTI†

Binuclear compounds of platinum(II) of the type $[Pt_2Cl_4L_2]$, where L is a tertiary phosphine, are useful intermediates for the preparation of complexes containing mixed ligands. Heretofore the most useful method[1] for the preparation of these complexes consisted of melting together the mononuclear complex $[PtCl_2L_2]$ and platinum(II) chloride. This method is unsuccessful, however, if either the starting material or the product decomposes before melting. These difficulties are reported[2] to be minimized by employing a slurry of the reactants in a hydrocarbon solvent consisting of varying proportions of xylene and naphthalene, depending on the desired reflux tem-

* University of Cincinnati, Cincinnati, Ohio. Financial support of the National Science Foundation and the Petroleum Research Fund administered by the American Chemical Society is gratefully acknowledged. We also wish to thank Engelhard Industries, Inc., for a generous supply of potassium tetrachloroplatinate-(II).
† Department of Chemistry, University of Utah, Salt Lake City, Utah 84112; current address: Department of Chemistry, Emory University, Atlanta, Ga. 30322.

perature. The hydrocarbon solvents are then removed by light ligroin extraction.

In the following procedure for the preparation of binuclear platinum(II) complexes, tetrachloroethane is used as the solvent, thus eliminating the extraction process and obviating the necessity of adjusting the proportions of the hydrocarbons.

The method, although general, is illustrated by the preparation of di-μ-dichloro-1,3-dichloro-2,4-bis(cyclohexyldiphenyl-phosphine)diplatinum(II).* The product has the structure:

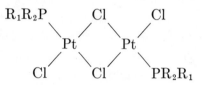

■ *Caution. Tetrachloroethane is a toxic liquid; contact with the skin and clothing as well as inhalation should be avoided.*

Procedure

To a pale yellow solution of 0.739 g. (0.92 mmole) of a mixture of *cis-* and *trans*-dichlorobis(cyclohexyldiphenylphosphine)-platinum(II)† in 25 ml. of tetrachloroethane (technical grade) is added 0.292 g. (1.10 mmoles) of platinum(II) chloride. The mixture is gently refluxed (b.p. 146°C.) with continuous stirring for 60 minutes under an atmosphere of nitrogen. The mixture

---

* The checker substituted triphenylphosphine for cyclohexyldiphenylphosphine in this preparation.   Yield: 70%.

† The mixture of cis and trans complex was prepared by a method analogous to that used by Jensen[3] to obtain dichloro-bis(triphenylphosphine)platinum(II). If the starting phosphine is appreciably water-soluble, the Jensen procedure as modified by Kauffman and Teeter[4] may be used. [See *Inorganic Syntheses*, **7**, 245 (1963).]   Our phosphine, m.p. 62–63°C., from ethanol, is readily prepared according to the standard procedure[5] by reacting chlorodiphenylphosphine (available commercially from Stauffer Chemical Co.) with cyclohexylmagnesium bromide.   The particular sample used in the preparation was a gift from Dr. J. Feldman, U.S.I. Division of National Distillers, Inc.

is then filtered to remove unreacted platinum(II) chloride, yielding a light orange filtrate. (A dark coloration of this filtrate, which may appear with other phosphine complexes, can be removed by treatment with decolorizing carbon.) About 20 ml. of the tetrachloroethane is then removed by evaporation under reduced pressure. The slow addition of 20–25 ml. of pentane to the vigorously stirred solution induces precipitation of the product. The solution is filtered after cooling to 0°C. The product is then purified by dissolving it in 20 ml. of chloroform, filtering, removing approximately 15 ml. of the chloroform by evaporation under reduced pressure, slowly adding 20 ml. of pentane, cooling the solution, and filtering the precipitated product. The yield is 0.79 g. [80% based on dichlorobis-(cyclohexyldiphenylphosphine)platinum(II)]. *Anal.* Calcd. for $[(C_6H_{11})(C_6H_5)_2P]_2Pt_2Cl_4$: C, 40.46; H, 3.96. Found: C, 40.17; H, 4.02%. Melting point, 248–251°C. (decomp.).

This procedure has also been utilized for the preparation of the dimeric complexes of this type with the following tertiary phosphines: tri-*n*-butylphosphine, dimethylphenylphosphine, dicyclohexylphenylphosphine, and triphenylphosphine. Satisfactory yields ranging from 50 to 80% were obtained in all cases.

Properties

The compound is a yellow-orange powder soluble in chloroform, methylene chloride, and dimethylformamide, and insoluble in practically all other common solvents.

References

1. J. Chatt and L. M. Venanzi, *J. Chem. Soc.*, 2787 (1955).
2. R. J. Goodfellow and L. M. Venanzi, *ibid.*, 7533 (1965).
3. K. A. Jensen, *Z. Anorg. Allgem. Chem.*, **299**, 225 (1936).
4. G. B. Kauffman and L. A. Teeter, *Inorganic Syntheses*, **7**, 245 (1963).
5. C. Steube, W. M. LeSuer, and G. R. Norman, *J. Am. Chem. Soc.*, **77**, 3526 (1955).

42. PENTAAMMINEIRIDIUM(III) COMPLEXES

Submitted by HANS-HERBERT SCHMIDTKE*
Checked by C. W. BRADFORD† and M. J. CLEARE†

A. PENTAAMMINECHLOROIRIDIUM(III) CHLORIDE

$$IrCl_6{}^{3-} + 5NH_3 \rightarrow [Ir(NH_3)_5Cl]^{2+} + 5Cl^-$$

This compound was first obtained by C. Claus[1] as a light pink crystalline powder; later it was described by W. Palmaer[2] as pale yellow octahedral crystals. Palmaer's preparation starts from an iridium(III) chloride solution which is treated with concentrated aqueous ammonia. By carrying out the reaction under pressure in a sealed tube, the yield is substantially increased (up to 70%).

An easier procedure, not requiring the use of a sealed tube, was described by Basolo and Hammaker.[3] A modification of their procedure is given here. The time required for the synthesis is about 8 hours.

Procedure

A solution of $K_3[IrCl_6] \cdot 3H_2O$‡ (3 g.) in 40 ml. of water is placed in a 150-ml. flask equipped with a standard-taper joint and a reflux condenser. To this solution is added NH_4Cl (20 g.) and $(NH_4)_2CO_3$ (30 g.). The system is then refluxed (steam bath). During initial heating, CO_2 is evolved, and the color changes gradually (about first 30 minutes) from greenish-black to light yellow after CO_2 evolution. It is desirable to complete the reaction under slight ammonia pressure in order to maximize the yield of pentaammine. Toward this end the condenser is fastened to a suitably adjusted mercury valve so that a pressure some 50–100 torr above atmospheric can be maintained in the flask. The system is then refluxed under pressure for an

---

* Institut für physikalische Chemie der Universität Frankfurt/Main, Germany.
† Johnson-Matthey Research Laboratories, Wembley, Mdx., England.
‡ $K_3[IrCl_6] \cdot 3H_2O$ is obtainable from Johnson Matthey Ltd., Wembley, Mdx., England, or from Alfa Inorganics, Inc., P.O. Box 159, Beverly, Mass. 01915.

additional 5-5½ hours. The brown-yellow precipitate which is slowly formed during the reaction is collected on a filter and extracted with 40 ml. of 3 M HCl to remove the tetraammine complex, $[Ir(NH_3)_4Cl_2]Cl$. The remaining solid material (1.51 g.) is dissolved in hot water (about 60 ml.). After a brown insoluble residue (0.15 g.) is removed by filtration, the solution is concentrated to 40-ml. volume. The product is precipitated from the warm (60°C.) solution by adding about 25 ml. of 12 M hydrochloric acid. The yellow or in some cases pink crystals (1.28 g.) are recrystallized from 40 ml. of warm water by precipitation with 20 ml. of 12 M hydrochloric acid. The pale yellow crystals are washed with 6 M HCl and alcohol and dried over silica in a desiccator. A yield of 1.18 g. (59%) is obtained. *Anal.* Calcd. for $[Ir(NH_3)_5Cl]Cl_2$: Ir, 50.05; N, 18.25; H, 3.94; Cl, 27.76. Found: Ir, 49.59; N, 18.46; H, 4.36; Cl, 27.65.

Properties

The color of the crystals is pale yellow to almost white. A pink shade shows the presence of minor impurities. The quality of the product is conveniently judged from the ultraviolet spectrum. The bands at wave numbers 35,000 and 44,100 cm.$^{-1}$ with molar extinction coefficients of 73 and 333, respectively, are typical of the chloropentaammine. Spectra of pink crystals show less resolved bands. These products are, however, sufficiently pure to serve as starting materials for further reactions. The molar conductivity in aqueous solution is 253 cm.^{2} ohm^{-1} mole^{-1}. The crystals are soluble in water but insoluble in HCl or in common organic solvents.

B. AQUOPENTAAMMINEIRIDIUM(III) PERCHLORATE

$$[Ir(NH_3)_5Cl]^{2+} + H_2O \xrightarrow{NaOH} [Ir(NH_3)_5OH]^{2+} \xrightarrow{HClO_4} [Ir(NH_3)_5(H_2O)](ClO_4)_3$$

For the preparation of the acidopentaammine series the appropriate intermediate complex is the aquo compound, which is conveniently made from the chloro complex in a way which was described by Palmaer.[2] The time required for this synthesis is about 9 hours. ■ *Warning*. *Perchlorates in contact with organic materials present a constant explosion hazard.*

Procedure

The chloride [Ir(NH$_3$)$_5$Cl]Cl$_2$ (3 g.) is dissolved in a solution of sodium hydroxide (100 ml. of a 1 M solution) and refluxed for 8 hours. The red-brown solution is filtered in order to remove impurities. To the filtrate 10–12 ml. of 60% perchloric acid is carefully added; white crystals form. The product is collected on a filter washed with 1 M HClO$_4$ and with alcohol. For purification the crystals are dissolved in about 40 ml. of warm (40°C.) water. After filtration from some gelatinous residue, the product is recovered by precipitation with 2 ml. of 60% HClO$_4$. The white crystals are washed by dilute HClO$_4$ and alcohol and dried in a desiccator over silica gel. Yield is 3.2–3.4 g. (69–73%). *Anal.* Calcd. for [Ir(NH$_3$)$_5$-(H$_2$O)](ClO$_4$)$_3$: Ir, 32.4; N, 11.8; H, 2.89; Cl, 17.91. Found: Ir, 32.23; N, 11.9; H, 3.02; Cl, 18.13.

Properties

Typical absorption bands in the ultraviolet are 38,800 and 47,000 cm.$^{-1}$ with molar extinction coefficients of 86 and 128, respectively. The compound is soluble in H$_2$O and almost insoluble in ethanol, acetone, or ether.

C. PENTAAMMINEISOTHIOCYANATOIRIDIUM(III) PERCHLORATE—PENTAAMMINEACIDOIRIDIUM(III) SALTS

$$\xrightarrow{\text{100–105°C.}} [\text{Ir(NH}_3)_5\text{X]X}_2 + \text{H}_2\text{O}$$

$$[\text{Ir(NH}_3)_5\text{OH}]^{2+} + 3\text{HX} \rightarrow [\text{Ir(NH}_3)_5\text{H}_2\text{O}]\text{X}_3 + 2\text{H}^+$$
$$[\text{Ir(NH}_3)_5\text{H}_2\text{O}]^{3+} + \text{X}^- \rightarrow [\text{Ir(NH}_3)_5\text{X}]^{2+} + \text{H}_2\text{O}$$

The acido complexes [Ir(NH$_3$)$_5$X]$^{2+}$ (X = I$^-$, Br$^-$, CHOO$^-$, CH$_3$COO$^-$, NO$_3^-$) for which the corresponding acids are easily available can be prepared according to the procedure given for the aquo compound. Instead of using HClO$_4$, the aquo salts are precipitated by the respective acid. The corresponding acidopentaammines are then obtained by heating the crystals in a dry-box at 105°C. or by boiling their aqueous solutions:[2,4]

$$[Ir(NH_3)_5H_2O]X_3 \xrightarrow{100-105°C.} [Ir(NH_3)_5X]X_2 + H_2O$$

The anion X can be substituted, either by repeated precipitation with a different acid HY or by passing through an anion exchanger.[4]

Other acidopentaammines (X = SCN^-, N_3^-, OH^-, NO_2^-) are prepared from the aquo salt by adding an excess of the corresponding anions to their solutions. For the nitro compound, the starting material is the chloride salt in order to avoid reaction with the perchlorate ion. As a representative of this substitution reaction, the preparation of the thiocyanato compound is described. The time necessary for the synthesis of this compound is about 24 hours. The specific equation for synthesis of the isothiocyanto complex is:

$$[Ir(NH_3)_5(H_2O)] + NaNCS \rightarrow [Ir(NH_3)_5(NCS)](ClO_4)_2\downarrow$$
$$+ NaClO_4$$

Procedure

A solution of the perchlorate $[Ir(NH_3)_5(H_2O)](ClO_4)_3$ (0.8 g.) and sodium thiocyanate (0.87 g., an eightfold excess) in 50 ml. of water is refluxed for 8 hours. The almost colorless solution is filtered while hot to remove impurities. It yields crystals which can be filtered off after standing for 15 hours. The product is collected on a filter, washed with alcohol, and dissolved for purification in 50 ml. of hot water. The crystals are recovered by adding 1 ml. of 60% perchloric acid. On cooling, white crystals of the isothiocyanato complex (*N*-bonded linkage isomer) are formed. Yield is 0.46 g. (65%). *Anal.* Calcd. for $[Ir(NH_3)_5(NCS)](ClO_4)_2$: Ir, 35.96; N, 15.72; C, 2.25; H, 2.83; S, 6.00; Cl, 13.26. Found: Ir, 35.73; N, 15.93; C, 2.46; H, 3.06; S, 5.93; Cl, 13.09.

From the filtrate of the above preparation the other linkage isomer (*S*-bonded) can be obtained in small quantities[4] by evaporation.

Properties

The compound is soluble in H_2O, slightly soluble in acetone, and insoluble in ethanol or ether. The ultraviolet absorption spectrum shows a band at 38,500 cm.$^{-1}$ with a molar extinction coefficient of 560. The conductivity in aqueous solution is 217 cm.^{2} ohm^{-1} mole^{-1}. The linkage properties are determined from the infrared spectrum[5] which shows a C—S stretching frequency at 825 cm.$^{-1}$ for the N-bonded isomer.

References

1. C. Claus, "Beiträge zur Chemie der Platinmetalle," Festschrift Universität Kasan, p. 92, Dorpat, Estonia, 1854.
2. W. Palmaer, *Z. Anorg. Allgem. Chem.*, **10**, 320 (1895).
3. F. Basolo and G. S. Hammaker, *Inorg. Chem.*, **1**, 1 (1962).
4. H. H. Schmidtke, *ibid.*, **5**, 1682 (1966).
5. A. Turco and C. Pecile, *Nature*, **191**, 66 (1961).

43. DICYANOBIS(1,10-PHENANTHROLINE)IRON(II) AND DICYANOBIS(2,2'-BIPYRIDINE)IRON(II)

Submitted by ALFRED A. SCHILT*
Checked by PETER A. RUSSO† and AARON WOLD†

The synthesis of dicyanobis(1,10-phenanthroline)iron(II) is best achieved by displacement of 1,10-phenanthroline from tris(1,10-phenanthroline)iron(II) cations by using cyanide ions[1-3] as the attacking reagent. The opposite approach, involving displacement of cyanide ligands from ferrocyanide by 1,10-phenanthroline, is slower and less satisfactory. Direct combination, by adding iron(II) ions to a solution of equimolar amounts of potassium cyanide and 1,10-phenanthroline, yields a mixture of complexes that rearrange too slowly to the desired product by ligand exchange to be practical.

* Northern Illinois University, DeKalb, Ill. 60115.
† Brown University, Providence, R.I. 02912.

Dicyanobis(2,2′-bipyridine)iron(II) can be prepared in a manner analogous to that of the 1,10-phenanthroline derivative, since 2,2′-bipyridine and 1,10-phenanthroline behave similarly. The complexes are useful as indicators for titrating weak bases in nonaqueous solvents and also for oxidation-reduction and aromatic diazotization titrations.[4,5]

A. DICYANOBIS(1,10-PHENANTHROLINE)IRON(II) DIHYDRATE

$$Fe(NH_4)_2(SO_4)_2 \cdot 6H_2O + 3(phen \cdot H_2O) \rightarrow$$
$$[Fe(phen)_3]SO_4 + (NH_4)_2SO_4 + 9H_2O$$
$$[Fe(phen)_3]SO_4 + 2KCN + 3H_2O \rightarrow$$
$$[Fe(phen)_2(CN)_2] \cdot 2H_2O + K_2SO_4 + (phen \cdot H_2O)$$
(phen is 1,10-phenanthroline.)

Procedure

■ *Caution. This synthesis should be performed in the hood. Potassium cyanide hydrolyzes under appropriate conditions to generate hydrogen cyanide.*

The dark-red solution obtained on dissolving 6.0 g. (0.030 mole) of 1,10-phenanthroline monohydrate together with 3.9 g. (0.010 mole) of ferrous ammonium sulfate hexahydrate in 400 ml. of water is heated to just below the boiling point. A freshly prepared solution of 10 g. of potassium cyanide in 20 ml. of water is added all at once to the hot solution. After thorough stirring, the hot mixture is set aside to cool at room temperature. Dark violet, almost black, dense, fine crystals develop rapidly. Precipitation is complete within several minutes; however, an aging period of an hour or longer is considered beneficial. The yield of crude product, after filtration and generous washing with water, is approximately 5 g. The filtrate contains about 2 g. of free 1,10-phenanthroline that can be recovered by extraction with either benzene or diethyl ether.

For purification, the crude product is dissolved in 30 ml. of concentrated sulfuric acid. To the vigorously stirred solution 1 l. of water is slowly added. The recrystallized product is collected by suction filtration and washed thoroughly with small portions of water until free of sulfuric acid. A final rinse with acetone facilitates drying. The product is vacuum-dried at room temperature. The final yield ranges from 4.0 to 4.9 g. (80–98% based on the iron(II) salt).* *Anal.* Calcd. for $[Fe(C_{12}H_8N_2)_2(CN)_2]\cdot2H_2O$: C, 61.92; H, 4.00; N, 16.67; H_2O, 7.1; formula weight, 504.3. Found: C, 63.7; H, 4.05; N, 17.0; H_2O, 6.8; formula weight, 505.

B. DICYANOBIS(2,2'-BIPYRIDINE)IRON(II) TRIHYDRATE

$Fe(NH_4)_2(SO_4)_2\cdot6H_2O + 3(bipy) \rightarrow$
$$[Fe(bipy)_3]SO_4 + (NH_4)_2SO_4 + 6H_2O$$
$[Fe(bipy)_3]SO_4 + 2KCN + 3H_2O \rightarrow$
$$[(Fe(bipy)_2(CN)_2]\cdot3H_2O + K_2SO_4 + (bipy)$$
(bipy is 2,2'-bipyridine.)

Procedure

The bipyridine complex $Fe(bipy)_2(CN)_2\cdot3H_2O$ can be prepared by using 4.7 g. of 2,2'-bipyridine in place of 6.0 g. of 1,10-phenanthroline. Excess, free 2,2'-bipyridine in the filtrate can be recovered by extraction with either benzene or diethyl ether.

The crude dicyanobis(2,2'-bipyridine)iron(II) can be purified by dissolving it in 25 ml. of concentrated sulfuric acid and diluting it slowly with 800 ml. of water while stirring. The recrystallized product is collected by suction filtration, washed with small portions of water until free of sulfuric acid, sucked dry, rinsed with acetone, and vacuum-dried at room temperature. The final yield is 3.5 g.† (75%). *Anal.* Calcd. for $[Fe(C_{10}H_8N_2)_2(CN)_2]\cdot3H_2O$: C, 55.71; H, 4.68; N, 17.72; H_2O, 11.4; formula weight, 474.3. Found: C, 55.8; H, 4.84; N, 17.9; H_2O, 11.6; formula weight, 472.

Properties

The dark violet, nonionic compounds possess appreciable dibasic character.[6,7] They display various colors in different

---

* The product is sufficiently soluble in water to give some loss during washing.

† The checkers reported a yield of 2.0 g. rather than 3.5 g.; they noted that solubility problems demand care in the washing of the precipitate.

solutions: violet, blue, red, orange, and yellow, depending on the nature of the solvent. They are most soluble in strongly acidic solvents. Concentrated sulfuric acid dissolves large amounts, producing yellow diprotonated species. Dilution of the concentrated sulfuric acid solutions with water causes precipitation of the yellow forms of the complexes, which on further dilution undergo gradual changes in hue through orange and red until the final dark violet of the original solid is attained. The compounds can be successfully titrated with perchloric acid in anhydrous acetic acid solvent.[6] ■ *Caution. Perchloric acid in contact with organic acids always poses a potential explosion hazard. Caution is advised.* Either spectrophotometric or potentiometric detection of the first equivalence point, corresponding to the formation of the orange monoprotonated species, is feasible. The second equivalence point, corresponding to the formation of the yellow diprotonated forms, is less distinct and is best detected spectrophotometrically.

Dicyanobis(1,10-phenanthroline)iron(II) dissolves sparingly in water to give a pale orange solution; it is moderately soluble in ethyl alcohol, giving a deep red solution, and moderately soluble in chloroform to give a deep violet color. Dicyanobis(2,2'-bipyridine)iron(II) dissolves in water to give a pale red solution; in ethyl alcohol it gives a deep red-violet liquid, and in chloroform it gives a deep blue color. Solubility in basic solvents is nil or very slight, with few exceptions.

Both complexes are readily oxidized to the corresponding ion(III) compounds by strong oxidants in acid solution. They can be successfully titrated with cerium(IV) sulfate in sulfuric acid.[4] The equivalence point is best detected potentiometrically (Pt vs. calomel electrode).

The complexes are diamagnetic.[3,8] Details of their electronic,[3,6,8] vibrational,[9,10] and Mössbauer spectra[11] have been reported. On the basis of infrared data[9,10] and stereochemical considerations,[12] both complexes are assigned a cis configuration.

References

1. G. A. Barbieri, *Atti Accad. Lincei*, **20**, 273 (1934).
2. A. A. Schilt, *J. Am. Chem. Soc.*, **79**, 5421 (1957).
3. A. A. Schilt, *ibid.*, **82**, 3000 (1960).
4. A. A. Schilt, *Anal. Chim. Acta*, **26**, 134 (1962).
5. A. A. Schilt and J. W. Sutherland, *Anal. Chem.*, **36**, 1805 (1964).
6. A. A. Schilt, *J. Am. Chem. Soc.*, **82**, 5779 (1960).
7. A. A. Schilt, *ibid.*, **85**, 904 (1963).
8. K. Madeja and E. König, *J. Inorg. Nucl. Chem.*, **25**, 377 (1963).
9. N. K. Hamer and L. E. Orgel, *Nature*, **190**, 439 (1961).
10. A. A. Schilt, *Inorg. Chem.*, **3**, 1323 (1964).
11. R. R. Berrett and B. W. Fitzsimmons, *J. Chem. Soc.*, A, 525 (1967).
12. K. Madeja, *Chem. Zvesti*, **19**, 186 (1965).

44. DIISOTHIOCYANATOTETRAPYRIDINE AND DIISOTHIOCYANATODIPYRIDINE COMPLEXES OF DIPOSITIVE FIRST-TRANSITION-METAL IONS (MANGANESE, IRON, COBALT, NICKEL, COPPER, AND ZINC)

$$MX_2 \cdot nH_2O + 4C_5H_5N + 2NH_4SCN \rightarrow$$
$$M(C_5H_5N_4)(NCS)_2 + 2NH_4X + nH_2O$$
$$(M = Mn, Fe, Co, or Ni)$$
$$MX_2 \cdot nH_2O + 2C_5H_5N + 2NH_4SCN \rightarrow$$
$$M(C_5H_5N)_2(NCS)_2 + 2NH_4X + nH_2O$$
$$(M = Cu or Zn)$$

Submitted by GEORGE B. KAUFFMAN,* RICHARD A. ALBERS,* and FRED L. HARLAN*
Checked by R. SCOTT STEPHENS† and RONALD O. RAGSDALE†

Complexes formed by heterocyclic amines with halides and pseudohalides of dipositive first-transition-series metal ions are usually prepared by the reaction between a metallic salt, a soluble thiocyanate, and pyridine in aqueous solution. Although the number of coordinated pyridine molecules varies,

* California State College at Fresno, Fresno, Calif. Financial support of the donors of the Petroleum Research Fund administered by the American Chemical Society (Grant 1152-B) and the California State College at Fresno Research Committee is gratefully acknowledged.
† University of Utah, Salt Lake City, Utah 84112.

tetrapyridine and dipyridine compounds are particularly numerous, and complexes of this type have been widely used in analytical chemistry.[1,2] Selective clathration by the diisothiocyanatotetrapyridines of cobalt(II), nickel(II), and copper(II) has been employed to separate and purify organic compounds, especially certain hydrocarbons.[3]

Procedure

Five-hundredths mole* of a manganese(II), iron(II),† cobalt(II), nickel(II), copper(II), or zinc(II) salt‡ is dissolved in 300 ml. of distilled water containing 20 ml. of pyridine. To this solution is added slowly with stirring a solution of 8.00 g. of ammonium thiocyanate in 100 ml. of distilled water, and the resulting mixture is allowed to stand in an ice bath until precipitation appears complete (*ca.* an hour).§ The precipitate is collected on an 8-cm. Büchner funnel and is washed with three 10-ml. portions of an ethanol–pyridine (9:1) solution. The product is air-dried for not longer than 10 minutes. It is powdered and then dried in a desiccator over potassium hydroxide until constant weight has been attained (*ca.* 24 hours). In the cases of copper(II) and zinc(II), the dipyridine compound is obtained. Details of separate syntheses are given in Table I.

Analysis

Pyridine[4]

A 0.5-g. sample is dissolved in 50 ml. of standard 0.1 *M* hydrochloric acid, and the solution is titrated with 0.1 *M* sodium hydroxide to a pH of 3.00.

---

* The checkers scaled down the preparation of all compounds by a factor of 10.
† In the case of iron(II), *ca.* 100 mg. of ascorbic acid is added to prevent oxidation by the atmosphere.
‡ The nitrates, sulfates, or chlorides are satisfactory.
§ In the case of zinc, the ammonium thiocyanate solution is first added to the solution of the zinc salt and then the pyridine. The addition of pyridine to a solution of zinc salts produces a white gelatinous precipitate.

TABLE I

Compound	Starting material	Yield	Analyses				Color[6-8]	Stability range,[2] °C.	Reference
			% Pyridine		% Isothiocyanate				
			Calcd.	Found	Calcd.	Found			
$Mn(C_5H_5N)_4(NCS)_2$	$MnSO_4 \cdot H_2O$ (8.45 g.)	17.56–20.72 g. (72–85%) 67%*	64.90	64.97 65.70*	23.83	23.91 23.26*	White	<40	2, 6 18,19
$Fe(C_5H_5N)_4(NCS)_2$	$FeSO_4 \cdot 7H_2O$ (13.90 g.)	22.00–24.20 g. (90–99%) 92%*	64.78	64.54 65.14*	23.78	24.03 23.90*	Yellow†	....	6, 18 20–28
$Co(C_5H_5N)_4(NCS)_2$	$Co(NO_3)_2 \cdot 6H_2O$ (14.55 g.)	22.60–24.31 g. (92–99%) 95%*	64.37	64.32 65.22*	23.63	23.79 24.16*	Pink	....	6, 18 29,30
$Ni(C_5H_5N)_4(NCS)_2$	$Ni(NO_3)_2 \cdot 6H_2O$ (14.54 g.)	24.31 g. (99%) 94%*	64.40	64.22 64.20*	23.64	23.52 23.80*	Blue	<63	6, 18 31–33
$Cu(C_5H_5N)_2(NCS)_2$	$CuSO_4 \cdot 5H_2O$ (12.49 g.)	15.5 g. (92.8%) 96%*	46.81	46.50	34.38	34.02 34.08*	Green	<46	2, 6 7, 34
$Zn(C_5H_5N)_2(NCS)_2$	....	....	46.57	46.41 47.36*	34.21	.... 33.92*	White	<70	2, 6

* Values obtained by checkers.
† Easily oxidized to the violet form. (See Properties.)

253

Isothiocyanate (Reverse Volhard Method)[5]

A 0.5-g. sample is dissolved in 50 ml. of 3 M aqueous ammonia, and the solution is made slightly acidic to litmus paper with 6 M nitric acid. Fifty milliliters of standard 0.1 M silver nitrate, 3 ml. of nitrobenzene, and 10 drops of ferric alum indicator are added, and the excess Ag^+ is titrated with 0.1 M potassium thiocyanate to a brownish-red end point.

Properties

The colors of the diisothiocyanatotetrapyridine and diisothiocyanatodipyridine compounds prepared according to the present synthesis, together with their temperature stability ranges, are given in Table I. They are soluble to varying degrees in chloroform. Davis et al.[6,7] measured dissociation pressures for these compounds, and Rogers et al. studied them by differential thermal analysis[9] and infrared absorption.[10] Stability trends conforming to the Irving-Williams series were found in most cases.[11]

Recent infrared absorption and x-ray diffraction studies have shown that metals of the first transition series normally form M—N bonds with the thiocyanate group.[12-14] X-ray diffraction studies have shown that the isothiocyanato groups are trans to each other in the tetrapyridine complexes of iron, cobalt, and nickel;[15-17] this configuration has been confirmed for the iron compound by electronic spectra.[8]

$Mn(C_5H_5N)_4(NCS)_2$[6,18] has been recommended for the gravimetric determination of manganese[19] even though it loses pyridine above 40°C.

$Fe(C_5H_5N)_4(NCS)_2$[18] was for many years believed to exist in two geometrically isomeric forms—a yellow form (α) and a violet form (β)—and the composition and structure of the violet form have long been a subject of controversy.[21-25] The yellow form has finally been identified as the trans octahedral compound, whereas the violet form is merely the yellow form contaminated with traces of iron(III).[26,27]   Infrared spectra have shown Fe—N bonding for the yellow compound.[28]   [Co(C$_5$H$_5$N)$_2$-(NCS)$_2$] has been reported to exist in two forms, violet and blue,[29] but later studies showed only one form to be stable.[30]

[Ni(C$_5$H$_5$N)$_4$(NCS)$_2$][6,18,31] has been recommended for the thermogravimetric determination of nickel or thiocyanate.[32,33]   The emerald-green chloroform-soluble copper complex [Cu(C$_5$H$_5$N)$_2$-(NCS)$_2$] has been suggested for the detection of copper, pyridine, or thiocyanate.[34]

References

1. A. I. Vogel, "A Textbook of Quantitative Inorganic Analysis Including Elementary Instrumental Analysis," p. 132, John Wiley & Sons, Inc., New York, 1961.
2. C. Duval, "Inorganic Thermogravimetric Analysis," 2d rev. ed., Elsevier Publishing Co., Amsterdam, 1963.
3. W. D. Schaeffer, W. S. Dorsey, D. S. Skinner, and C. G. Christian, *J. Am. Chem. Soc.*, **79**, 5870 (1957).
4. D. A. Skoog and C. M. West, "Fundamentals of Analytical Chemistry," pp. 262–263, Holt, Rinehart and Winston, Inc., New York, 1963.
5. K. Kodama, "Methods of Quantitative Inorganic Analysis," p. 186, Interscience Publishers, a division of John Wiley & Sons, Inc., New York, 1963.
6. T. L. Davis and H. R. Batchelder, *J. Am. Chem. Soc.*, **52**, 4069 (1930).
7. T. L. Davis, *ibid.*, **58**, 2155 (1936).
8. D. M. L. Goodgame, M. Goodgame, M. A. Hitchman, and M. J. Weeks, *Inorg. Chem.*, **5**, 635 (1966).
9. P. B. Bowman and L. B. Rogers, *J. Inorg. Nucl. Chem.*, **28**, 2215 (1966).
10. C. W. Frank and L. B. Rogers, *Inorg. Chem.*, **5**, 615 (1966).
11. H. Irving and R. J. P. Williams, *J. Chem. Soc.*, 3192 (1953).
12. J. Lewis, R. S. Nyholm, and P. W. Smith, *ibid.*, 4590 (1961).
13. K. Nakamoto, "Infrared Spectra of Inorganic and Coordination Compounds," p. 175, John Wiley & Sons, Inc., New York, 1963.
14. D. P. Graddon, R. Schulz, E. C. Watton, and D. G. Weeden, *Nature*, **198**, 1299 (1963).
15. M. A. Porai-Koshits, *Tr. Inst. Kristallogr. Akad. Nauk SSSR*, **10**, 117 (1954).
16. M. A. Porai-Koshits and A. S. Antsishkina, *Kristallografiya*, **3**, 386 (1958).
17. A. S. Antsishkina and M. A. Porai-Koshits, *ibid.*, **3**, 676 (1958).
18. H. Grossmann, *Ber.*, **37**, 559 (1904).
19. G. Spacu and J. Dick, *Z. Anal. Chem.*, **74**, 188 (1928); *Bull. Soc. Stiinte Cluj*, **4**, 431 (1929).
20. G. Spacu, *Ann. Sci. Univ. Jassy*, **8**, 162 (1914); **9**, 337 (1915).
21. A. Rosenheim, E. Roehrich, and L. Trewendt, *Z. Anorg. Allgem. Chem.*, **207**, 97 (1932).

22. G. Spacu, *ibid.*, **216**, 165 (1933).
23. R. Asmussen, *ibid.*, **218**, 425 (1934).
24. O. Binder and P. Spacu, *Compt. Rend.*, **200**, 1405 (1935).
25. P. Spacu, M. Teodorescu, and C. Lepadatu, *Rev. Roumaine Chim.*, **9**, 39 (1964).
26. N. E. Erickson and N. Sutin, *Inorg. Chem.*, **5**, 1834 (1966).
27. C. D. Burbridge, M. J. Cleare, and D. M. L. Goodgame, *J. Chem. Soc., Sec. A, Inorg. Phys. Theo. Chem.* 1968 (1966).
28. J. F. Duncan and K. F. Mok, *Australian J. Chem.*, **19**, 701 (1966).
29. A. Hantzsch, *Z. Anorg. Chem.*, **166**, 237 (1927).
30. N. S. Gill, R. S. Nyholm, G. A. Barclay, T. I. Christie, and P. J. Pauling, *J. Inorg. Nucl. Chem.*, **18**, 88 (1961).
31. G. Spacu, *Bull. Soc. Stiinte Cluj*, **1**, 314 (1922).
32. C. Duval and N. D. Xuong, *Anal. Chim. Acta*, **5**, 506 (1951).
33. R. Duval and C. Duval, *ibid.*, **5**, 71 (1951).
34. G. Spacu, *Bull. Soc. Stiinte Cluj*, **1**, 284 (1922).

45. $\alpha,\beta,\gamma,\delta$-TETRA(4-PYRIDYL)PORPHINEZINC

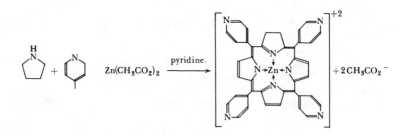

Submitted by HOWARD J. BUTCHER* and EVERLY B. FLEISCHER†
Checked by E. L. MUETTERTIES,‡ E. J. MUETTERTIES,§ and M. C. MUETTERTIES§

A class of synthetic metalloporphyrins which are conveniently studied in aqueous solution derives from coordination compounds of the quadridentate ligand $\alpha,\beta,\gamma,\delta$-tetra(4-pyridyl)-porphine. Although a method of synthesis for the zinc complex has been published,[1] the yield is often poor and the method unreliable. If the same reaction is allowed to proceed under

* Washington Research Center, W. R. Grace and Co., Clarksville, Md.
† Department of Chemistry, University of Chicago, Chicago, Ill.
‡ Central Research Department, Experimental Station, E. I. du Pont de Nemours & Company, Wilmington, Del. 19898.
§ 1137 South Concord Road, West Chester, Pa.

more strictly controlled conditions, the yield is consistently satisfactory. The working time is short, but maximum yield requires 48 hours of reaction time on the scale described here.

Procedure

Fifty milliliters (0.625 mole) of reagent-grade pyridine is saturated with reagent-grade zinc acetate dihydrate. The solution is dehydrated over molecular sieves. A quantity of technical-grade pyrrole is distilled in the air just prior to use, with molecular sieves added to the receiver. A *freshly opened* bottle of technical-grade 4-pyridine aldehyde is also treated with molecular sieves. The zinc acetate solution, 50 ml. (0.71 mole) of the redistilled pyrrole, and 35 ml. (0.37 mole) of the dry aldehyde are filtered into a 250-ml. pressure bottle* containing about 5 g. of molecular sieves. After 10 minutes of deaeration with oil-pumped nitrogen, the bottle is sealed with a new neoprene gasket. The sealed bottle is immersed to the liquid line in a preheated oil bath at 130–150°C. for 48 hours (or less for a small yield). The bottle is removed from the bath, cooled, then opened, and its contents warmed to 80°C. The hot, viscous reaction mixture is suction-filtered through a medium-porosity glass frit. The purple crystals of porphinezinc on the frit are washed with absolute ethanol and air-dried. Complete separation from the molecular sieves can be achieved by extraction with chloroform in a Soxhlet apparatus. The zinc complex may be purified by passing it down an alumina column in chloroform (0.5% ethanol stabilized) and eluting with a 5% ethanol solution in chloroform.

Properties

The porphinezinc dissolves only sparingly in chloroform to give an intensely colored solution. The visible absorption

---

* The checkers used a Carius tube of 350-ml. capacity. Reactants were cooled to liquid-nitrogen temperatures, and the tube evacuated and sealed.

spectrum in chloroform contains principal peaks at 424 nm., $a = 512 \times 10^3$ M^{-1} cm.$^{-1}$, and 559 nm., $a = 19.4 \times 10^3$ M^{-1} cm.$^{-1}$. The x-ray diffraction powder pattern generated with copper radiation through a nickel filter contains major peaks at 2θ values of 17.65, 19.52, 30.40, and 39.70°. The free ligand can be generated by the methods described in reference 1. A large variety of metal salts each reacts with the free ligand in glacial acetic acid to form metalloporphyrins.[2] The free ligand and its metal complexes may all be dissolved in aqueous acids since the pyridyl groups are readily protonated.

References

1. E. B. Fleischer, *Inorg. Chem.*, **1**, 493 (1962).
2. J. E. Falk, "Porphyrins and Metalloporphyrins," American Elsevier Publishing Co., Inc., New York, 1964.

46. POLY{DI-μ-(DIPHENYLPHOSPHINATO)}AQUO- HYDROXYCHROMIUM(III)

$$CrCl_2 + 2KOP(C_6H_5)_2O + H_2O \xrightarrow{CH_3OH} [Cr\{OP(C_6H_5)_2O\}_2H_2O] + 2KCl$$

$$4[Cr\{OP(C_6H_5)_2O\}_2H_2O] \xrightarrow[C_4H_8O-H_2O]{\text{oxygen of air}} \frac{4}{x}[Cr(H_2O)(OH)\{(OP(C_6H_5)_2O\}_2]_x$$

Submitted by KEITH D. MAGUIRE*
Checked by EDWARD E. FLAGG† and RICHARD E. RIDENOUR†

Poly{di-μ-(diphenylphosphinato)}aquohydroxychromium(III) has been prepared starting with chromium(II) acetate and an aqueous solution of potassium diphenylphosphinate. The chromium(II) diphenylphosphinate produced was then oxidized by exposure of an aqueous suspension to the air.[1] The polymers

* Technological Center, Pennwalt Chemicals Corp., King of Prussia, Pa.
† The Dow Chemical Co., Midland, Mich.

produced by this procedure can have acetate groups incorporated, and the oxidation in aqueous suspension gives rise to a mixture of products. The following modified procedure based upon chromium(II) chloride[2,3] and oxidation in a tetrahydrofuran–water mixture avoids these problems.[4]

Procedure

All operations involving chromium(II) compounds must be carried out in the complete absence of atmospheric oxygen.*

The apparatus is assembled as shown in Fig. 21.† All the stopcocks and ball joints are fitted with retaining clips to prevent accidental opening when the apparatus contains nitrogen pressure in excess of 1 atmosphere.‡

The gum-rubber tubing connections between vessels *A*, *B*, and *C* should be of sufficient length to allow these vessels to be inverted without producing undue strain on the assembly. The reaction flask *B* is a 2-l. three-necked flask to which a 4-mm. stopcock has been attached. The filter tube *C* can be readily constructed from a 600-ml. coarse-fritted-glass filter funnel.

Fifty grams of highly purified chromium metal shot§ (99.999% Cr) is placed in flask *A*, and the whole apparatus is then thoroughly purged with nitrogen to remove all the air. At least 20 minutes with a continuous nitrogen flow (nitrogen pressure

* This is a particularly important point since Cr(II) is oxidized very easily.

† Regular 24/40 S.T., etc., joints may be substituted for the ball joints, but because of the positive nitrogen pressures employed, they must be firmly secured by spring clips or very strong rubber bands.

‡ The checkers reported that, by using ball joints (with stopcocks) at the bottom of both flasks, an *all-glass* reaction column including vessels *A*, *B*, and *C* could be assembled. A mechanical paddle stirrer was used for agitation in vessel *B* (3-l. three-necked flask). This stirrer minimized problems in transferring the Cr(II) precipitate. The funnels were equipped with pressure-equalizing tubes, thus simplifying the addition of solutions.

§ Chromium may be obtained as 99.999% Cr shot approximately 3–5 mm. in diameter from United Mineral and Chemical Corporation, 129 Hudson St., New York, N.Y. 10013. The checkers reported that it is very important to use the high-purity chromium if a pure product is desired.

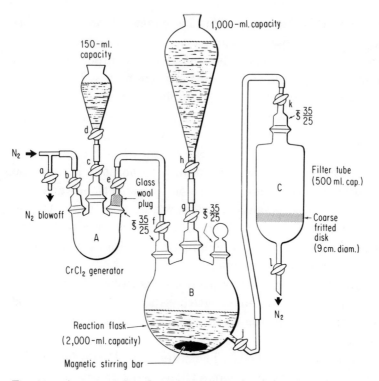

Fig. 21. **Apparatus for the preparation of poly{di-μ-(diphenyl-phosphinato)}aquohydroxychromium(III).**

regulated to 10 p.s.i.g. maximum) is required. When purging is completed, all stopcocks (except *a*) are closed, and a small nitrogen flow is allowed to vent throughout the reaction via the blowoff. One hundred and ten milliliters of 2 *M* hydrochloric acid (0.22 mole) is placed in the funnel above flask *A* and deaerated by blowing a stream of nitrogen through it for 15 minutes. It is allowed to enter flask *A* by stopcocks *c* and *d*. Care is required to prevent entry of air into the flask.

Reaction usually commences immediately, and the hydrogen which is evolved is vented into a hood via stopcock *b* while stopcock *c* is closed. After most of the hydrogen evolution has ceased (approximately 30 minutes), a heating mantle is placed around the flask, and the solution is heated to 85–90°C. for another 30 minutes to complete the reaction.   During this time stopcock *b* should remain open, and a small nitrogen flow is maintained.   The chromium(II) chloride solution should be bright blue; gray-blue or green-blue solutions are partially oxidized and are useless.   The addition funnel is removed from flask *A*, and the flask is slightly pressurized with nitrogen.   The chromium(II) chloride is forced into the reaction flask *B* by inverting flask *A* and then opening stopcocks *e* and *f*.   The glass-wool plug serves to prevent the excess chromium shot from clogging the stopcock or entering the reaction flask.   If some chromium(II) solution remains in the generator, the pressurizing process is repeated to make the transfer as complete as possible.

A solution of 51.2 g. (0.2 mole) of potassium diphenylphosphinate in 1500 ml. of methanol is prepared as follows: A suspension of 43.64 g. (0.2 mole) of diphenylphosphinic acid* in 500 ml. of methanol is slowly neutralized by adding a solution of 16.3 g. (0.25 mole) of potassium hydroxide (assay 86.0% KOH) in 500 ml. of methanol until a pH of 6.5–7.0 is reached. The solution is then diluted to 1500 ml. with methanol and deaerated by passing a stream of nitrogen for 15 minutes through a gas dispersion tube immersed in the solution.

The large dropping funnel attached to reaction flask *B* is filled with the potassium diphenylphosphinate solution.   The solution is then allowed to enter the reaction flask slowly via stopcocks *g* and *h* while the contents of the flask are stirred continuously with the magnetic stirrer.   The complete addition of the diphenylphosphinate solution should take approximately an hour.   The blue-gray precipitate which forms rapidly at first assumes the bright pink color of chromium(II) diphenylphosphinate.   Toward the end of the addition, the flow should be interrupted briefly and the supernatant liquid examined visually for the presence of excess chromium(II).   A small excess of chromium(II) (blue color) should be present at the

---

* See *Inorganic Syntheses*, **8**, 71 (1966).   Also obtainable from Aldrich Chemical Co., Milwaukee, Wis.

end of the addition to minimize co-precipitation of excess diphenylphosphinate which is detrimental to the final polymer. When the addition is complete, the reaction mixture is stirred for 30 minutes and then allowed to settle.

The reaction product is collected in filter tube C by pressurizing the reaction flask with nitrogen and allowing the pressure to force the slurry through stopcocks j and k into filter tube C. The solid chromium(II) diphenylphosphinate collects on the coarse glass frit, and the solution exits via stopcock l. In order to get as complete a transfer as possible, the reaction flask has to be tilted and swirled. The final precipitate transfer can be aided by a small amount of deaerated methanol, added via the dropping funnel.

Most of the mother liquor adhering to the precipitate can be removed by passing a stream of nitrogen through it for about 5 minutes. The filter tube is pressurized with nitrogen (10 p.s.i.g. maximum), stopcocks l and k are closed, and filter tube C is disconnected from the rest of the apparatus. A deaerated wash solution (1:1 v/v methanol:water) is placed in a dropping funnel connected to the top of filter tube C with a *short* piece of gum-rubber tubing. Opening stopcock k allows nitrogen to escape upward and out of the top of the dropping funnel, thus preventing access of air to the filter tube. One hundred milliliters of wash solution is allowed to enter the filter tube via stopcock k (l closed), and then k is closed and the filter tube disconnected and thoroughly shaken. Application of nitrogen pressure via stopcock k forces the wash liquid out via stopcock l. This washing procedure is repeated until the washings are virtually chloride-free. Four to six washings are usually required, accompanied by sufficient shaking to break up the filter cake each time.

The wet filter cake is transferred to an open beaker containing a mixture of 700 ml. of tetrahydrofuran and 300 ml. of water. The transfer is readily accomplished by adding approximately

75 ml. of the tetrahydrofuran–water mixture to the filter tube in a manner similar to the one used for the wash addition. The filter tube is then shaken vigorously until the filter cake has become a fine suspension. The filter tube is then pressurized with nitrogen, inverted over the beaker of tetrahydrofuran–water mixture, and stopcock k is opened, allowing the nitrogen pressure to force the suspension out into the beaker. The suspension is then allowed to oxidize by stirring it while exposed to the air. The oxidation is complete when a clear, bright green solution has formed (3–4 hours). The green solution is then poured into 4 l. of stirred, ice-cold sodium chloride solution (10 g. NaCl/l.), whereupon the polymer precipitates as a green solid. The precipitate is allowed to settle (usually overnight) and then collected on a Büchner filter, washed chloride-free with water, and dried in a vacuum oven at 65–75°C. for 8 hours. A yield of 42–44 g. or about 80–85% based upon the potassium diphenylphosphinate is usually obtained. *Anal.* Calcd. for $[Cr(H_2O)(OH)\{OP(C_6H_5)_2O\}_2]_x$; $C_{24}H_{23}CrO_6P_2$: C, 55.29; H, 4.45; Cr, 9.97; P, 11.88. Found: C, 55.9; H, 4.57; Cr, 10.02; P, 12.16.

Properties

The $[Cr(H_2O)(OH)\{OP(C_6H_5)_2O\}_2]_x$ polymer is a green solid which is readily soluble in chloroform, benzene, and tetrahydrofuran but is insoluble in water and diethyl ether. It does not melt before decomposing; thermogravimetric analysis indicates decomposition starting at 365°C. A freshly prepared solution in chloroform has an intrinsic viscosity ranging from 0.03 to 0.04 dl./g. The intrinsic viscosity increases slowly when solutions in organic solvents (for example, 1 g./100 ml. in chloroform) are allowed to stand at temperatures of approximately 55°C., and values of 0.6–0.8 dl./g. are common after a number of days. A sample with an intrinsic viscosity of 0.04 dl./g.

has a number average molecular weight of 6160 (measured by vapor-pressure osmometer); a sample of intrinsic viscosity 0.86 dl./g. has a number average molecular weight of 180,000 (measured by membrane osmometer). The infrared spectrum[1] has absorptions at 3600 cm.$^{-1}$ (sharp) and 3300–3450 cm.$^{-1}$ (broad, weak) assigned as coordinated OH and H_2O, respectively.

References

1. A. J. Saraceno and B. P. Block, *Inorg. Chem.*, **3**, 1699 (1964).
2. J. P. Fackler, Jr., and D. G. Holah, *ibid.*, **4**, 954 (1965).
3. H. Lux and G. Illmann, *Ber.*, **91**, 2143 (1958).
4. K. D. Maguire and B. P. Block, unpublished results.

47. DIHALOGENODINITROSYLMOLYBDENUM AND DIHALOGENODINITROSYLTUNGSTEN

$$M(CO)_6 + 2NOX \rightarrow M(NO)_2X_2 + 6CO \qquad (X = Cl, Br)$$

Submitted by B. F. G. JOHNSON* and K. H. AL-OBADI*
Checked by ROBERT T. PAINE†

Dichlorodinitrosylmolybdenum and tungsten have been prepared by the reaction of the appropriate hexacarbonyl with nitrosyl chloride in dichloromethane.[1] The corresponding dibromo derivatives have been prepared by a similar reaction with nitrosyl bromide.[2] These are the only synthetic routes to these compounds; they are rapid (3 hours) and convenient. The corresponding chromium compounds are unknown. This method is a general route to transition-metal nitrosyl halide

* University College, London, England.
† Department of Chemistry, University of Michigan, Ann Arbor, Mich. 48104.

compounds.[3-5]   Dihalogenodinitrosylmolybdenum and -tung-
sten are valuable starting materials in the synthesis of NO-
containing organometallic compounds.[5,6]   The most probable
structure is:

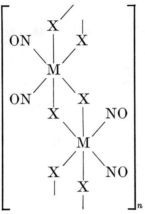

Procedure

The experiment described herein should be carried out in a
well-ventilated hood, and the normal safety measures adopted
for use of metal carbonyls should be carefully observed.   All
apparatus should be dried thoroughly before use.

The reaction is carried out in a 250-ml. three-necked flask
which is swept out with dry, oxygen-free nitrogen.   The metal
carbonyl* (5 g.) and degassed dichloromethane (50 ml.) are
introduced.   A solution of freshly distilled nitrosyl chloride (5
ml.) in dichloromethane is added dropwise and the reaction
mixture stirred vigorously at 20°C.   As the reaction takes
place, carbon monoxide is evolved, and a green solid is produced.
After stirring for 2 hours, the solvent is removed under vacuum
and any unreacted carbonyl sublimed into a trap at −196°C.
The pure compound is obtained as a green powder.

* Molybdenum and tungsten hexacarbonyls were obtained from Climax Molybdenum Company, 1270 Avenue of the Americas, New York, N.Y. 10020.

The preparation of the-corresponding dibromo derivatives is carried out in a similar manner, using nitrosyl bromide. The yields are *ca.* 90% in all cases.

These preparations may be carried out on a larger scale without difficulty.

Anal. Calcd. for $[Mo(NO)_2Cl_2]_n$: Mo, 42.3; N, 12.4; Cl, 31.3. Found: Mo, 41.9; N, 12.7; Cl, 32.3. ν_{NO} (Nujol mull) cm.$^{-1}$: 1805(s); 1690(s); 1600(w).

Calcd. for $[Mo(NO)_2Br_2]_n$: Mo, 30.3; N, 8.9; Br, 50.6. Found: Mo, 30.1; N, 8.4; Br, 51.6. ν_{NO}: 1805(s), 1695(s), 1595(w).

Calcd. for $[W(NO)_2Cl_2]_n$: W, 58.9; N, 8.9; Cl, 22.6. Found: W, 58.4; N, 8.3; Cl, 21.6. ν_{NO}: 1800(s); 1680(s); 1600(w).

Calcd. for $[W(NO)_2Br_2]_n$: W, 45.5; N, 6.9; Br, 39.6. Found: W, 44.9; N, 6.7; Br, 40.3. ν_{NO}: 1800(s); 1690(s); 1600(w).

Properties

These have been discussed previously.[1,2] The compounds are green solids, very hygroscopic, and must be kept under vacuum or in an atmosphere of dry nitrogen. They dissolve in alcohols, acetone, acetonitrile, or tetrahydrofuran without gas evolution to give green solutions but are insoluble in nondonor solvents such as benzene, carbon tetrachloride, chloroform, dichloromethane, and alkanes. They are thermally stable and do not decompose below 100°C. The compounds have a characteristic three-band spectrum in the infrared. Chemical and infrared evidence are consistent with a polymeric linked chain structure involving bridging halogens between the metal atoms.[1,2]

References

1. F. A. Cotton and B. F. G. Johnson, *Inorg. Chem.*, **3**, 1609 (1964).
2. B. F. G. Johnson, *J. Chem. Soc.*, **A**, 475 (1967).
3. W. P. Griffith, J. Lewis, and G. Wilkinson, *ibid.*, 775 (1961).
4. C. C. Addison and B. F. G. Johnson, *Proc. Chem. Soc.*, 305 (1962).
5. G. Crookes, B. F. G. Johnson, and K. H. Al-Obadi, *J. Chem. Soc.*, in press.
6. R. B. King, *Inorg., Chem.*, **7**, 90 (1968).

48. SODIUM (+)-ARSENYL TARTRATE

$$As_2O_3 + 2H_2(+)C_4H_4O_6 \rightarrow 2H(+)AsOC_4H_4O_6 + H_2O$$
$$2NaOH + 2H(+)AsOC_4H_4O_6 \rightarrow 2Na(+)AsOC_4H_4O_6 + 2H_2O$$

Submitted by G. SCHLESSINGER*
Checked by A. E. GEBALA† and M. M. JONES†

Sodium (+)-arsenyl tartrate, a useful resolving agent,[1] is not commercially available. It has been made by the reaction of arsenic(III) oxide with sodium hydrogen (+)-tartrate.[2] The following synthesis is an adaptation of this method.

Procedure

■ *Caution. The product is intensely poisonous like its antimony analog; the lethal dose is probably of the order of 100 mg.[3]* All work should be conducted in the hood with adequate safety equipment.

Twenty grams (0.10 mole) of pure arsenic(III) oxide is added to a mixture of 30 g. (0.20 mole) of (+)-tartaric acid and 25 ml. of water. After heating just to the boiling point, a solution of 8 g. of sodium hydroxide (0.2 mole) in 20 ml. of water is carefully added dropwise with stirring. The suspension is digested over a small flame until only a small amount of undissolved oxide (*ca.* 0.2 g.) remains. After cooling, 25 ml. of water is added and the residue filtered off.

Since the above solutions are rather concentrated and losses in this preparation are largely mechanical, small volumes of water should be used to rinse out all funnels and other vessels.

The filtrate is evaporated to a syrup on a steam bath and then heated for an additional 1–2 hours at 105–110°C.; any vented

* Department of Chemistry, Newark College of Engineering, Newark, N.J. 07102.
† Department of Chemistry, Vanderbilt University, Nashville, Tenn. 37203.

oven is suitable for this second step.   On cooling, with occasional stirring, a semisolid crystalline mass is obtained consisting of the $2\frac{1}{2}$ hydrate of the product.   This is triturated to a smooth slurry with 100 ml. of absolute ethanol, cooled in ice, and filtered. The hydrated material is heated to constant weight at 105–110°C.; at this temperature the anhydrous compound results. The yield is 46–50 g. (88–95%).   This crude product is satisfactory for most purposes; it may be purified further by recrystallization.   The compound is recrystallized from a 50:50 (v/v) ethanol–water mixture, using between 3 and 4 ml. of solvent per gram of complex.   It is then washed with absolute ethanol and dried.   Yield is about 85% of the crude product used.

Anal. Calcd. for $NaAsC_4H_4O_7$: C, 18.33; H, 1.54; As, 28.59. Found: C, 18.48; H, 1.76; As, 28.44.*   Checker obtained C, 18.54; H, 1.37; As, 28.44.

Properties

The anhydrous salt is very soluble in water and insoluble in ethanol.   The pH of both a 1 and 2% aqueous solution is 3.7; at about 250°C. the material begins to darken with progressive deepening of color as the temperature increases.   A solution of 2.50 g. of the anhydrous salt in 25.0 ml. of water, in a 2.0-dm tube, had a rotation of +17.11° at 25.0°C.   This corresponds to a specific rotation of +85.5°.

References

1. I. K. Reid and A. M. Sargeson, *Inorganic Syntheses*, **9**, 167 (1967).
2. G. G. Henderson and A. R. Ewing, *J. Chem. Soc.*, **67**, 107 (1895).

* Arsenic may be determined iodometrically.   A dried sample of 150–200 mg. is mixed with 7 ml. of 15 N nitric acid and the solution boiled gently for 5–10 minutes.   Three milliliters of 2 N sodium bromate is then added and the mixture evaporated *gently* to dryness.   The residue is taken up in 150 ml. of 4 M hydrochloric acid in an iodine flask; then 1 g. of sodium hydrogen carbonate is cautiously mixed in to displace the air, followed by 1 g. of potassium iodide.   The flask is then stoppered and allowed to stand 5 minutes.   The contents are rapidly titrated with 0.05 N sodium thiosulfate to the disappearance of the iodine color.

3. S. Moeschlin, "Klinik und Therapie der Vergiftungen," p. 186, Georg Thieme Verlag, Stuttgart, 1964.
4. H. W. Riemann, J. D. Neuss, and B. Naiman, "Quantitative Analysis," 3d ed., p. 449, McGraw-Hill Book Company, New York, 1951.

49. RESOLUTION OF TRIS(ETHYLENEDIAMINE)-RHODIUM(III) AND -CHROMIUM(III) CHLORIDE BY MEANS OF (+)-TARTRATE

SUBMITTED BY F. GALSBØL*
Checked by N. AHMAD,† R. HALLIDAY,† and S. KIRSCHNER†

Methods for the preparation of tris(ethylenediamine)rhodium-(III) chloride hydrate and for its resolution by means of the diastereoisomer $Li\{(-)_D\text{-}[Rh(en)_3]\}\{(+)\text{-tart}\}_2\cdot 3H_2O$ are presented. The resolution of tris(ethylenediamine)chromium(III) chloride by means of an analogous diastereoisomeric compound is also described.

A. TRIS(ETHYLENEDIAMINE)RHODIUM(III) CHLORIDE TRIHYDRATE

$$RhCl_3\cdot 3H_2O + 3en \rightarrow [Rh(en)_3]Cl_3\cdot 3H_2O$$

Tris(ethylenediamine)rhodium(III) chloride, first reported by Werner,[1] was prepared by the reaction of sodium hexachlororhodate(III) dodecahydrate with ethylenediamine monohydrate. The reaction product, however, contained sodium chloride, which could not be removed by recrystallization,[1] probably because of the formation of a double salt.‡

Jaeger[2,3] prepared the compound from rhodium(III) chloride trihydrate and 50% aqueous ethylenediamine. Gillard, Osborn, and Wilkinson[4] carried out the reaction of rhodium(III) chloride

* Chemistry Department I, H. C. Ørsted Institute, Universitetsparken 5, 2100, Copenhagen, Denmark.
† Department of Chemistry, Wayne State University, Detroit, Mich. 48202.
‡ See Properties subsection on page 277.

trihydrate with ethylenediamine monohydrate in ethanolic solution. By these two methods, however, oily by-products are formed, often in large quantity, and yields are not reproducible. In the synthesis reported here, rhodium(III) chloride trihydrate is allowed to react with aqueous ethylenediamine, and the complex is precipitated with ethanol. The noncrystalline residue obtained by evaporation of the filtrate is converted into the desired product by a heat treatment during which ethylenediamine is given off. The yield is nearly quantitative. The compounds [Rh{(−)pn}₃]Cl₃·aq, [Rh{(+)pn}₃]Cl₃·aq,[24] [Rh(tn)₃]Cl₃·aq, and [Rh{(−)chxn}₃]Cl₃·aq (pn = 1,2-diaminopropane; tn = 1,3-diaminopropane; chxn = *trans*-1,2-diaminocyclohexane) have been prepared with minor modifications of the procedure described below.

Procedure

Rhodium(III) chloride trihydrate (26.3 g., 0.10 mole) is placed in a 100-ml. round-bottomed flask which is then fitted with a condenser. The flask is immersed in an ice bath, and an ice-cold mixture of 27 ml. (0.33 mole) of ethylenediamine monohydrate and 25 ml. of water is added through the condenser. The mixture should be added very carefully at first, and the whole addition should take at least 5 minutes as the initial reaction is very vigorous.* The ice bath is removed (but kept at hand for controlling the reaction, if necessary). The reaction mixture is heated very carefully and finally refluxed until the rhodium chloride is dissolved (10–15 minutes). The solution is cooled for 15 minutes before 25 ml. of ethanol is added through the condenser. The mixture is boiled until the precipitate formed has redissolved, and the yellow solution is then allowed to stand at room temperature for 2 hours. During standing, white, needlelike crystals separate. Finally the mix-

---

* Rather large crystals of rhodium(III) chloride are preferred. The violence of the initial reaction is probably due to crystal powder and to smaller crystals accompanying the bigger crystals.

ture is cooled in ice. The crystals are collected on a filter, washed successively with 20-ml. portions of 60, 80, and 96% ethanol, and dried in air. The yield is approximately 29 g.

The combined filtrate and washings are evaporated to complete dryness.* The residue is placed in an oven at 120°C. for 4 hours and is then dissolved in 15 ml. of hot water. Twelve milliliters of absolute ethanol is added to the solution, and the mixture is boiled until a clear solution has formed. The brownish solution is allowed to stand for 2 hours before cooling in ice. The precipitate is filtered, washed successively with 10-ml. portions of 60, 80, and 96% ethanol, and dried in air. The yield is approximately 9 g.

A third crop of crystals is obtained in the following way: The combined filtrate and washings are evaporated to complete dryness.* The residue is placed in an oven at 220°C. for 15 minutes† and is then dissolved in 12 ml. of hot water. The solution is filtered.‡ During heating, 25 ml. of absolute ethanol is added to the filtrate, and the mixture is boiled until a clear solution has formed. The grayish solution is allowed to stand for one hour before cooling in ice. The precipitate is filtered, washed successively with 5-ml. portions of 60, 80, and 96% ethanol, and dried in air. The yield is approximately 6 g.

The combined crude products are dissolved in 45 ml. of hot water, and the solution is filtered.‡ During heating, 55 ml. of absolute ethanol is added to the filtrate, and the mixture is boiled until a clear solution has formed. Five drops of ethylenediamine monohydrate is added, and the solution is allowed to stand for 3 hours before it is cooled in ice. The crystals are filtered, washed first with two 20-ml. portions of 80% ethanol,

---

* Preferably in a rotating vacuum evaporator to avoid splashing of the syrupy concentrate.

† If the heat treatment is prolonged too much, colloidal rhodium is produced. If the heat treatment is shortened, the amorphous residue is not converted into a crystalline one. Fifteen minutes at 220°C. is a suitable compromise.

‡ A filter with extremely high retention, e.g., Whatman No. 50, is needed in order to get a pure product.

then with 20 ml. of 96% ethanol, and dried in air. The yield is 40 g. (90%). *Anal.* Calcd. for $[Rh(en)_3]Cl_3 \cdot 3H_2O$:* C, 16.25; H, 6.82; N, 18.95; Cl, 23.98; Rh, 23.20. Found: C, 16.42; H, 6.74; N, 18.64; Cl, 24.13; Rh, 23.25.

B. RESOLUTION OF TRIS(ETHYLENEDIAMINE)RHODIUM(III) CHLORIDE

$$2[Rh(en)_3]Cl_3 \cdot 3H_2O + 2H_2(+)\text{-tart} + 4LiOH \cdot H_2O \rightarrow$$
$$Li\{(-)_D\text{-}[Rh(en)_3]\}\{(+)\text{-tart}\}_2 \cdot 3H_2O +$$
$$(+)_D\text{-}[Rh(en)_3]Cl_3 \cdot 3H_2O + 3LiCl + 8H_2O$$
$$Li\{(-)_D\text{-}[Rh(en)_3]\}\{(+)\text{-tart}\}_2 \cdot 3H_2O + 4HCl \rightarrow$$
$$(-)_D\text{-}[Rh(en)_3]Cl_3 \cdot 3H_2O + 2H_2(+)\text{-tart} + LiCl$$

(en = ethylenediamine; (+)-tart = (+)-tartrate anion;
D = sodium D line)

Werner[1] resolved the tris(ethylenediamine)rhodium(III) ion both with sodium-3-nitro-(+)-camphor and with (+)-tartrate as resolving agents. Both methods, however, suffer from disadvantages. The solubility of the diastereoisomer $\{(-)_D\text{-}[Rh(en)_3]\}$-$\{(3\text{-nitro-}(+)\text{-camphor})_3\}$ precipitated by the first method is so low that purification by recrystallization is impossible. On the other hand, the solubilities of the diastereoisomeric chloride (+)-tartrates used in the second method are of the same order of magnitude, so that the solution has to be seeded with a crystal of $\{(-)_D\text{-}[Rh(en)_3]\}Cl\{(+)\text{-tart}\}\cdot aq$. The (+)$_D$ form of the chloride (+)-tartrate is not obtained in an optically pure condition.[3] Furthermore, the yields of both methods are low because of the many operations needed for obtaining the stereoisomers of the chloride. By the present method $(-)_D\text{-}[Rh(en)_3]^{3+}$ is precipitated as the double salt $Li\{(-)_D\text{-}[Rh(en)_3]\}\{(+)\text{-tart}\}_2\cdot 3H_2O$, which is purified before conversion to the chloride. The (+)$_D$-chloride is precipitated from the mother liquor by the addition of hydrochloric acid and ethanol.

---

* The content of water of crystallization varies according to the history. The products prepared by this method (air-drying in about 50% relative humidity) contain approximately 3 moles of water of crystallization.

Procedure

Tris(ethylenediamine)rhodium(III) chloride trihydrate (11.10 g., 0.025 mole) and 5.63 g. (0.0375 mole, 50% excess) of (+)-tartaric acid and 3.15 g. (0.075 mole) of LiOH·H$_2$O are placed in a 100-ml. conical flask and dissolved in 25 ml. of boiling water. When a clear solution has formed, the solution is allowed to cool somewhat by standing for 5 minutes. Fifteen milliliters of ethanol is then added, and the solution is allowed to stand for 3 hours at room temperature with occasional shaking.* During standing, white crystals of the tartrate double salt separate. The precipitate is filtered, washed with three 10-ml. portions of a 1:1 mixture of ethanol and water, and dried in air. The yield is approximately 8.3 g.

The filtrate is transferred to a 250-ml. flask, 8 ml. (0.1 mole) of 12 *M* HCl is added, and the solution is heated to boiling. After standing for 5 minutes, 75 ml. of absolute ethanol is added, and the solution is allowed to stand for 2 hours at room temperature. The crystal cake is broken up, and the mixture is cooled in ice before filtering. The crystals are washed, first with 9 ml. of 80% and then with two 9-ml. portions of 96% ethanol, and dried in air. The yield is 3.5 g. (65%). *Anal.* Calcd. for (+)$_D$-[Rh(en)$_3$]Cl$_3$·2.3H$_2$O†: C, 16.72; H, 6.69; N, 19.50; Cl, 24.68; Rh, 23.88. Found: C, 16.93; H, 6.24; N, 19.17; Cl, 26.66; Rh, 23.74.

The crude tartrate double salt is dissolved in 25 ml. of boiling water, and a solution of 5.4 g. of LiCl in 9 ml. of boiling water is added. The solution is filtered. The filtrate is heated to boiling and allowed to stand for 5 minutes. Thirty-five milliliters of absolute ethanol is added slowly with vigorous stirring, and the solution is allowed to cool by standing for 3 hours at room temperature. The crystal cake is broken up, and the mixture is cooled in ice. The precipitate is filtered, washed

* By cooling in ice, Li$_2$(+)-tart·H$_2$O is co-precipitated.
† Water content depends on method of drying.

successively with 9-ml. portions of 60, 80, and 96% ethanol, and dried in air.* The yield is 7.3 g. (92%). *Anal.* Calcd. for Li{(−)$_D$-[Rh(en)$_3$]}{(+)-tart}$_2$·3H$_2$O: C, 26.26; H, 5.98; N, 13.13; Li, 1.08. Found: C, 26.18; H, 5.92; N, 13.00; Li, 1.10.

The recrystallized tartrate double salt (7.0 g., 0.011 mole) is dissolved in 10 ml. (0.12 mole) of boiling 12 M HCl, and 40 ml. of absolute ethanol is added. A large white crystalline precipitate is formed. The mixture is heated to boiling, and the suspension is allowed to stand for one hour before cooling in ice. The precipitate is filtered, washed with 10 ml. of 80% and two 10-ml. portions of 96% ethanol, and dried in air. The yield is 4.3 g. (91%). *Anal.* Calcd. for (−)$_D$-[Rh(en)$_3$]Cl$_3$·2.3H$_2$O: C, 16.72; H, 6.69; N, 19.50; Cl, 24.68; Rh, 23.88. Found: C, 16.96; H, 6.36; N, 19.17; Cl, 26.66; Rh, 23.73.

C. RESOLUTION OF TRIS(ETHYLENEDIAMINE)CHROMIUM(III) CHLORIDE

$$2[Cr(en)_3]Cl_3 \cdot 3H_2O + 2H_2(+)\text{-tart} + 4LiOH \cdot H_2O \rightarrow$$
$$Li\{(+)_D\text{-}[Cr(en)_3]\}\{(+)\text{-tart}\}_2 \cdot 3H_2O +$$
$$(-)_D\text{-}[Cr(en)_3]Cl_3 \cdot 3H_2O + 3LiCl + 8H_2O$$
$$Li\{(+)_D\text{-}[Cr(en)_3]\}\{(+)\text{-tart}\}_2 \cdot 3H_2O + 4HCl \rightarrow$$
$$(+)_D\text{-}[Cr(en)_3]Cl_3 \cdot 3H_2O + 2H_2(+)\text{-tart} + LiCl$$

(en = ethylenediamine; (+)-tart = (+)-tartrate anion;
D = sodium D line)

The tris(ethylenediamine)chromium(III) ion was first resolved by Werner[5] by means of sodium 3-nitro-(+)-camphor. What has been said concerning the resolution of the corresponding rhodium compound holds true of the chromium compound, except that for the chromium compound the solubility difference of the diastereoisomeric chloride (+)-tartrates is so small that a resolution via these diastereoisomers has not been achieved.[5,6] The method reported here is essentially the same as the one described for the rhodium complex but with minor alterations

---

* At this point the optical purity should be ascertained because the optical isomers of the chloride are much more soluble than the corresponding racemic salt. It is therefore difficult, or, when it is present in large quantities, impossible, to remove racemic salt by recrystallization of the stereoisomers of the chloride.

because of the decreased robustness and thereby decreased stability toward racemization of the chromium complex.

Procedure

Tris(ethylenediamine)chromium(III)   chloride   trihydrate* (39.3 g., 0.10 mole) and 22.5 g. (0.15 mole) of (+)-tartaric acid, and a magnetic stirring bar, are placed in a 400-ml. beaker. A 100-ml. volume of water is added, and the mixture is dissolved with stirring.   Immediately after the dissolution, 12.6 g. (0.30 mole) of $LiOH \cdot H_2O$ is added with continued stirring.   The hydroxide dissolves, and shortly after a yellow crystalline precipitate is formed.   About 5 minutes after the addition of the hydroxide, 100 ml. of ethanol is added dropwise with continued stirring during a 15-minute period.   The precipitate is filtered, and the filtrate is run directly into a mixture of 15 ml. (0.18 mole) of 12 M HCl and 600 ml. of absolute ethanol.   From this system a finely crystalline precipitate of impure $(-)_D$-chloride separates out.   The filter flask is replaced, and the crystals are washed with three 50-ml. portions of a 1:1 mixture of ethanol and water and dried in air.   (Yield is 27–28 g.)   The precipitate of $(-)_D$-chloride from the filtrate is filtered, washed with three 50-ml. portions of ethanol, and dried in air.   (Yield is 13–14 g.)

The crude tartrate double salt is dissolved in 500 ml. of water, and the solution is filtered.   A filtered solution of 21.2 g. (0.50 mole) of LiCl in 500 ml. of ethanol is added dropwise to the stirred filtrate during a period of approximately an hour, and the stirring is continued for 15 minutes.   The precipitate is filtered, washed with three 30-ml. portions of a 1:1 mixture of ethanol and water, and dried in air.†   The yield is 23–24 g.

* If tris(ethylenediamine)chromium(III) chloride is prepared by the method of Rollinson and Bailar,[7] it is worthy of note that if the dehydration of the hydrated chromium(III) sulfate is performed in a vacuum oven (2–3 mm. Hg, 70–80°C.) the anhydrous sulfate is obtained in a much more reactive form so that the reaction with ethylenediamine, when started by local heating, proceeds by itself and is finished within an hour. *Editor's note:* For an alternative preparation, see *Inorganic Syntheses,* **10,** 35 (1967); also ref. 26.

† See footnote * on p. 274.

(80%).   *Anal.* Calcd. for $Li\{(+)_D\text{-}[Cr(en)_3]\}\{(+)\text{-tart}\}_2 \cdot 3H_2O$:
C, 28.53; H, 6.50; N, 14.26; Li, 1.18; Cr, 8.82.   Found: C,
28.93; H, 6.06; N, 14.44; Li, 1.18; Cr, 8.82.

To the recrystallized double salt (approximately 23 g. or
approximately 0.04 mole) is added 30 ml. (0.36 mole) of 12 *M*
HCl.   A 300-ml. quantity of methanol is added dropwise with
stirring during a period of approximately 30 minutes, and the
stirring is continued for 5 minutes.   The precipitate is filtered,
washed with three 25-ml. portions of methanol, and dried in
air.   Yield is approximately 13 g. of $(+)_D$-chloride.

The crude $(+)_D$-chloride is dissolved in 23 ml. of 4 *M* HCl,
and the solution is filtered.   To the stirred filtrate is added
300 ml. of methanol during a period of approximately 30 minutes,
and the stirring is continued for 5 minutes.   The precipitate is
filtered, washed with three 25-ml. portions of methanol, and
dried in air.   The yield is 11 g. (60%).   *Anal.* Calcd. for
$(+)_D\text{-}[Cr(en)_3]Cl_3 \cdot 1.6H_2O^*$: C, 19.61; H, 7.46; N, 22.87; Cl,
28.94; Cr, 14.15.   Found: C, 19.98; H, 7.42; N, 22.38; Cl, 28.63;
Cr, 14.00.

The crude $(-)_D$-chloride is treated with 38 ml. of 4 *M* HCl
by stirring for a few minutes, and the residue of racemic salt
is filtered as fast as possible.   To the stirred filtrate is added
450 ml. of methanol during a period of approximately 30 minutes.
The precipitate is filtered, washed with three 15-ml. portions of
methanol, and dried in air.   The yield is 9–10 g. (50%).   *Anal.*
Calcd. for $(-)_D\text{-}[Cr(en)_3]Cl_3 \cdot 1.6H_2O^*$: C, 19.61; H, 7.46; N,
22.87; Cl, 28.94; Cr, 14.15.   Found: C, 21.24; H, 7.30; N,
22.70; Cl, 28.59; Cr, 14.15.

Properties

Tris(ethylenediamine)rhodium(III) chloride is obtained as
white needlelike crystals, and $Li\{(-)_D\text{-}[Rh(en)_3]\}\{(+)\text{-tart}\}_2 \cdot$
$3H_2O$ and $(-)_D$- and $(+)_D\text{-}[Rh(en)_3]Cl_3 \cdot aq$ as small white
crystals. $Li\{(+)_D\text{-}[Cr(en)_3]\}\{(+)\text{-tart}\}_2 \cdot 3H_2O$ and $(+)_D$- and

---

* The content of water of crystallization varies according to the history.

$(-)_D$-[Cr(en)$_3$]Cl$_3$·aq are isolated as small yellow crystals.   The tartrate double salts are sparingly soluble in cold water and are probably not congruently soluble, whereas the stereoisomeric chlorides are highly soluble and have a solubility much higher than the corresponding racemic salts.   The tris(ethylenediamine)rhodium(III) ion is very robust, and the stability toward racemization is so high that an aqueous solution can be evaporated to dryness without loss of optical activity.[1]   The robustness of [Cr(en)$_3$]$^{3+}$ is much lower.[7]

X-ray powder diagrams obtained by the Guinier method indicate great structural similarities between Li$\{(-)_D$-[Rh(en)$_3$]$\}$-$\{(+)$-tart$\}_2$·3H$_2$O and Li$\{(+)_D$-[Cr(en)$_3$]$\}\{(+)$-tart$\}_2$·3H$_2$O and between [Co(en)$_3$]Cl$_3$·3H$_2$O, 2[Co(en)$_3$]Cl$_3$·NaCl·6H$_2$O, 2$(+)_D$-[Co(en)$_3$]Cl$_3$·NaCl·6H$_2$O, [Cr(en)$_3$]Cl$_3$·3H$_2$O, 2[Cr(en)$_3$]Cl$_3$·KCl·6H$_2$O, 2[Cr(en)$_3$]Cl$_3$·RbCl·6H$_2$O, 2$(+)_D$-[Cr(en)$_3$]Cl$_3$·RbCl·6H$_2$O, [Rh(en)$_3$]Cl$_3$·3H$_2$O, and 2[Rh(en)$_3$]Cl$_3$·NaCl·6H$_2$O, whereas $(+)_D$-[Co(en)$_3$]Cl$_3$·aq, $(+)_D$-[Cr(en)$_3$]Cl$_3$·aq, and $(+)_D$-[Rh(en)$_3$]Cl$_3$·aq do not show such structural similarities, either among themselves or with the above-mentioned series of compounds.[24]   It is noteworthy that [Co(en)$_3$]Cl$_3$·3H$_2$O[9] and 2$(+)_D$-[Co(en)$_3$]Cl$_3$·NaCl·6H$_2$O[10] are closely related in structure even though their space groups are P$\bar{3}$c1 and P3, respectively.   It has been found[8] that the racemic salts [Cr(en)$_3$]Cl$_3$·3H$_2$O and 2[Cr(en)$_3$]Cl$_3$·KCl·6H$_2$O also belong to the space group P$\bar{3}$c1. The unit-cell dimensions are given in Table I.

In [Co(en)$_3$]Cl$_3$·3H$_2$O and 2$(+)_D$-[Co(en)$_3$]Cl$_3$·NaCl·6H$_2$O the water molecules are enclosed in hollow channels in the structure. On the basis of the structural similarities,[10] it is probably

TABLE I   Unit-cell Dimensions for Racemic Salts Belonging to Space Group P$\bar{3}$c1

	a	c
[Co(en)$_3$]Cl$_3$·3H$_2$O[9]	11.50 A.	15.52 A.
2[Cr(en)$_3$]Cl$_3$·KCl·6H$_2$O[8]	11.60 A.	15.34 A.
[Cr(en)$_3$]Cl$_3$·3H$_2$O[8]	11.56 A.	15.40 A.

Inorganic Syntheses

allowable to draw the conclusion that the same is true of the racemic salts [Cr(en)$_3$]Cl$_3$·3H$_2$O and [Rh(en)$_3$]Cl$_3$·3H$_2$O. The experiments reported here further seem to indicate that a similar zeolitic behavior characterizes the optically active tris(ethylenediamine)chromium(III) and -rhodium(III) chlorides.

Some optical properties determined in aqueous solution on the compounds prepared and purified according to the present procedures are given in Tables II and III.

TABLE II   Absorption and CD Spectra

	Absorption		Circular dichroism	
	λ_{max}, nm.	ϵ_{max}, l./mole-cm.	λ_{max}, nm.	$(\epsilon_l - \epsilon_r)_{max}$, l./mole-cm.
Li{(−)$_D$-[Rh(en)$_3$]}{(+)-tart}$_2$·3H$_2$O $c \sim 3 \times 10^{-3}\,M$	301 255	243 194	320 288 257	1.88 −0.33 0.77
(+)$_D$-[Rh(en)$_3$]Cl$_3$·2.3H$_2$O $c \sim 4 \times 10^{-3}\,M$			319 287 257	−2.02 0.10 −0.78
	301 255	238* 191		
(−)$_D$-[Rh(en)$_3$]Cl$_3$·2.3H$_2$O $c \sim 4 \times 10^{-3}\,M$			319 287 257	1.97† −0.11 0.73
Li{(+)$_D$-[Cr(en)$_3$]}{(+)-tart}$_2$·3H$_2$O $c \sim 7 \times 10^{-3}\,M$	457 351	76.2 61.0	457 ~350 330	1.54 $\gtrsim$0 0.071
(+)$_D$-[Cr(en)$_3$]Cl$_3$·1.6H$_2$O $c \sim 13 \times 10^{-3}\,M$(abs.); $9 \times 10^{-3}\,M$(CD)			457 ~350 330	1.69§ −0.004 0.069
	457 351	75.0‡ 60.6		
(−)$_D$-[Cr(en)$_3$]Cl$_3$·1.6H$_2$O $c \sim 13 \times 10^{-3}\,M$(abs.); $9 \times 10^{-3}\,M$(CD)			457 ~350 330	−1.62 0.004 −0.068

* Literature,[11] λ = 301 (ϵ = 210), 253 (190); literature,[12] 302 (238); literature,[13] 306 (251), 257 (246); literature,[4] 299 (101), 253 (107).

† Literature,[14] λ = 310 ($\Delta\epsilon$ = 1.4); literature,[13] 320 (2.0), 287 (−0.1), 258 (0.8).

‡ Literature,[13] 460 (74), 353 (65); literature,[15] 457 (76.5), 351 (60.7).

§ Literature,[14] 460 (1.7); literature,[13] 456 (1.36), 350 (−0.05).

TABLE III Specific and Molar Rotation at 20°C.

		313 nm.	364 nm.	436 nm.	546 nm.	578 nm.	589 nm.
Li{(−)$_D$-[Rh(en)$_3$]}{(+)-tart}$_2$·3H$_2$O $c \sim 4.7 \times 10^{-3}$ M	[α] deg. ml./g. dm.	25	25	−58.7	−42.3	−38.0	−37.1
	[M] = [α]M/100	160	160	−376	−271	−243	−238
(+)$_D$-[Rh(en)$_3$]Cl$_3$·2.3H$_2$O $c \sim 5.8 \times 10^{-3}$ M	[α] deg. ml./g. dm.	....	7.4	134	94.7	84.8	81.7
	[M] = [α]M/100	....	32	575	408	374	352*
(−)$_D$-[Rh(en)$_3$]Cl$_3$·2.3H$_2$O $c \sim 5.8 \times 10^{-3}$ M	[α] deg. ml./g. dm.	....	−5.2	−126	−88.6	−79.7	−75.1
	[M] = [α]M/100	....	−22	−541	−382	−344	−324†
Li{(+)$_D$-[Cr(en)$_3$]}{(+)-tart}$_2$·3H$_2$O $c \sim 3.5 \times 10^{-3}$ M	[α] deg. ml./g. dm.	−447	−389	−530	149	89.8	77.1
	[M] = [α]M/100	−2630	−2290	2650	876	529	454
(+)$_D$[Cr(en)$_3$]Cl$_3$·1.6H$_2$O $c \sim 4.9 \times 10^{-3}$ M	[α] deg. ml./g. dm.	−865	−829	....	223	124	103
	[M] = [α]M/100	−3180	−3050	....	818	456	380
(−)$_D$-[Cr(en)$_3$]Cl$_3$·1.6H$_2$O $c \sim 4.9 \times 10^{-3}$ M	[α] deg. ml./g. dm.	852	810	....	−218	−122	−101
	[M] = [α]M/100	3130	2980	....	802	−447	−370

* Literature,[1] 338.9.
† Literature,[1] −347.6.

The ions $(+)_D$-$[Cr(en)_3]^{3+}$ and $(-)_D$-$[Rh(en)_3]^{3+}$ have been assigned the absolute configuration Λ, the same* as that of $(+)_D$-$[Co(en)_3]^{3+}$ for which the configuration was determined by the anomalous diffraction of x-rays.[17, 25] The assignments for the Cr and Rh compounds were deduced from the method of active racemates[18, 24] and from conformational analysis on tris complexes of optically active diamine ligands[19, 20] together with comparison of Cotton effects for related electronic transitions,[13, 14, 21−23] and both are in full agreement with Werner's criterion of least soluble diastereoisomers.[1]

References

1. A. Werner, *Ber.*, **45**, 1228 (1912).
2. F. M. Jaeger, *Proc. Koninkl. Akad. Amsterdam*, **20**, 244 (1917).
3. F. M. Jaeger, *Rec. Trav. Chim.*, **38**, 171 (1919).
4. R. D. Gillard, J. A. Osborn, and G. Wilkinson, *J. Chem. Soc.*, 1951 (1965).
5. A. Werner, *Ber.*, **45**, 865 (1912).
6. F. M. Jaeger, *Bull. Soc. Chim. France*, [5], **4**, 1201 (1937).
7. C. L. Rollinson and J. C. Bailar, Jr., *Inorganic Syntheses*, **2**, 196 (1946).
8. P. J. McCarthy, private communication.
9. K. Nakatsu, Y. Saito, and H. Kuroya, *Bull. Chem. Soc. Japan*, **29**, 428 (1956).
10. K. Nakatsu, M. Shiro, Y. Saito, and H. Kuroya, *ibid.*, **30**, 158 (1957).
11. C. K. Jørgensen, *Acta Chem. Scand.*, **10**, 500 (1956).
12. H.-H. Schmidtke, *Z. Physik. Chem.*, Neue Folge, **38**, 170 (1963).
13. A. J. McCaffery, S. F. Mason, and R. E. Ballard, *J. Chem. Soc.*, 2883 (1965).
14. J. H. Dunlop, R. D. Gillard, and G. Wilkinson, *ibid.*, 3160 (1964).
15. F. Woldbye, *Acta Chem. Scand.*, **12**, 1079 (1958).
16. IUPAC Information Bulletin 33, p. 68, 1968.
17. K. Nakatsu, *Bull. Chem. Soc. Japan*, **35**, 832 (1962).
18. M. Delépine and R. Charonnat, *Bull. Soc. Mineral.*, **53**, 73 (1930).
19. E. J. Corey and J. C. Bailar, Jr., *J. Am. Chem. Soc.*, **81**, 2620 (1959).
20. A. M. Sargeson, in R. L. Carlin, "Transition Metal Chemistry," Vol. 3, Marcel Dekker, Inc., New York, 1966.
21. J. H. Dunlop and R. D. Gillard, *J. Inorg. Nucl. Chem.*, **27**, 361 (1965).
22. R. D. Gillard and G. Wilkinson, *J. Chem. Soc.*, 1368 (1964).
23. S. F. Mason and B. J. Norman, *Chem. Commun.*, 73 (1965).
24. P. Andersen, F. Galsbøl, and S. E. Harnung, *Acta Chem. Scand.*, **23**, 3027 (1969).
25. Y. Saito, K. Nakatsu, M. Shiro, and H. Kuroya, *Bull. Chem. Soc. Japan*, **30**, 795 (1957).
26. E. Pedersen, *J. Sci. Inst.*, [2], **1**, 1013 (1968).

* Λ is the symbol proposed by IUPAC.[16]

Chapter Six

NONMETAL SYSTEMS

I. PHOSPHORUS COMPOUNDS

50. DIFLUOROPHOSPHINES

Submitted by LOUIS CENTOFANTI* and R. W. RUDOLPH†
Checked by MAX LUSTIG‡

The compounds F_2POPF_2, F_2PPF_2, and F_2PH are examples of difluorophosphines which can be conveniently prepared from various reactions involving the labile P—halogen bonds in F_2PI and F_2PBr.§ A metathetical reaction is used for the preparation of μ-oxo-bis(difluorophosphine), F_2POPF_2, whereas reductive coupling effects the syntheses of tetrafluorodiphosphine, F_2PPF_2, and difluorophosphine, F_2PH. Recent studies in our laboratories and others[2-3] indicate that the general methods outlined here can be extended to prepare a host of substituted difluorophosphines.

■ *Caution. Since many of the fluoro derivatives of the oxygen acids of phosphorus are known[5] to be extremely toxic and since*

* Department of Chemistry, Emory University, Atlanta, Ga. 30322.
† Department of Chemistry, University of Michigan, Ann Arbor, Mich. 48104.
‡ Department of Chemistry, Memphis State University, Memphis, Tenn. 38111.
§ The preparation of pure halodifluorophosphines, F_2PX, is described in *Inorganic Syntheses*, **10**, 147 (1967).

detailed and reliable toxicity data on all the compounds considered here are not available, all these fluorophosphine derivatives should be handled as though they were highly toxic materials.

A. TETRAFLUORODIPHOSPHINE, P_2F_4

$$2F_2PI + 2Hg \rightarrow F_2PPF_2 + Hg_2I_2$$

Procedure

Standard high-vacuum techniques[6,7] are used since some of the reagents and products are volatile, pyrophoric, and/or toxic. Since F_2PI and F_2PBr attack mercury, a mercury float-valve system cannot be used, and manometers should be protected with stopcocks until pressure measurements are to be made.*

A 1000-ml. bulb, equipped with a stopcock and containing 2 ml. of triply distilled mercury, is attached to the vacuum system by a standard-taper connection. The bulb is thoroughly evacuated. A quantity of F_2PI (4.80 mmoles) is condensed into the bulb and the stopcock closed. The bulb is then removed from the vacuum system, warmed to 25°C., and shaken. After 6 hours of shaking, the bulb is attached to the vacuum manifold and its volatile contents removed through U-tubes maintained at −111°C. (CCl_3F slush), −126°C. (methylcyclohexane slush), and −196°C. (liquid nitrogen). Unreacted F_2PI (~0.50 mmole) is retained at −111°C. The desired P_2F_4 (2.00 mmole) slowly passes through the −111°C. trap and is held at −126°C. The −196°C. fraction is PF_3 (0.20 mmole). The yield is 93% based on the amount of F_2PI which reacted. If the initial pressure of F_2PI in the reaction bulb is too high, low yields result. It is recommended that the calculated initial pressure of F_2PI approximate 100 mm.

---

* Kel-F-10 oil may be used to protect the manometers, although it is not entirely satisfactory.

Properties

The vapor-pressure data for P_2F_4 follow the equation $\log P(\text{mm.}) = -1290/T + 7.716.$[8] A vapor pressure of 114 mm. is observed at $-45.2°C$. (chlorobenzene slush). In the region from 4000–200 cm.$^{-1}$ the infrared spectrum of P_2F_4 shows absorptions at 847(vvs), 839(vvs), 834(vvs), 828(vvs), 365(m), and 361(m) cm.$^{-1}$.[2] The ^{19}F and ^{31}P n.m.r. spectra are reported elsewhere.[4] Tetrafluorodiphosphine can be stored in clean glass at $-78°C$. (Dry Ice) with little decomposition; however, it is best kept at $-196°C$.

B. DIFLUOROPHOSPHINE

$$F_2PI + HI + 2Hg \rightarrow F_2PH + Hg_2I_2$$

Procedure I

A 70-ml. reaction tube equipped with a stopcock is charged with 2 ml. of triply distilled mercury. The reaction tube is attached to the vacuum line through a standard-taper connection and is thoroughly evacuated. Equal amounts (3.36 mmoles) of F_2PI and HI* are condensed into the tube at $-196°C$. The stopcock is then closed; the tube is removed from the vacuum system, warmed to 25°C., and shaken for 2 hours. The tube and its contents are then frozen with liquid nitrogen. Recovery is effected by allowing the tube to warm to 25°C. while pumping the volatile products through *two* traps held at $-196°C$.† The contents of the two $-196°C$. traps are then combined. Separation is effected by fractional distillation through traps at $-140°C$. (cooled 30–60°C. petro-

---

* Hydrogen iodide can be obtained from Matheson Co., Inc., 932 Paterson Plank Road, East Rutherford, N.J. 07073, or prepared as outlined by C. J. Hoffman, *Inorganic Syntheses*, **7**, 180 (1963).

† The second trap recovers any product which is entrained in the small amount of H_2 (noncondensable at $-196°C$.) which is formed during the reaction.

leum ether), $-160°$C. (isopentane slush), and $-196°$C. An unstable material thought to be $PHF_2 \cdot HI$ is held at $-140°$C. The desired F_2PH (1.85 mmoles) is retained at $-160°$C. The $-196°$C. trap contains PF_3 (0.72 mmole). The yield is 55% based on the F_2PI taken.

Procedure II

$$2F_2PI + 2Hg + PH_3 \rightarrow 2F_2PH + Hg_2I_2 + \frac{1}{n}[PH]_n$$

If PH_3 is available,* F_2PH is obtained in very good yield from the reduction of F_2PI by PH_3 in the presence of mercury.

In this procedure, a 500-ml. reaction bulb equipped with a stopcock and containing about 2 ml. of triply distilled mercury is attached to the vacuum system and thoroughly evacuated. Samples of F_2PI (2.36 mmoles) and PH_3 (3.40 mmoles) are condensed into the bulb at $-196°$C. The bulb is then removed from the vacuum system, warmed to $25°$C., and shaken for 15 hours. The product is recovered by condensation at $-160°$C. (isopentane slush), while unreacted PH_3 (2.34 mmoles) and PF_3 (0.15 mmole) slowly pass through the $-160°$C. trap and are held at $-196°$C. The yield is 2.11 mmoles of F_2PH, or 90% based on the amount of F_2PI taken.

Properties

Liquid F_2PH displays a vapor pressure of 208 mm. at $-83.6°$C. (ethyl acetate slush). The vapor-pressure equation is $\log P$(mm.) $= -1126/T + 8.280$.[9] The infrared spectrum[9] of the gas displays absorptions at 2251(s), 2240(vs), 2233(s), 1016(vs), 1008(s and br), 958(vw), 838(vs), 851(vvs), 825(vvs),

* Phosphine is conveniently prepared as outlined by S. D. Gokhale and W. L. Jolly, *Inorganic Syntheses*, **9**, 56 (1967).

367(w), 348(w) cm.$^{-1}$ in the 4000–200 cm.$^{-1}$ region. The proton n.m.r. spectrum[9] of the liquid* consists of a doublet of 1:2:1 triplets centered −7.65 p.p.m. from TMS internal standard, $J_{PH} = 182$ and $J_{FPH} = 42$ Hz. To ensure no decomposition, difluorophosphine is best stored at −196°C.; however, little decomposition is observed after 1 week at −78°C. in a clean glass tube equipped with a stopcock.

C. μ-OXO-BIS(DIFLUOROPHOSPHINE)

$$2F_2PX + M_2O \rightarrow F_2POPF_2 + 2MX$$

Procedure I, Using Copper(I) Oxide

In a typical experiment, a 40-ml. reaction tube equipped with a stopcock and standard-taper joint is charged with 0.5 g. of Cu_2O (3.5 mmoles), attached to the vacuum system, and evacuated for one hour in order to ensure removal of trace amounts of water. Then F_2PI (2.14 mmoles) is condensed into the tube with liquid nitrogen. The contents of the tube are allowed to warm to 25°C. and cooled to −196°C. several times. Reaction is evidenced by the formation of tan-colored copper(I) iodide. Finally, the volatile components are removed from the vessel by pumping them through a series of traps held at −112°C. (CS_2 slush), −145°C. (cooled 30–60°C. petroleum ether), and −196°C. It is necessary to heat the reaction tube to ~150°C. with a hot-air gun to remove the last trace of product. This procedure gives 0.78 mmole of F_2POPF_2 in the −145°C. trap, corresponding to a 73% yield based on the amount of F_2PI taken. Trace amounts of an unidentified gas and PF_3 are held in the −112°C. and −196°C. traps, respectively.

---

* If the spectrum is determined at the ambient temperature of the probe (usually +35°C.), the sample should be contained in a heavy-walled tube such as the semimicro n.m.r. tubes available from NMR Specialties, Inc., New Kensington, Pa.

Procedure II, Using Tri-*n*-butyltin(IV) Oxide

$$2F_2PBr + [(C_4H_9)_3Sn]_2O \rightarrow 2(C_4H_9)_3SnBr + F_2POPF_2$$

When larger amounts of F_2POPF_2 are desired, the reaction of F_2PBr with tri-*n*-butyltin oxide is preferable.

In a typical run tri-*n*-butyltin oxide (4.41 g., 7.41 mmoles) is added to a 250-ml. flask equipped with a stopcock and standard-taper joint. The flask is attached to the vacuum manifold and evacuated. After the tip of the flask is frozen with liquid nitrogen, F_2PBr (14.10 mmoles) is condensed into the vessel and the stopcock closed. After repeated warming to 25°C. and cooling to -196°C., the desired product is separated by fractional condensation as described above. With this procedure 4.79 mmoles of F_2POPF_2 is retained at -145°C. The unidentified difluorophosphites ($ROPF_2$, 0.68 mmole) and PF_3 (2.2 mmoles) found in the -112°C. and -196°C. traps, respectively, are discarded. The yield based on the amount of F_2PBr taken is about 60–70%. Optimum yields result if a slight stoichiometric excess of F_2PBr is used. However, F_2PBr and F_2POPF_2 cannot be separated by fractional condensation; hence, if pure F_2POPF_2 is desired, it is necessary to adhere strictly to a mole ratio for F_2PBr to $[(C_4H_9)_3Sn]_2O$ of 1.9:1.0.

Properties

F_2POPF_2 is a water-white liquid having a vapor pressure of 61.0 mm. at -63.6°C. (chloroform slush). The vapor-pressure equation is

$$\log P(\text{mm.}) = \frac{-1300}{T} + 7.981^2$$

The infrared spectrum of the vapor shows absorptions at 1077(w), 976(vvs and br), 863, 853, 842(vvs and br), 682(m), 519, 515(m), 460(w), 359(w) cm.$^{-1}$ in the 4000–200 cm.$^{-1}$ region. The material can be stored in clean glass at 25°C. and saturation pressure with less than 1% decomposition after a day. However, it is most conveniently kept at $-78°C$. (Dry Ice) in a glass tube equipped with a stopcock and standard-taper joint. The ^{31}P and ^{19}F n.m.r. spectra are described in the literature.[2]

References

1. J. G. Morse, K. Cohn, R. W. Rudolph, and R. W. Parry, *Inorganic Syntheses*, **10**, 147 (1967).
2. R. W. Rudolph, R. C. Taylor, and R. W. Parry, *J. Am. Chem. Soc.*, **88**, 3729 (1966).
3. R. G. Cavell and R. C. Dobbie, *J. Chem. Soc.*, 1308 (1967).
4. F. A. Johnson and R. W. Rudolph, *J. Chem. Phys.*, **47**, 5449 (1967).
5. M. Pianka, *J. Appl. Chem. (London)*, **5**, 109 (1955).
6. R. T. Sanderson, "Vacuum Manipulation of Volatile Compounds," John Wiley & Sons, Inc., New York, 1948.
7. W. L. Jolly, "Synthetic Inorganic Chemistry," Prentice-Hall, Inc., Englewood Cliffs, N.J., 1960.
8. M. Lustig, J. K. Ruff, and C. B. Coburn, *J. Am. Chem. Soc.*, **88**, 3875 (1966).
9. R. W. Rudolph and R. W. Parry, *Inorg. Chem.*, **4**, 1339 (1965).

51. DIMETHYLTHIOPHOSPHINIC BROMIDE
(*Dimethylphosphinothioic Bromide*)

$$(CH_3)_2P(S)\text{-}P(S)(CH_3)_2 + Br_2 \rightarrow 2(CH_3)_2P(S)Br$$

Submitted by R. SCHMUTZLER*
Checked by L. F. CENTOFANTI†

Dimethylthiophosphinic bromide has been prepared by addition of elemental bromine to tetramethyldiphosphine disulfide suspended in carbon tetrachloride.[1-3] The reaction is rather general in that alkyl or aryl groups other than methyl can be attached to the phosphorus atom to give products of the type

* Lehrstuhl B für Anorganische Chemie der Technischen Universität, Pockelsstrasse 4, 33 Braunschweig, Germany.
† Department of Chemistry, University of Utah, Salt Lake City, Utah 84112; current address: Emory University, Atlanta, Ga. 30322.

RR'P(S)Br.[2,4,5] Dimethylthiophosphinic bromide is also formed by the addition of elemental sulfur to dimethylbromophosphine, $(CH_3)_2PBr$, or to the product $(CH_3)_2PBr_3$ obtained in the reaction of methyl bromide with red phosphorus.[6]

Dimethylthiophosphinic bromide and its analogs, RR'P(S)Br, are useful in the introduction of the $(CH_3)_2P(S)$- group into other molecules, wherever two hydrocarbon groups bonded to phosphorus are desired. The following reactions, for instance, have been described for $R_2P(S)Br$[7]

$$R_2P(S)Br + R_2P(S)SNa \rightarrow R_2P(S)\!-\!S\!-\!P(S)R_2 + NaBr$$
$$2R_2P(S)Br + Ag_2O \rightarrow R_2P(S)\!-\!O\!-\!P(S)R_2 + 2AgBr$$

Reaction of dimethylthiophosphinic bromide with Grignard reagents gives rise to formation of tertiary phosphine sulfides[3,8]:

$$(CH_3)_2P(S)Br + RMgX \rightarrow (CH_3)_2P(S)R + MgXBr$$
$$(R = CH_3, C_6H_5)$$

Upon desulfurization of $(CH_3)_2(P(S)Br$ with tributylphosphine, dimethylbromophosphine, $(CH_3)_2PBr$, is obtained in high yield.[2,6]

In general, $(CH_3)_2P(S)Br$ (and its analogs) will react in the same way as $(CH_3)_2P(S)Cl$ or, more generally, RR'P(S)Cl. The bromides, however, are preferred to the chlorides, since their preparation, involving the liquid bromine instead of gaseous chlorine, is more convenient.

Procedure*

A 1-l., four-necked, round-bottomed flask is equipped with a mechanical stirrer, a 250-ml. dropping funnel with sidearm, a thermometer reaching to the bottom of the flask, and a reflux condenser, topped by a drying tube. The system is flushed with dry nitrogen. In a countercurrent of nitrogen, 84 g.

* The checker carried out the reaction on $\frac{1}{5}$ scale with a yield of 82%.

(0.45 mole) of tetramethyldiphosphine disulfide[9a,b] is suspended in 500 ml. of dry carbon tetrachloride, while a solution of 73.5 g. (0.46 mole) of bromine in 200 ml. of carbon tetrachloride is placed in the dropping funnel. The bromine is added dropwise with stirring over a period of one hour. A mildly exothermic reaction, accompanied by an immediate discharge of the bromine color and gradual dissolution of the phosphine sulfide, takes place. The temperature is kept below 30°C. by occasional cooling with ice water. After stirring at room temperature for an additional 15 hours, the mixture is filtered rapidly through a folded paper filter. Carbon tetrachloride is distilled off at atmospheric pressure through a 12-in. Vigreux column. The product is obtained by fractionation of the residue *in vacuo* as a colorless liquid which rapidly solidifies in the receiver. Yield is 132–144 g. (85–92%).

Properties

Dimethylthiophosphinic bromide is a deliquescent solid, m.p. 34°C.[1] (32–34°C.[2]); b.p. 205–207°C./718 mm.[2] (90°/13 mm.;[1] 87–88°C./14 mm.[3]); refractive index, $n_D^{35} = 1.5482$.[2]

The ^{31}P chemical shift in chloroform is −63.2 p.p.m., relative to 85% phosphoric acid.[10] A 1:1 doublet is observed in the [1]H n.m.r. spectrum, $J_{H-P} = 13.2$ Hz., $\delta_H = -2.15$ p.p.m. (in cyclohexane solution), relative to an internal $Si(CH_3)_4$ reference.[11]

References

1. R. Cölln and G. Schrader (to Farbenfabriken Bayer A. G.): German patent 1,054,453 (April 9, 1959).
2. L. Maier, *Chem. Ber.*, **94**, 3051 (1961).
3. H. J. Harwood and K. A. Pollart, *J. Org. Chem.*, **28**, 3430 (1963).
4. W. Kuchen and H. Buchwald, *Angew. Chem.*, **71**, 162 (1959).
5. W. Kuchen, H. Buchwald, K. Strolenberg, and J. Metten, *Ann.*, **652**, 28 (1962).
6. L. Maier, *Helv. Chim. Acta*, **46**, 2026 (1963).
7. W. Kuchen, K. Strolenberg, and H. Buchwald, *Chem. Ber.*, **95**, 1703 (1962).

8. H. J. Harwood and K. A. Pollart (to Monsanto Chemical Co.), U.S. patent 3,053,900 (Sept. 11, 1962).
9a. H. Reinhardt, D. Bianchi, and D. Mölle, *Chem. Ber.*, **90**, 1656 (1957).
9b. G. W. Parshall, *Org. Syn.*, **45**, 102 (1965).
10. K. Moedritzer, L. Maier, and L. C. D. Groenweghe, *J. Chem. Eng. Data*, **7**, 307 (1962).
11. J. F. Nixon and R. Schmutzler, *Spectrochim. Acta*, **22**, 565 (1966).

52. (TRICHLOROMETHYL)DICHLOROPHOSPHINE
(*Trichloromethylphosphonous Dichloride*)

$$[CCl_3PCl_3][AlCl_4] + CH_3OPCl_2 \rightarrow CCl_3PCl_2 + \cdots$$

Submitted by R. SCHMUTZLER* and M. FILD*
Checked by STUART GOLINGER† and ROBERT R. HOLMES†

(Trichloromethyl)dichlorophosphine can be prepared by the reduction of the corresponding tetrachlorophosphorane, CCl_3PCl_4, with white phosphorus in a carbon tetrachloride medium.[1] Methyl dichlorophosphite, CH_3OPCl_2, was also employed as the reducing agent in an alternative method of synthesis of CCl_3PCl_2 from CCl_3PCl_4.[2] A particularly convenient starting material for the preparation of CCl_3PCl_2 is provided in the ternary complex between carbon tetrachloride, phosphorus trichloride, and aluminum chloride.[3,4] Reducing agents such as white phosphorus,[5] methyl dichlorophosphite,[6] or phenyldichlorophosphine[7] have been employed. The photochemical or thermal reaction between elemental phosphorus and carbon tetrachloride,[8] or exposure to gamma[8] or neutron irradiation[9] of the same reactants, also leads to formation of (trichloromethyl)dichlorophosphine. The reduction by methyl dichlorophosphite of the ternary complex formed by phosphorus trichloride, carbon tetrachloride, and aluminum chloride is described in the following procedure.

* Lehrstuhl B für Anorganische Chemie der Technischen Universitat, Pockelsstrasse 4, 33 Braunschweig, Germany.
† Department of Chemistry, University of Massachusetts, Amherst, Mass. 01002.

Procedure

The complex $[CCl_3PCl_3][AlCl_4]$ is conveniently made by the direct combination of its constituents; carbon tetrachloride, phosphorus trichloride, and aluminum chloride[3,4] are used.*

The reduction is conducted in a 500-ml. four-necked flask, fitted with an efficient reflux condenser topped by a drying tube, a well-sealed mechanical stirrer, a thermometer, and a 200-ml. dropping funnel with sidearm. The system is evacuated repeatedly through a stopcock adapter, temporarily placed on top of the reflux condenser. The evacuated system is filled with dry nitrogen and is maintained under slight positive nitrogen pressure by exhausting the nitrogen through a mercury bubbler.

To the flask, maintained under a dry nitrogen atmosphere, are added 1.80 moles of PCl_3 (240 g. or 152 ml.), 0.44 mole of CCl_4 (68.0 g. or 43 ml.), and 0.22 mole of Al_2Cl_6 (58.5 g.). The suspension is heated to a temperature of 75°C. while stirring constantly. Then 58 g. of methyl dichlorophosphite (0.445 mole)[10] is added slowly (dropwise) to the stirred suspension over a $\frac{1}{2}$-hour period. The stirring is continued for 4 hours while the temperature is held at 75 ± 5°C.

Dichloromethane† (methylene chloride) (200 ml.) is placed in the dropping funnel and added to the stirred solution over a 5-minute period. The resulting suspension is filtered through a sintered-glass filter tube, using one of the previously described techniques (page 10) for filtering in the absence of air. The residue is washed with 200 ml. of dry dichloromethane. The filtrate is distilled in a 12-in. Vigreux column to remove the volatile products. The residue‡ thus left is transferred under

* It was found that the ternary complex could also be prepared conveniently and in nearly quantitative yield in a stainless-steel shaker tube (1 hour/70°C.), instead of an open system as described in reference 4.

† A normal reagent-grade product which has been dried over calcium chloride and distilled is used.

‡ The checkers noted that this residue is mostly solid at room temperature but may be liquid if the material is warm. They noted that a short condenser or none at all should be used since the product solidifies in the vacuum distillation (m.p. 50°C.).

a nitrogen blanket to a 50-ml. flask and is distilled *in vacuo* through a 6-in. Vigreux column. (Trichloromethyl)dichloro-phosphine is obtained as a liquid of b.p. 55°C./13 mm. which readily solidifies in the condenser. The yield is 44–50 g. (45–51%). (Checker reported a yield of only 20–25%.)

Properties

(Trichloromethyl)dichlorophosphine is a volatile low-melting solid which is exceedingly sensitive to air. Boiling points between 65–67°C./20 mm.[7] and 51°C./13 mm.[6] have been reported. According to the earliest reference on CCl_3PCl_2, the compound distilled with decomposition at 82–83°C./7 mm.[1] but a boiling point of 171–172°C. at atmospheric pressure was subsequently reported.[2] Reported melting points for CCl_3PCl_2 range between 40 to 50°C., the recent value 49.6–50°C.[11] being most likely. The ^{31}P chemical shift of CCl_3PCl_2 in carbon tetrachloride, relative to an external 85% H_3PO_4 reference, is -148.6 p.p.m.[12]

(Trichloromethyl)dichlorophosphine, like other halophos-phines, is an important reactive intermediate in phosphorus chemistry. Its reactions with various nucleophilic reagents have been described. Thus hydrolysis under controlled con-ditions gives rise to formation of (trichloromethyl)phosphonous acid,[11] $[CCl_3P(O)H(OH)]_3$. Reaction of CCl_3PCl_2 with ethanol in the presence of a tertiary base forms $CCl_3P(OC_2H_5)_2$.[13] Fluorination of (trichloromethyl)dichlorophosphine with anti-mony trifluoride yields (trichloromethyl)difluorophosphine, CCl_3PF_2.[14,15] This compound is of importance both as a precursor to penta-coordinate phosphorus compounds, con-taining the —CCl_3 group, and, especially, as a versatile, novel ligand in transition-metal coordination chemistry.[16,17]

References

1. V. A. Ginsburg and A. Ya. Yakubovich, *Zh. Obshch. Khim.*, **28**, 728 (1958).
2. L. D. Quin and C. H. Rolston, *J. Org. Chem.*, **23**, 1693 (1958).

3. A. M. Kinnear and E. A. Perren, *J. Chem. Soc.*, 3437 (1952).
4. K. C. Kennard and C. S. Hamilton, *Org. Syn.*, Coll. Vol. **4**, 950 (1963).
5. J. L. Van Winkle, S. C. Bell, and R. C. Morris (to Shell Development Co.), U.S. patent 2,875,224 (Feb. 24, 1959); *C.A.*, **53**, 13055 (1959).
6. V. P. Davydova and M. G. Voronkov, U.S.S.R. patent 135,485 (Feb. 15, 1961); *C.A.*, **55**, 15350 (1961).
7. S. Z. Ivin, L. E. Dimitrieva, and K. V. Karavanov, *Zh. Obshch. Khim.*, **36**, 950 (1966).
8. D. Perner and A. Henglein, *Z. Naturforsch.*, **17b**, 703 (1962).
9. A. Henglein, H. Drawe, and D. Perner, *Radiochim. Acta*, **2**, 19 (1963).
10. E. J. Malowan, D. R. Martin, and P. J. Pizzolato, *Inorganic Syntheses*, **4**, 63 (1953).
11. J. F. Nixon, *J. Chem. Soc.*, 2471 (1964).
12. R. Schmutzler, unpublished work.
13. R. E. Atkinson, J. I. G. Cadogan, and J. Dyson, *J. Chem. Soc.*, **C**, 2542 (1967).
14. J. F. Nixon, *Chem. Ind.* (*London*), 1555 (1963).
15. J. F. Nixon, *J. Inorg. Nucl. Chem.*, **27**, 1281 (1965).
16. J. F. Nixon, *J. Chem. Soc.*, **A**, 1136 (1967).
17. C. G. Barlow, J. F. Nixon, and M. Webster, *ibid.*, **A**, 2216 (1968).

53. PHENYL-SUBSTITUTED PHOSPHONITRILE FLUORIDE TRIMERS

Submitted by CHRISTOPHER W. ALLEN* and THERALD MOELLER|
Checked by JAMES E. DUNNING‡

The phosphonitrile fluoride trimers can be represented as

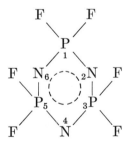

The fluorine atoms are readily replaced by phenyl or *o*-, *m*-, and *p*-tolyl groups on treatment with the corresponding lithium

* Department of Chemistry, University of Vermont, Burlington, Vt. 05401.
† Department of Chemistry, University of Illinois, Urbana, Ill. 61801; present address: Department of Chemistry, Arizona State University, Tempe, Ariz. 85281.
‡ Department of Chemistry, University of Michigan, Ann Arbor, Mich. 48104.

reagents.[1,2] When two aryl groups are introduced, the products are primarily nongeminally substituted.[2] Although the replacement is stepwise and can be controlled to yield primarily products of any degree of substitution, mixtures of positional and stereo isomers result whenever this is possible.[1,2] In combination with the Friedel-Crafts reaction, however, the aryllithium reaction permits the syntheses of a number of isomeric products.[3] For phenyl-substituted compounds, the monophenyl compound, $N_3P_3F_5(C_6H_5)$, is useful as a starting material (Sec. A). The conversion of this compound to the geminal diphenyl derivative is described in Sec. B.

A. (MONOPHENYL)PHOSPHONITRILE FLUORIDE TRIMER

(1-Phenyl-1,3,3,5,5-pentafluorocyclotriphosphazatriene)

$$N_3P_3F_6 + C_6H_5Li \rightarrow N_3P_3F_5(C_6H_5) + LiF$$

Procedure

Approximately 100 ml. of diethyl ether is distilled from elemental sodium directly into a dry 500-ml., three-necked, round-bottomed flask which is equipped with a nitrogen inlet tube and a reflux condenser attached to a mercury-bubbler outlet. An additional 100 ml. of ether is distilled into a pressure-equalizing dropping funnel. Then 2.94 g. (0.42 mole) of lithium metal, cut in the form of thin 1×1-cm. squares, is added to the flask, together with a magnetic stirring bar.* The dropping funnel is attached immediately to the flask and charged with 31.4 g. (0.2 mole) of dry bromobenzene. The flask is flushed for 5 minutes with dry nitrogen. The flow of nitrogen is stopped, the stirrer is started, and *ca.* 5 ml. of the solution of bromobenzene is added. After the reaction has started, as indicated by a light gray coloration in the reaction mixture, the remaining bromobenzene solution is added at such a rate as to maintain constant reflux. Stirring is continued until reflux

---

* Stirring bars used in this preparation must *not* be Teflon-coated.

ceases, and the mixture is allowed to cool. The concentration of phenyllithium* is then determined by titrating a 1-ml. aliquot of the solution with 0.1 M hydrochloric acid, using methyl red as indicator.

A volume† of the phenyllithium solution corresponding to 0.1 mole of the solute is transferred to a pressure-equalizing dropping funnel. This transfer is accomplished by connecting the reaction flask and funnel (with standard-taper joints) through a 105° glass tube containing a small filter of glass wool.

The phenyllithium solution is poured through the filter into the funnel. The dropping funnel containing the phenyllithium solution is then attached to a 500-ml., three-necked, round-bottomed flask containing 24.9 g. (0.1 mole) of trimeric phosphonitrile fluoride,[4,5] 100 ml. of diethyl ether freshly distilled from sodium, and a magnetic stirring bar. The flask is equipped as indicated above, and the system is flushed with dry nitrogen for 5 minutes. The flask is then placed in an ice-water bath, the flow of nitrogen is stopped, and the stirrer is started. The phenyllithium solution is added over a period of 1–2 hours. The ice-water bath is then removed, and the reaction mixture is heated at reflux for one hour. The solvent is removed by means of a rotary evaporator, and the resulting oily mixture is treated with 100 ml. of low-boiling (30–60°C.) petroleum ether. The precipitated lithium fluoride is removed by one or more filtrations through infusorial earth.

The resulting clear filtrate is passed through a 30 × 700-mm. chromatographic column which is one-third filled with silica gel.‡ With low-boiling petroleum ether as the eluting agent and a rapid flow rate, 200-ml. fractions are collected until no

---

* Commercial phenyllithium does not give satisfactory results. Phenyllithium solutions prepared as described herein should be used within 24 hours. The major impurities that appear in the product are phenol and biphenyl which result from decomposition and coupling reactions of the lithium reagent.

† Use of a calibrated dropping funnel simplifies measurement of this volume. The checker added a standard volume of the phenyllithium solution from a pipet and adjusted the amount of trimeric phosphonitrile fluoride added in the next step.

‡ Complete removal of lithium fluoride is essential to rapid flow.

more product is removed from the column. Phenol is retained. After solvent removal, the resulting liquid fractions are combined and distilled through a 15-cm. Vigreux column at 1 mm. Hg* at an oil-bath temperature of 85–110°C. The product distills at 50–52°C. Yield, based upon trimeric phosphonitrile fluoride, is 19.1 g. (62.5%). *Anal.* Calcd. for $N_3P_3F_5(C_6H_5)$: C, 23.45; H, 1.63; N, 13.68; mol. wt., 307. Found: C, 23.74; H, 1.72; N, 13.72; mol. wt., 319.

Properties

(Monophenyl)phosphonitrile fluoride trimer is a hydrolytically stable, colorless liquid that boils at 183–184°C. with slight decomposition. It is soluble in petroleum ether, benzene, diethyl ether, and chloroform but insoluble in either cold or boiling water. The infrared spectrum has a strong phosphorus-nitrogen stretching vibration at 1270 cm.$^{-1}$. Strong bands associated, respectively, with asymmetric and symmetric phosphorus-fluorine stretching modes appear at 940 and 850, 830 cm.$^{-1}$

B. (1,1-DIPHENYL)PHOSPHONITRILE FLUORIDE TRIMER

(1,1-Diphenyl-3,3,5,5-tetrafluorocyclotriphosphazatriene)

$$N_3P_3F_5(C_6H_5) + C_6H_6 + (C_2H_5)_3N \xrightarrow{Al_2Cl_6} 1,1\text{-}N_3P_3F_4(C_6H_5)_2 + (C_2H_5)_3NHF$$

The compound $1,1\text{-}N_3P_3F_4(C_6H_5)_2$ can be prepared by the reaction between (monophenyl)phosphonitrile fluoride trimer and benzene in the presence of anhydrous aluminum chloride and triethylamine.[2] In the absence of triethylamine, the product is contaminated with difficultly separable biphenyl. Triethylamine also improves the conversion of phosphonitrile chloride trimer to the geminally substituted diphenyl derivative[6] by the Friedel-Crafts procedure.[7] The Friedel-Crafts reaction can be used also for the preparation of the geminally substituted

---

* At lower pressures, removal of biphenyl is incomplete.

tetra- and hexaphenyl derivatives from trimeric phosphonitrile chloride.[7]   A similar reaction converts the mixture of geminal, cis, and trans (diphenyl)phosphonitrile fluoride isomers obtained by the phenyllithium reaction to 1,1,3,3-tetraphenylphosphonitrile fluoride trimer.[2]   Unlike the aryllithium reaction, the Friedel-Crafts reaction favors the formation of geminally substituted aryl derivatives in the fluoride system.[2]

Procedure

To a 500-ml., three-necked, round-bottomed flask, fitted with two pressure-equalizing dropping funnels and a mechanical stirrer, are added 175 ml. of benzene, freshly distilled from sodium, and 60 g. (0.225 mole) of anhydrous aluminum chloride. To one dropping funnel are added 25 ml. of freshly distilled benzene and 7.6 g. (0.075 mole) of triethylamine; to the other, 50 ml. of freshly distilled benzene and 15.35 g. (0.05 mole) of (monophenyl)phosphonitrile fluoride trimer (Sec. A).   The triethylamine solution is added slowly with stirring,* and the empty dropping funnel is then replaced by a reflux condenser bearing a phosphorus(V) oxide drying tube.   The contents of the flask are brought to reflux temperature and maintained at that temperature for 30 minutes, after which time the (monophenyl)phosphonitrile fluoride solution is added over a period of 30 minutes.

The reaction is allowed to continue for 24 hours.   The stirrer is then stopped, and the reaction mixture is allowed to cool to room temperature.   The contents of the flask are then hydrolyzed by pouring into a 1000-ml. beaker which is half-filled with ice and contains 5 ml. of 12 M hydrochloric acid.†   After hydrolysis is complete, the liquid layers are separated, and the aqueous layer is washed once with benzene.   The benzene layer and benzene washings are combined and washed

---

* The reaction is exothermic.

† This operation should be carried out in a hood because of the quantities of hydrogen chloride that are evolved.

alternately three times with water and saturated sodium hydrogen carbonate solution and finally twice with water.

The washed benzene solution is dried over anhydrous magnesium sulfate and decolorized with charcoal. The benzene is then removed by means of a rotary evaporator. The resulting yellow oil is dissolved in a minimum volume of *n*-pentane and the solution cooled in a Dry Ice–acetone bath. The crystals which are formed are removed and purified by sublimation at 70°C. and 0.025 mm. Hg. Yield, based upon (monophenyl)-phosphonitrile fluoride trimer, is 13 g. (71%). *Anal.* Calcd. for $N_3P_3F_4(C_6H_5)_2$: C, 39.45; H, 2.74; N, 11.51; mol. wt., 365. Found: C, 39.85; H, 2.76; N, 11.58; mol. wt., 359.

Properties

The (1,1-diphenyl)phosphonitrile fluoride trimer is a colorless crystalline solid that melts at 68.5–69.5°C. It can be recrystallized from *n*-pentane, *n*-heptane, petroleum ether, or absolute methanol. It is also soluble in diethyl ether, carbon disulfide, and chloroform, but it is insoluble in and not attacked by water. The infrared spectrum shows a strong phosphorus-nitrogen stretching mode at 1250–1265 cm.$^{-1}$. Strong bands at 914–920, 900–906 cm.$^{-1}$ and at 812–820 cm.$^{-1}$ are associated, respectively, with phosphorus-fluorine asymmetric and symmetric stretching modes.

This compound crystallizes in the orthorhombic system with $a = 14.803(4)$, $b = 12.571(5)$, and $c = 16.732(8)$ A.; $Z = 8$; space group = Pnma.[8,9] The phenyl-substituted phosphorus atom is out of plane with the five other ring atoms by 0.20 A. There are three sets of phosphorus-nitrogen bonds with mean lengths 1.618(5), 1.558(4), and 1.539(5) A. The exocyclic bond angles are C—P—C, 107.9(3)°, and F—P—F, 96.9(2)°. The endocyclic bond angles are N—P—N, 115.5(3)° and 120.6(3)°, and P—N—P, 120.5(2)°.

The compound $1,1\text{-}N_3P_3F_4(C_6H_5)_2$ reacts with an equimolar quantity of phenyllithium to form $1,1,3\text{-}N_3P_3F_3(C_6H_5)_3$.[2]

References

1. T. Moeller and F. Tsang, *Chem. Ind. (London)*, 361 (1962).
2. C. W. Allen and T. Moeller, *Inorg. Chem.*, **7**, 2177 (1968).
3. C. W. Allen, F. Y. Tsang, and T. Moeller, *ibid.*, **7**, 2183 (1968).
4. T. Moeller, K. John, and F. Tsang, *Chem. Ind. (London)*, 347 (1961).
5. T. Moeller, R. Schmutzler, and F. Tsang, *Inorganic Syntheses*, **9**, 75 (1967).
6. E. T. McBee, K. Okuhara, and C. J. Morton, *Inorg. Chem.*, **4**, 1672 (1965).
7. R. A. Shaw, R. Keat, and C. Hewlett, in "Preparative Inorganic Reactions," W. L. Jolly (ed.), Vol. 2, pp. 74–75, Interscience Publishers, a division of John Wiley & Sons, Inc., New York, 1965.
8. C. W. Allen, J. B. Faught, T. Moeller, and I. C. Paul, *Inorg. Chem.*, **8**, 1719 (1969).

II. FLUORINE COMPOUNDS

■ *Warning.* *Many of the nitrogen-fluorine compounds discussed in this section are very hazardous materials. Their syntheses should not be undertaken without adequate shielding equipment and a prior knowledge of techniques for the handling of explosive substances.* Compounds of this type are labeled extra-hazardous or hazardous, depending upon properties of the material.

The editor is happy to acknowledge the assistance of Prof. John K. Ruff of the University of Georgia, who assisted in the preparation of this section.

54. DIFLUOROAMIDO COMPOUNDS— EXTRA-HAZARDOUS MATERIALS

Submitted by G. W. FRASER,* J. M. SHREEVE,† MAX LUSTIG,‡
and CARL L. BUMGARDNER§
Checked by CLAUDE I. MERRILL∥

Difluoroamidocarbonyl fluoride, difluoroamidosulfuryl fluoride, difluoroaminofluorosulfate, and difluoroamidosulfur penta-

* Mount LaSalle Novitiate, Christian Brothers, Napa, Calif.
† Department of Chemistry, University of Idaho, Moscow, Idaho 83843.
‡ Department of Chemistry, Memphis State University, Memphis, Tenn. 38111.
§ Department of Chemistry, North Carolina State University, Raleigh, N.C. 27607.
∥ Chemical Research Laboratory, Edwards Air Force Base, Calif. 93534.

fluoride are formed in the photolytic decomposition of tetra-fluorohydrazine, N_2F_4, at room temperature in the presence of excess carbon monoxide,[1] sulfur dioxide,[2,3] sulfur trioxide,[4] and sulfur tetrafluoride,[5] respectively.

A standard Pyrex-glass apparatus with a mercury manometer covered with Kel-F-10 oil is employed for the transfer of gaseous materials. During the preparation and subsequent handling of the above difluoroamido compounds, all apparatus must be scrupulously dry, and as much work as possible should be performed under conditions of high vacuum to minimize hydrolysis. ■ *Caution. All starting materials and most of the products of these syntheses are highly toxic gases;* also, the tetrafluorohydrazine, dinitrogen difluoride (N_2F_2; also called difluorodiazine), and difluoroamido derivatives are oxidizing agents and must be kept free from hydrocarbon material or other reducing agents. *Workers should use appropriate shielding and wear gloves when manipulating the nitrogen fluoride reactants and products.*

A. DIFLUOROAMIDOCARBONYL FLUORIDE

(Difluorocarbamoyl Fluoride)

$$2N_2F_4 + 2CO \overset{h\nu}{\rightarrow} 2NF_2C(O)F + N_2F_2$$

Procedure

The reaction vessel is a 3-l. round Pyrex bulb equipped with a 2-mm. stopcock and a quartz tube insert (20 cm. in length and 2.5-cm. i.d.) which is sealed into the bulb by means of 45/50 S.T. joints.* Fluorocarbon grease is used for lubricant. The bulb is evacuated and charged with tetrafluorohydrazine† (100-mm. pressure, 16 mmoles), which is used without purifica-

---

* A diagram of the photolysis apparatus used by the checker for all syntheses listed is shown in Fig. 22.

† Air Products and Chemicals, Inc., 733 W. Broad St., Emmaus, Pa. 18049.

tion, and carbon monoxide* (200-mm. pressure, 32 mmoles), which is then taken from a sample condensed in liquid nitrogen at −196°C. to ensure removal of condensable impurities. The reaction mixture is then irradiated with a water-cooled, 70-watt, Hanovia type 81 high-pressure mercury arc lamp for 2–5 hours† through the quartz insert. After irradiation, the contents of the bulb are slowly pumped out through a series of two U-traps cooled in liquid nitrogen. The condensable fraction is about 20 mmoles. There is no significant change in pressure in the bulb during irradiation.

The product mixture is separated by fractional co-distillation[6] (batch size about 1.5 mmoles) using a 12-ft. fractionating column of $\frac{1}{8}$-in.-o.d. unpacked aluminum tubing which has been coiled to fit into a half-pint Dewar flask. The helium from the tank is passed through a U-trap cooled in liquid nitrogen to remove any condensable impurities and is adjusted to about 4 ml. per minute. The major components of the mixture are, in the order of appearance during the distillation, *cis*-dinitrogen difluoride, nitrous oxide, carbonyl fluoride, carbon dioxide, tetrafluorohydrazine (which is typically removed at a column temperature of about −75°C.), difluoroamidocarbonyl fluoride, and difluoroamine (fluoroimide). The fraction of crude difluoroamidocarbonyl fluoride (2–3 mmoles) is reseparated in three or four batches to remove low- and high-boiling impurities, primarily tetrafluorohydrazine and difluoroamine, respectively. The yield of purified difluoroamidocarbonyl fluoride is about 0.15–0.25 g. (1.5–2.5 mmoles, 10–15% of theory).

Purity of the product may be checked by vapor-density molecular weight (99.01) or by analysis. Oxidizing power of

* Matheson Co., Inc., 932 Paterson Plank Rd., East Rutherford, N.J. 07073.

† The irradiation period depends on the time the lamp has been in service. Two hours suffices for a new lamp. Extended irradiation will result in complete decomposition of the desired product.

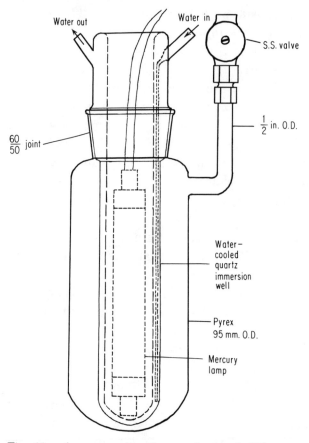

Fig. 22. *Apparatus for the synthesis of difluoro-amido compounds.*

the sample is easily determined by hydrolyzing a weighed sample in an excess of triiodide-free hydriodic acid at room temperature and titrating the liberated iodine with standard thiosulfate solution (0.0404 equiv./g.). The amount of difluoro-amine in the product, usually the major impurity, is easily estimated by examining a high-pressure infrared spectrum of the gas in the region of 11–11.5 μ where difluoroamine absorbs strongly.

Properties

Difluoroamidocarbonyl fluoride is a colorless gas which condenses to a colorless mobile liquid at its normal boiling point, $-52°C$. The liquid has a vapor pressure and density of approximately 160 mm. and 1.63 g./ml., respectively, at the temperature of Dry Ice; it forms a colorless glass below $-130°C$. The gas is hydrolyzed instantly to carbon dioxide and difluoroamine by water, to carbonate and (primarily) dinitrogen difluoride by sodium hydroxide solutions, and is reduced to ammonium ion by hydriodic acid solutions. Clean Pyrex glass and mercury are not attacked at room temperature, but ultraviolet light causes slow decomposition to carbonyl fluoride and dinitrogen difluoride, N_2F_2. Extended storage is preferably under high vacuum at the temperature of liquid nitrogen. Its infrared spectrum consists of the following bands (expressed in cm.$^{-1}$): 1900(s), 1212(s), 1035(m), 960(m), 845(s), 835(w, triplet), and 766(m, triplet).

B. DIFLUOROAMIDOSULFURYL FLUORIDE

(Difluorosulfamoyl Fluoride)

$$2N_2F_4 + 2SO_2 \xrightarrow{h\nu} 2NF_2SO_2F + N_2F_2$$

This compound may be obtained also by heating the two reactants to about 120°C.,[3] but temperature control is vital and yields are considerably lower than when the photolytic method is used.

Procedure

The reaction vessel is a 550-ml. Pyrex cylindrical flask with a Hanovia internally sealed cold-cathode low-pressure mercury resonance coil lamp (2537 A.). The lamp is fitted to the reactor through a 29/42 S.T. joint. Tetrafluorohydrazine and sulfur dioxide (10 mmoles each) are admitted to the reactor. Irradiation is continued for $2\frac{1}{2}$ hours or until no further pressure decrease occurs.

The products, according to mass spectral analysis, are NF_2SO_2F (8.9 mmoles, 89% yield), N_2F_2 (3 mmoles), and smaller amounts of N_2O, NO, SiF_4, SO_2, and SO_2F_2. The NF_2SO_2F is obtained pure by passing the gaseous mixture at 25°C. through a 20-ft. × 0.25-in. column packed with Dow 710 silicone on Chromosorb W.

Properties

Difluoroamidosulfuryl fluoride is a colorless gas which condenses to a colorless liquid at its boiling point, −18°C. The liquid has a vapor pressure of 33.6 mm. at −79.2°C. and forms a colorless solid at −111°C. Liquid densities are given by the equation $d = 1.545 - 0.00308t(°C.)$. Its infrared spectrum consists of bands at 1488(s), 1250(s), 1000(w), 920(s), 845(s), 685(w), 604(m), 530(w), 483(w), and 467(w)μ cm.$^{-1}$. The compound undergoes hydrolysis upon contact with aqueous alkaline solutions to evolve nitrogen, nitrogen(I) oxide, tetrafluorohydrazine, and dinitrogen difluoride. Purity may also be checked by molecular weight (135) and elemental analysis (N, 10.36%; S, 22.2%) using the Dumas method and alkaline hydrolysis, respectively.

C. DIFLUOROAMINOFLUOROSULFATE

$$N_2F_4 + 2SO_3 \xrightarrow{h\nu} 2NF_2OSO_2F + N_2F_2$$

This compound can be made in almost quantitative amounts by the reaction of peroxydisulfuryl difluoride, $S_2O_6F_2$, and tetrafluorohydrazine.[7] However, since the procedure below avoids the necessity of a two-step synthesis, it is more convenient and also does not require the special reactor for making $S_2O_6F_2$.

Procedure

Ten millimoles each of SO_3* and N_2F_4 are introduced into the apparatus described in Sec. B. The mixture is photolyzed for 10 minutes. Purification of NF_2OSO_2F is carried out by vacuum-line fractionation of the crude mixture through a cold trap at $-126°C$. The less volatile NF_2OSO_2F is retained (55% yield), and the other products, including SO_2F_2, N_2O, NF_3, N_2F_2, N_2, and NO, pass through to a trap at $-195°C$.

Properties

Difluoroaminofluorosulfate is readily hydrolyzed by dilute alkaline solution to nitrate, fluorosulfate, and fluoride ions; this provides an easy route to decomposition for elemental analysis (N, 9.27; S, 21.2). It is a colorless liquid which boils at $-2.5°C$. and condenses to a glass at about $-129°C$. Its density at $25°C$. is 1.498 g./cc. Purity can be readily determined from molecular weight (151.1) and its infrared spectrum which has bands at 1492, 1250, 1032, 913, 848, and 748 cm.$^{-1}$.

D. DIFLUOROAMIDOSULFUR PENTAFLUORIDE

$$2N_2F_4 + 2SF_4 \xrightarrow{\Delta} 2NF_2SF_5 + N_2F_2$$

Difluoroamidosulfur pentafluoride results from the photolysis of N_2F_4 and SF_5Cl[5] or from the thermolysis of N_2F_4 with either S[5] or S_2F_{10}[8,9] and also from heating a mixture of N_2F_4, N_2F_2, and SF_4.[10] A photochemical reaction of N_2F_4 and SF_4[5] is the most convenient method.

---

* Sulfan B (General Chemical Division, Allied Chemical Corporation, P.O. Box 405, Morristown, N.J. 07960).

Procedure

A 12-l. nickel tank with a calcium fluoride window is charged with SF_4 (25 g., 0.25 mole) and N_2F_4 (26 g., 0.25 mole).* The mixture is irradiated for 80 hours with a low-pressure mercury resonance lamp. Gas chromatographic as well as mass spectrometric analyses indicate a yield of ~30%. Other products are N_2F_2, NF_3, N_2, and SF_6. Some unreacted starting materials are also present. Purification of SF_5NF_2 is achieved by passing the crude mixture through 5% NaOH solution and then through cold traps at −45, −126, and −196°C. The trap at −126°C. retains the SF_5NF_2.

Properties

Difluoroamidosulfur pentafluoride is a colorless liquid that boils at $−7.5 \pm 0.50°C$. It can be stored in stainless steel at room temperature. Its infrared spectrum shows major bands at 885, 913, and 945 cm.$^{-1}$. The purity may be checked by molecular weight (179) and elemental analysis (N, 7.82; S, 17.9).

References

1. G. W. Fraser and J. M. Shreeve, *Inorg. Chem.*, **4**, 1497 (1965).
2. C. L. Bumgardner and M. Lustig, *ibid.*, **2**, 662 (1963).
3. M. Lustig, C. L. Bumgardner, F. A. Johnson, and J. K. Ruff, *ibid.*, **3**, 1165 (1964).
4. M. Lustig, C. L. Bumgardner, and J. K. Ruff, *ibid.*, **3**, 917 (1964).
5. A. L. Logothetis, G. N. Sausen, and R. J. Shozda, *ibid.*, **2**, 173 (1963).
6. G. H. Cady and D. P. Siegwarth, *Anal. Chem.*, **31**, 618 (1959).
7. M. Lustig and G. H. Cady, *Inorg. Chem.*, **2**, 388 (1963).
8. G. H. Cady, D. F. Eggers, and B. Tittle, *Proc. Chem. Soc.*, 65 (1963).
9. E. C. Stump, Jr., C. D. Padgett, and W. S. Brey, *Inorg. Chem.*, **2**, 648 (1963).
10. M. Lustig, *ibid.*, **4**, 104 (1965).

* The checker obtained a yield of 70% SF_5NF_2 by using apparatus and conditions similar to those employed for the synthesis of difluoroamidosulfuryl fluoride. Use of scrupulously dry glassware (dried in a 500°C. oven and used as soon as possible following removal from the oven or flamed out under vacuum) and a few grams of sodium fluoride within the reaction vessel to absorb traces of hydrogen fluoride served to reduce the etching of the glassware to an invisible level.

55. 1,1-DIFLUOROUREA SOLUTIONS AND DIFLUORO-AMINE—EXTRA-HAZARDOUS MATERIALS

$$H_2N-\overset{\overset{\displaystyle O}{\|}}{C}-NH_2 + 2F_2 \xrightarrow{H_2O} H_2N-\overset{\overset{\displaystyle O}{\|}}{C}-NF_2 + 2HF$$

$$H_2N-\overset{\overset{\displaystyle O}{\|}}{C}-NF_2 + 2HF + H_2O \xrightarrow{\Delta} HNF_2 + CO_2 + NH_4F{\cdot}HF$$

Submitted by C. O. PARKER* and J. P. FREEMAN†
Checked by WILLIAM GRAHAM* and MAX LUSTIG‡

Difluoroamine (fluorimide) was discovered by Kennedy and Colburn as a by-product accompanying tetrafluorohydrazine in the reduction of NF_3 by arsenic[1] and arsine. It was found to be a product of hydrolysis of fluorinated urea[2a] and was detected in trace amounts among the products of fluorination of ammonia.[3] Reactions to be considered for preparative purposes include the following: (1) reduction of tetrafluoro-hydrazine,[4] (2) hydrolysis of difluorourea,[5] (3) protonation of trityldifluoroamine,[6] (4) hydrolysis of difluorosulfamide,[7] and (5) hydrolysis of isopropyl difluorocarbamate.[8]

Trityldifluoroamine and isopropyl difluorocarbamate are reported to be stable reagents which permit the generation of difluoroamine directly with a minimum of preliminary effort. Tetrafluorohydrazine also can be converted to difluoroamine with ordinary laboratory equipment. If these difluoroamino compounds are not available or if expense is a factor to be considered, the straightforward fluorination of urea or sulfamide in water followed by hydrolysis *in situ* permits the preparation of usable quantities of difluoroamine in a simple manner.

* Rohm and Haas Company, Redstone Research Laboratories, Huntsville, Ala. 35807.

† Department of Chemistry, University of Notre Dame, Notre Dame, Ind. 46556.

‡ Department of Chemistry, Memphis State University, Memphis, Tenn. 38111.

Fluorination of urea in aqueous solution at 0–10°C. proceeds efficiently and selectively.   With solutions about 2 M in urea, the progress of reaction[9] can be followed by the ^{19}F n.m.r. spectra[9] of aqueous samples, whereby monofluorourea, difluorourea, and difluoroamine can be distinguished as the only oxidizing species present.   The point of stoichiometric equivalence between urea and 2 moles of fluorine corresponds to a maximum in the iodimetric titer of the reaction mixture, to the predominance of difluorourea among fluorinated species detected by n.m.r., and to an 80–90% yield based on urea. Continued fluorination of difluorourea solutions results in little absorption of fluorine and gradual decomposition of difluorourea. Carbon dioxide, nitrogen oxides, NF_3, FNO_2, $FONO_2$, FNO, and SiF_4 can be detected as by-products.

Acidification of the aqueous difluorourea solution with a nonoxidizing mineral acid results in the smooth liberation of difluoroamine as well as carbon dioxide.

A. 1,1-DIFLUOROUREA SOLUTIONS

Procedure

A solution of 30 g. of urea in 350 ml. of water is placed in a 1-l., three-necked, creased Pyrex flask and cooled to 0°C. with an ice bath.   A gas inlet tube made of 2-mm., heavy-wall, glass tubing is attached to the end of the fluorine supply line by means of a Tygon sleeve in such a manner that no Tygon is exposed.   It is fitted into one of the necks of the flask so that the tip is under the surface of the solution.   A thermometer and a gas outlet are attached to the other two necks.   It is convenient to use a bubbler on the gas outlet containing urea, potassium iodide, and a little acid.   This not only will remove the excess fluorine but will serve to indicate when the fluorination is complete.

The equipment for handling tank fluorine has been described in Volume XI of this series.[10]   The fluorine tank is opened, and the concentration of fluorine is adjusted to approximately 30% in helium by means of a dual rotameter system and allowed to bubble into the reaction flask. The rate is controlled so that the temperature of the flask does not rise above 10°C. When the fluorination is complete, as evidenced by the very dark color of aqueous triiodide ion in the bubbler, the fluorine tank is turned off and the apparatus is swept for 20 minutes with nitrogen.

The solution may then be removed from the reactor and stored in capped polyethylene bottles at −20 to −80°C. ■ *Caution. Concentrated difluorourea solutions decompose above −20°C. and commonly reek of tetrafluorohydrazine and difluoroamine. Therefore storage facilities should be well vented.*

The solution may be conveniently analyzed by the addition of excess potassium iodide followed by an iodimetric titration. The stoichiometry of the reaction is:

$$H_2N\overset{\displaystyle O}{\overset{\displaystyle \|}{C}}NF_2 + H_2O + 4HI \rightarrow 2I_2 + 2NH_4F + CO_2$$

In the case of the solution prepared above, it was found to be 1.21 M in difluorourea. The checker obtained a solution which was 0.99 M. Yields of 80–90% are generally obtained. For disposal, concentrated difluorourea solutions should be heavily diluted with water before being neutralized with sodium hydroxide.

Properties

The course of the fluorination may be followed by ^{19}F n.m.r. Figure 23 shows the variation during fluorination of urea to monofluorourea (MFU) and difluorourea (DFU). Only trace amounts of MFU were found at the end of the fluorination (at 110 p.p.m. relative to CCl_3F). In addition to the DFU (at −32 p.p.m. relative to CCl_3F) a relatively large amount of difluoroamine (at 8 p.p.m. relative to CCl_3F) was present.

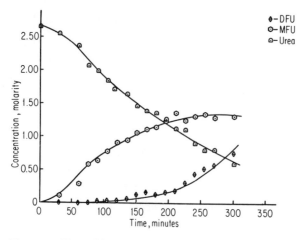

Fig. 23. *Urea, MFU, and DFU concentrations during fluorination reaction.*

The solutions decompose above $-20°$C. to yield difluoroamine but can be stored safely at $-80°$C.

B. DIFLUOROAMINE

Procedure

An amount of aqueous difluorourea solution necessary to give 10–20 mmoles of difluoroamine (calculated from the concentration of the solution as determined by iodimetric titer) is used. In the case of the solution prepared in Sec. A, 15 ml. of the aqueous difluorourea solution is placed in a 100-ml. three-necked flask equipped with a pressure-equalized dropping funnel, magnetic stirrer, and water-cooled reflux condenser. About 3 ml. of concentrated sulfuric acid is placed in the dropping funnel. The system is flushed with nitrogen admitted over the dropping funnel. The N_2 is conducted from the top of the water condenser through a 2.5×10-cm. drying tube containing Drierite and into a trap.' The trap contains stopcocks so that it can be isolated. The trap is cooled to $-78°$C. and the purging nitrogen stream is slowed to a trickle. ■ *Warning. Under no circumstances should liquid nitrogen be used to condense* difluoroamine. *Many detonations have been experienced when this is done.* Slow dropwise addition of the sulfuric acid is begun, and the temperature of the reaction flask is raised to 80°C. When gas evolution slows, a gentle nitrogen sweep is resumed while heating is continued for an hour. The solution is cooled; the trap is isolated from the system and moved to a vacuum line while still at −78°C. The solution in the reaction flask may be discarded. The contents of the trap are fractionated through cold traps at −78 and −130°C.; this serves to remove the carbon dioxide. A 14.6-mmole sample of difluoroamine was contained in the −130°C. trap for a yield of 81%. (The checkers reported a 67% yield.) The purity was 99.5% with CO_2 (0.2%) as the only impurity detectable by mass spectroscopy.

Properties

Difluoroamine is a colorless gas, b.p. −23°C., m.p. −116°C.,[1] which, although not decomposed immediately in air, is both easily oxidized and easily reduced.[1,2b,8] It is incompatible with many metals, particularly iron. Significant decomposition is incurred by conducting it through Tygon or rubber tubing.

The mass cracking pattern for difluoroamine is reproduced[1] in Table I for convenience in identification. Vapor pressure values are given by the equation:

$$\log P(\text{mm.}) = \frac{-1298}{T} + 8.072$$

TABLE I **Fragmentation Pattern**

m/e	Ion	Abundance, %
53	HNF_2^+	100
52	NF_2^+	8.7
34	HNF^+	99.5
33	NF^+	47.4
20	HF^+	4.0
19	F^+	5.2
15	NH^+	8.7
14	N^+	23.0

Recent studies of difluoroamine chemistry may be found in references 11 to 15.

References

1. A. Kennedy and C. B. Colburn, *J. Am. Chem. Soc.*, **81**, 2906 (1959).
2a. E. A. Lawton and J. Q. Weber, *ibid.*, **81**, 4755 (1959).
2b. E. A. Lawton, D. Pilipovitch, and R. D. Wilson, *Inorg. Chem.*, **4**, 118 (1965).
3. S. I. Morrow et al., Abstracts, 137th American Chemical Society Meeting, Cleveland, Ohio, April, 1960, p. 11M.
4. J. P. Freeman, A. Kennedy, and C. B. Colburn, *J. Am. Chem. Soc.*, **82**, 5304 (1960).
5. V. Grakauskas, Abstracts, 140th American Chemical Society Meeting, Chicago, Ill., September, 1961, p. 23M.
6. W. H. Graham and C. O. Parker, *J. Org. Chem.*, **29**, 850 (1963).
7. R. A. Wiesboeck and J. K. Ruff, *Inorg. Chem.*, **4**, 123 (1965).
8. D. L. Klopotek and B. G. Hobrock, *ibid.*, **6**, 1750 (1967).
9. F. A. Johnson, unpublished results.
10. J. K. Ruff, *Inorganic Syntheses*, **11**, 131 (1968).
11. W. T. Yap, A. D. Craig, and G. A. Ward, *J. Am. Chem. Soc.*, **89**, 3442 (1967).
12. C. B. Colburn, *Chemistry in Britain*, **2**, 336 (1966).
13. C. B. Colburn, NF Chemistry, in "Encyclopedia of Chemistry," George L. Clark (ed.), pp. 693–695, 2d ed., Reinhold Publishing Corp., N.Y., 1966.
14. C. B. Colburn, *Endeavor*, **24**(3) (September, 1965).
15. K. J. Martin, *J. Am. Chem. Soc.*, **87**, 394 (1965).

56. BIS(TRIFLUOROMETHYL) TRIOXIDE—HAZARDOUS MATERIAL

$$2F_2CO + OF_2 \xrightarrow{CsF} CF_3OOOCF_3$$

Submitted by L. R. ANDERSON,* D. E. GOULD,* and W. B. FOX*
Checked by JOHN K. RUFF†

Bis(trifluoromethyl) trioxide, CF_3OOOCF_3, was probably first prepared in low yield ($<5\%$) from the fluorination of sodium trifluoroacetate.[1] A substance of this composition was also

* Industrial Chemicals Division, Allied Chemical Corporation., P.O. Box 405, Morristown, N.J. 07960
† Department of Chemistry, University of Georgia, Athens, Ga. 30601.

reported to result from the ultraviolet irradiation of mixtures of hexafluoroazomethane and oxygen but was not fully characterized.[2]

The new preparation described below, which involves the cesium fluoride–catalyzed reaction between equimolar amounts of carbonyl fluoride and oxygen difluoride,[3] has the advantage over the methods cited above of producing high yields of the trioxide from easily obtained starting materials. The use of oxygen difluoride in excess of the stoichiometric amount required is not necessary but is helpful in ensuring that no carbonyl fluoride remains unreacted and thus facilitates the separation of the final products. The excess oxygen difluoride can be recovered and used in subsequent preparations.

Procedure

■ *Caution. Oxygen difluoride is a very strong oxidizing agent and also extremely toxic. Explosions and/or fire result if care is not taken to prevent it from coming into contact with organic or other reducing materials. It has a distinctive and disagreeable odor and has a threshold limit value (TLV) of 0.05 p.p.m.[4]*

Although no explosions have occurred with bis(trifluoromethyl) trioxide, the amount of material prepared in the synthesis has been limited in order to minimize the hazards. All reactions should be carried out behind adequate safety shields. The toxicity of bis(trifluoromethyl) trioxide is unknown but probably high.

In a dry-box operation, a 3-ounce Fischer-Porter* aerosol compatibility tube is loaded with approximately 10 g. of cesium fluoride and then equipped with a stainless-steel needle-valve adapter and a 323A Hoke valve.† After removal from the dry-box, Swagelok† connections are made from the valve to a metal standard-taper joint and the whole assembly attached to

---

* Fischer and Porter Co., County Line Road, Warminster, Pa. 18974.
† Hoke, Inc., 1 Tenakill Park, Cresskill, N.J. 07626.

a standard vacuum system and evacuated. A small furnace is placed around the tube, and the cesium fluoride is dried by heating to 160°C., with pumping, overnight.

The dried cesium fluoride is then ground in the dry-box with a mortar and pestle. (Higher yields are obtained if the cesium fluoride is ground in a nitrogen atmosphere for 5 hours with a Spex* Mixer/Mill No. 8000 using a tungsten carbide capsule and ball.) A 2.0-g. sample (13.2 mmoles) of the ground cesium fluoride is introduced into a 30-ml. stainless-steel Hoke cylinder, which is then fitted with a 323A Hoke valve. The assembly is removed from the dry-box and connected through Swagelok fittings and a metal standard-taper joint to the vacuum system.†

After evacuation, standard vacuum techniques are used to measure and condense equimolar amounts (20 mmoles each) of carbonyl fluoride and oxygen difluoride into the cylinder. The cylinder should be equipped with an explosion rupture assembly and should be in a shielded location before it is allowed to warm up. The valve is then closed, and the cylinder and its contents are allowed to warm to room temperature and remain at that temperature for 5 days. After this time, the cylinder contents are passed through a trap kept at approximately −140°C. (melting Skelly-Solve F or methylcyclopentane).

The CF_3OOOCF_3 is collected in the −140°C. trap, and the excess OF_2 along with traces of CF_3OF and any CF_3OOCF_3 that have been formed pass through the trap. (A more rapid fractionation with some loss of product can be accom-

---

* Spex Industries, Inc., 3880 Park Ave., Edison, N.J. 08840.

† The checker used the following procedure for preparing CsF. The CsF was dried and ground as described and then placed in a stainless-steel Hoke cylinder together with thirty $\frac{1}{8}$-in. stainless-steel balls. The cylinder was then closed and evacuated. Carbonyl fluoride or hexafluoroacetone (about 7 mmoles) was placed in the cylinder. The assembly was closed, removed from the vacuum line, and shaken on a wrist-action shaker overnight. The volatile contents were then removed from the bomb; the bomb was cooled to −196°C. and 10 mmoles of F_2 expanded into it. The cylinder was allowed to warm to about 25°C. and to stand for one hour. The contents of the bomb were removed by pumping through a −196°C. trap and a soda-lime trap. Following this treatment, the bomb was ready for use.

plished by using a $-130°C$. trap.) Yields of CF_3OOOCF_3 vary according to the preparation and previous use of the catalyst. With no previous use of CsF which has been ground by mortar and pestle, a yield of about 67% is obtained (based upon COF_2 introduced). A second use of this same catalyst leads to yields of about 74%. If CsF which has been ground with the Spex Mixer/Mill is used, the yields average about 84%. Subsequent use of CsF treated in this way has little effect on the yield.

Properties

Bis(trifluoromethyl) trioxide is a colorless gas which has a boiling point of $-16°C$. and a freezing point of $-138°C$. Its vapor pressure over the range -80 to $-23°C$. may be calculated from the equation:

$$\log P(\text{mm.}) = \frac{-1241}{T(°K)} + 7.705$$

It is most readily identified by its infrared spectrum in the gas phase which is as follows (in $cm.^{-1}$): 2600–2400(vw complex), 2123(vw), 1290(s), 1252(s), 1169(s), 1067(vw), 997(vw), 929(vw), 897(m), 773(m), 755(w shoulder), and 699(w).

The ^{19}F n.m.r. spectrum of pure CF_3OOOCF_3 at $-80°C$. is a single peak at $+72.4$ p.p.m. (relative to $CFCl_3$).

Bis(trifluoromethyl) trioxide is stable in glass at 25°C. with a half-life of many weeks. At 70°C. its decomposition is rapid and results in the formation of bis(trifluoromethyl) peroxide and molecular oxygen. It does not hydrolyze rapidly in pure water at 25°C. and is not destroyed rapidly by mercury although the mercury darkens.

References

1. P. G. Thompson, *J. Am. Chem. Soc.*, **89**, 4316 (1967).
2. V. A. Ginsburg et al., *Dokl. Akad. Nauk SSSR*, **149**, 97 (1963).
3. L. R. Anderson and W. B. Fox, *J. Am. Chem. Soc.*, **89**, 4313 (1967).
4. N. V. Steere, *J. Chem. Educ.*, **44**, A49 (1967).

Appendix

PURIFICATION OF TETRAHYDROFURAN

WARNING

It has been reported that serious explosions may occur when impure tetrahydrofuran is treated with solid potassium hydroxide or with concentrated aqueous potassium hydroxide, as has been recommended widely for the purification of tetrahydrofuran; see *Organic Syntheses*, Coll. **4**,* 474, 792 (1963); **40,** 94 (1960). There is evidence that the presence of peroxides in the tetrahydrofuran being purified was causal. It is strongly recommended, therefore, that this method not be used to dry tetrahydrofuran, if the presence of peroxides is indicated by a qualitative or quantitative test with acidic aqueous iodide solution. Traces of peroxide can be removed by treatment with copper(I) chloride (cuprous chloride) see *Organic Syntheses*, **45,** 57 (1965). The safety of this operation should be checked first on a small scale (1–5 ml.). It is recommended that tetrahydrofuran containing larger than trace amounts of peroxides be discarded by flushing down a drain with tap water. It must be kept in mind that mixtures of tetrahydrofuran vapor and air are easily ignitable and explosive; purification is best

* Collective Volume 4 of *Organic Syntheses* is a revised edition of Annual Volumes 30–39 published by John Wiley & Sons, Inc., New York.

carried out in a hood which is well exhausted and which does not contain an ignition source.

The best procedure for drying tetrahydrofuran appears to be distillation (under nitrogen) from lithium aluminum hydride. This operation should not be attempted until it is ascertained that the tetrahydrofuran is peroxide-free and also not grossly wet. A small-scale test can be carried out in which a small amount of lithium aluminum hydride is added to *ca.* 1 ml. of the tetrahydrofuran to determine whether a larger-scale drying operation with lithium aluminum hydride would be too vigorous for safe operation. Tetrahydrofuran so purified rapidly absorbs both oxygen and moisture from air. If not used immediately, the purified solvent should be kept under nitrogen in a bottle labeled with the date of purification. Storage for more than a few days is not advised unless 0.025% of 2,6-di-*t*-butyl-4-methylphenol is added as an antioxidant.

There are no indications that peroxide-free, but moisture-containing, tetrahydrofuran cannot safely be predried over potassium hydroxide. However, even this operation should be attempted only after a test-tube-scale experiment to make sure that a vigorous reaction does not occur.

A peroxide-free grade of anhydrous tetrahydrofuran (stabilized by 0.025% of 2,6-di-*t*-butyl-4-methylphenol) in 1-lb. bottles is available currently (1970) from Fisher Scientific Co. This product as obtained from freshly opened bottles has been found to be suitable for reactions, such as the formation of Grignard reagents, in which purity of solvent is critical (Du Pont Company, unpublished observations). It is standard practice in at least one laboratory to use only tetrahydrofuran (Fisher) from freshly opened bottles and to discard whatever material is not used within 2 to 3 days.

INDEXES

INDEX OF CONTRIBUTORS

SUBJECT INDEX

Names used in this cumulative Subject Index for Volumes XI and XII, as well as in the text, are based for the most part upon the "Definitive Rules for Nomenclature of Inorganic Chemistry," 1957 Report of the Commission on the Nomenclature of Inorganic Chemistry of the International Union of Pure and Applied Chemistry, Butterworths Scientific Publications, London, 1959; American version, *J. Am. Chem. Soc.*, **82,** 5523–5544 (1960); and the latest (unpublished) revisions, including "Coordination Compounds" (Tentative Rules, IUPAC, 1968, when definitive to replace Section 7 of the 1957 Rules); also on "Tentative Rules for Nomenclature of Organometallic Compounds," IUPAC, 1967; "D-4 Nomenclature of Organosilicon Compounds," IUPAC, 1969; and "The Nomenclature of Boron Compounds" [Committee on Inorganic Nomenclature, Division of Inorganic Chemistry, American Chemical Society, published in *Inorganic Chemistry*, **7,** 1945 (1968) as tentative rules following approval by the Council of the ACS]. All of these rules have been approved by the ACS Committee on Nomenclature. Conformity with approved organic usage is also one of the aims of the nomenclature used here.

In line, to some extent, with *Chemical Abstracts* practice, more or less inverted forms are used for many entries, with the substituents or ligands given in alphabetical order (even though they may not be in the text); for example, derivatives of arsine, phosphine, silane, germane, and the like; organic compounds; metal alkyls, aryls, 1,3-diketone and other derivatives and relatively simple specific neutral (nonelectrolyte) coordination complexes: *Iron, cyclopentadienyl-* (also at *Ferrocene); Cobalt(II), bis(2,4-pentanedionato)-* [instead of *Cobalt(II) acetylacetonate*]. In this way, or by the use of formulas, many entries beginning with numerical prefixes are avoided (as for halogeno, cyano, and a few other complexes); thus, *Chlorovanadate(III), tetra-*. Numerical and some other prefixes are also avoided by restricting entries to group headings where possible: *Sulfur imides,* with the formulas; *Molybdenum carbonyl,* $Mo(Co)_6$; both *Perxenate,* $HXeO_6^{3-}$, and *Xenate-(VIII),* $HXeO_6^{3-}$. In cases where the cation (or anion) is of little or no significance in comparison with the emphasis given to the anion (or cation), one ion has been omitted; e.g., also with less well-known complex anions (or cations): $CsB_{10}H_{12}CH$ is entered only as *Carbaundecaborate(1−), tridecahydro-* (and as $B_{10}CH_{13}^-$ in the Formula Index).

Under general headings such as *Cobalt(III) complexes* and *Ammines,* used for grouping coordination complexes of similar types having names considered unsuitable for individual headings, formulas or names of specific compounds are not usually given. Hence it is imperative to consult the Formula Index for entries for specific complexes.

Two entries are made for compounds having two cations and for addition compounds of two components, with extra entries or cross references for synonyms. Unsatisfactory or special trivial names that have been retained for want of better ones or as synonyms are placed in quotation marks.

Boldface type is used to indicate individual preparations described in detail, whether for numbered syntheses or for intermediate products (in the latter case, usually without stating the purpose of the preparation). Group headings, as *Xenon fluorides*, are in lightface type unless all the formulas under them are boldfaced.

As in *Chemical Abstracts* indexes, headings that are phrases are alphabetized straight through, letter by letter, not word by word, whereas inverted headings are alphabetized first as far as the comma and then by the inverted part of the name. Stock Roman numerals and Ewen-Bassett Arabic numbers with charges are ignored in alphabetizing unless two or more names are otherwise the same. Footnotes are indicated by *n.* following the page number.

FORMULA INDEX

The Formula Index, as well as the Subject Index, is a cumulative index for Volumes XI and XII. The chief aim of this index, like that of other formula indexes, is to help in locating specific compounds or ions, or even groups of compounds, that might not be easily found in the Subject Index, or in the case of many coordination complexes are to be found only as general entries in the Subject Index. *All* specific compounds, or in some cases ions, with definite formulas (or even a few less definite) are entered in this index or noted under a related compound, whether entered specifically in the Subject Index or not. As in the latter index, **boldface type** is used for formulas of compounds or ions whose preparations are described in detail, in at least one of the references cited for a given formula.

Wherever it seemed best, formulas have been entered in their usual form (*i.e.*, as used in the text) for easy recognition: Si_2H_6, XeO_3, NOBr. However, for the less simple compounds, including coordination complexes, the significant or central atom has been placed first in the formula in order to throw together as many related compounds as possible. This procedure often involves placing the cation last as being of relatively minor interest (*e.g.*, alkali and alkaline earth metals), or dropping it altogether: MnO_4Ba; $Mo(CN)_8K \cdot 2H_2O$; $Co(C_5H_7O_2)_3Na$; $B_{12}H_{12}^{2-}$. Where there may be almost equal interest in two or more parts of a formula, two or more entries have been made: Fe_2O_4Ni and $NiFe_2O_4$; $NH(SO_2F)_2$, $(SO_2F)_2NH$, and $(FSO_2)_2NH$ (halogens other than fluorine are entered only under the other elements or groups in most cases); $(B_{10}CH_{11})_2Ni^{2-}$ and $Ni(B_{10}CH_{11})_2^{2-}$.

Formulas for organic compounds are structural or semistructural so far as feasible: $CH_3COCH(NHCH_3)CH_3$. Consideration has been given to probable interest for inorganic chemists, *i.e.*, any element other than carbon, hydrogen, or oxygen in an organic molecule is given priority in the formula if only one entry is made, or equal rating if more than one entry: only $Co(C_5H_7O_2)_2$, but $AsO(+)$-C_4-H_4O_6Na and $(+)$-$C_4H_4O_6AsONa$. Names are given only where the formula for an organic compound, ligand, or radical may not be self-evident, but not for frequently occurring relatively simple ones like C_5H_5 (cyclopentadienyl), $C_5H_7O_2$ (2,4-pentanedionato), C_6H_{11}(cyclohexyl), C_5H_5N(pyridine). A few abbreviations for ligands used in the text are retained here for simplicity and are alphabetized as such: "en" (under "e") stands for ethylenediamine, "py" for pyridine, "bipy" for bipyridine, "pn" for 1,2-propanediamine (propylenediamine), "fod" for 1,1,1,2,2,3,3-heptafluoro-7,7-dimethyl-4,6-octanedionato, "thd" for 2,2,6,6-tetramethylheptane-

3,5-dionato, "DH" for dimethylglyoximato and "D" for the dianion, $(CH_3)_2C_2$-$N_2O_2{}^{2-}$.

The formulas are listed alphabetically by atoms or by groups (considered as units) and then according to the number of each in turn in the formula rather than by total number of atoms of each element. This system results in arrangements such as the following:

NHS_7
$(NH)_2S_6$ (instead of $N_2H_2S_6$)
$NH_3B_{10}CH_{12}$

$(FSO_2)_2NH$ (instead of $F_2S_2O_4NH$)
FSO_3H
F_2SO_3

$[Mo(CO)_3C_5H_5]K$
$[Mo(CO)_3C_5H_5CH_3]$

FNO

$[Cr(en)_3][Ni(CN)_5]$ ["en" instead of $(NH_2)_2C_2H_4$ or $N_2H_4C_2H_4$]
$[Cr(NH_3)_6][Ni(CN)_5]$

Footnotes are indicated by *n.* following the page number.